Processing, Properties, and Applications of Glass and Optical Materials

Processing, Properties, and Applications of Glass and Optical Materials

Ceramic Transactions, Volume 231

Proceedings of the 9th International Conference on Advances in the Fusion and Processing of Glass (AFPG9) and Symposium 15—Structure, Properties and Photonic Applications of Glasses held during PACRIM-9, Cairns, Australia, July 10–14, 2011

Edited by
Arun K. Varshneya
Helmut A. Schaeffer
Kathleen A. Richardson
Marlene Wightman
L. David Pye

A John Wiley & Sons, Inc., Publication

Published by John Wiley & Sons, Inc., Hoboken, New Jersey.
Published simultaneously in Canada.

Library of Congress Cataloging-in-Publication Data is available.

ISBN: 978-1-118-27374-6
ISSN: 1042-1122

Printed in the United States of America.

10 9 8 7 6 5 4 3 2 1

Contents

PART B: STRUCTURE, PROPERTIES, AND PHOTONIC APPLICATIONS OF GLASS

Preface

The proceedings at hand are a bound volume of some of the presentations made at both "The 9th International Conference on Advances in the Fusion and Processing of Glass" (AFPG9) as Part A and Symposium 15, "Structure, Properties and Photonic Applications of Glasses" as Part B.

The AFPG9 was held July 10-14, 2011, co-located in Cairns (Australia) with the 9th PACRIM and the annual meeting of the Australian Ceramic Society. The organizers of PACRIM-9, namely Dan Perera and Philip Walls, thought (wisely) that they needed to approach some of the undersigned members of the Glass & Optical Materials Division of The American Ceramic Society to organize the glass and related technical program portion of the PACRIM-9.

We thought that a clear opportunity existed to expand the glass technical program to include the 9th AFPG which hitherto had missed its target of having a meeting about once every three years.

The "AFPG" conferences have a strong history of bringing together highly skilled professionals who have been contributing scientifically, educationally or technologically in direct support of the world-wide glass industry. Beginning with the first conference in this series at Alfred University in 1988, the conferences have been organized in rotation between Alfred and the Deutsche Glastechnische Gesellschaft (DGG): Dusseldorf (1990), New Orleans (1992), Wurzburg (1995), Toronto (1997), Ulm (2000), Rochester (2003), Dresden (2006).

The AFPG meeting in Australia was therefore a departure from the tradition, yet an experiment worth trying. With China, Japan, Korea and India representing better than half the gross tonnage of glass production and Australia beginning to establish itself as an important new supplier of raw materials, it didn't take much convincing to bring the AFPG to where the action may be in the foreseeable future and co-locate with the PACRIM meeting in Australia.

New challenges were foreseen, however; in particular, how do we expect to draw attendance to a faraway part of the globe while the world economy was not doing well with unemployment rates hitting highs not seen since the Great Depression and the Australian dollar had appreciated nearly 60% over a matter of months!

No challenge is fun until it is met head-on. A sound technical program and megafunds are essentially the winning combination. That is just what the AFPG9 organization aimed to do. Technical programming focus, of course, was on new ways of glass fusion and raw materials that improve the efficiency of the fusion process and hence are energy conserving, sustainable, and leave a smaller carbon footprint. In addition, new thoughts were sought for container lightweighting, increasing the usable strength of the glass product. An equally important focus was on bringing the pharmaceutical glass packaging professionals to the attention of the glass industry. Because of the low tonnage involved, glass industry has barely paid attention to the needs of the pharma industry; however, because of the potential lifesaving issues, the pharmaceutical industry invests a great deal in solving its own problems without a significant interaction with the traditional glass professionals. AFPG9 sought to change that. To complete the technical program package, new research on the photonic applications of glass were also sought. Presentations on optical materials and applications were organized as "Symposium 15".

Some of us donned our fundraising caps and looked for financial support. Surprisingly, we did well. We would like to thank the following corporations for their generous financial support: Ivoclar Vivadent (Volker Rheinberger), NEG (Akihiko Sakamoto), Corning (Ivan Cornejo), Owens-Corning (Manoj Choudhary) and PPG (Mehran Arbab). Internal organizational support from the Australian Ceramic Society made it possible to support several invited speakers at least partially and underwrite the expense of these proceedings. Between AFPG9 and Symposium 15, a total of six keynote lectures, 75 oral and 13 poster presentations were scheduled. The following seven students were judged to receive awards up to $1500, plus registration, towards their travel: Fumitake Tada (Japan), Laura Adkins (USA), Sefina Ali (USA), Zhuoqi Tang (UK), Sebastian Krolikowski (Germany), J. David Musgraves (USA), and Rolf Weigand (Germany). These students received an opportunity to present posters as well as a brief oral presentation.

As it happens, the Cairns region was hit by the category 4 cyclone Yasi in early February 2011. For a brief moment in time, it appeared that the PACRIM/AFPG conferences may need to be relocated. Fortunately, early fears were unfounded. Yet, to add insult to injury, five weeks later, Japan was hit by an earthquake and a resulting nuclear catastrophe which triggered new concerns over attendance by Japanese professionals (more than what bird flu did to PACRIM 8). Undoubtedly, the attendance by glass professionals from China and India lacked luster. However, our friends from Japan and Korea made up for this shortage. In particular, we would like to thank Akio Makishima of Japan Advanced Institute of Science & Technology and Satoru Inoue of National Institute for Materials Science (NIMS) who encouraged the participation by the conglomerate of Japanese glass industry, university and government labs to bring their work on in-flight melting of glass to AFPG9. We would also like to thank Ron Iacocca of Eli Lilly to have understood the need for greater interaction between the traditional glass professional and the pharmaceutical glass packaging professional.

Special thanks go to our longtime friend, Fabiano Nicoletti of the Stevanato Group and president of the International Commission on Glass. Dr. Nicoletti delivered the AFPG9 Premiere Lunch lecture entitled, "The Global Role of the International Commission on Glass", Tuesday July 12, 2011, and presented the student awards.

Since the metropolitan Cairns had escaped the wrath of Yasi, the travel, transportation and sightseeing were not an issue. For many of us, it was the first time visiting Australia. Through the tireless efforts of our primary hosts, namely Phil Walls (President, The Australian Ceramic Society), Dan Perera (The Australian Ceramic Society), Nick Koerbin (Materials Australia), Hussein Hamka (Materials Australia), and Yi-Bing Cheng (Monash University), the conference was a roaring success. Their hospitality seems to have left no stones unturned.

Thanks to Dr. Doreen Edwards, Dean, School of Engineering, Alfred University for making the university infrastructure available for launching the conference. Endorsements by the Glass & Optical Materials Division of The American Ceramic Society and the International Commission on Glass did much to help; they are gratefully acknowledged.

To all of you, the members of the Organizing Committee extend their sincere thanks for your help, encouragement, and steadfast support during those times of uncertainty while planning AFPG 9.

Arun K. Varshneya, Chair
Alfred University and Saxon Glass Technologies, Inc., Alfred, New York, USA

Helmut A. Schaeffer
University of Erlangen and Deutsche Glastechniche Gesellschaft, Berlin, Germany

Kathleen A. Richardson
Clemson University, Clemson, South Carolina, USA

Marlene Wightman
Alfred University, Alfred, New York, USA

L. David Pye
Alfred University and Empire State Glassworks LLC, Little Falls, New York, USA

Endorsed By:

Acknowledgments

Thank you to our sponsors:

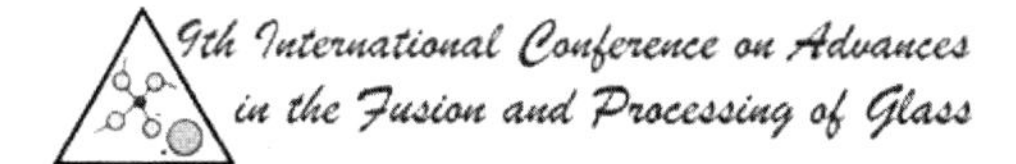

CORNING

Nippon Electric Glass Co., Ltd.

PART A: THE 9TH INTERNATIONAL CONFERENCE ON ADVANCES IN THE FUSION AND PROCESSING OF GLASS

Fusion of Glass

NEW CONCEPTS FOR ENERGY EFFICIENT & EMISSION FRIENDLY MELTING OF GLASS

Ruud Beerkens
TNO
Eindhoven, The Netherlands

ABSTRACT

The paper starts with a short analysis of the performance of currently applied glass melting processes/furnaces concerning their energy consumption levels (benchmarking), as well as NOx emissions. The best practice situations will be presented for few glass products (e.g. container glass). Potential improvements in furnace designs to obtain more intensified melting or smaller glass furnace sizes and improved-controlled process steps (melting-in, complete raw material (e.g. sand) digestion of the melt, fining, homogenization & conditioning) will be discussed. Important aspects per process step are control of the applied temperature level, finding optimum glass melt flow regimes and residence times in each compartment of the melting tank, dedicated for a specific process step. Controlled & intensified heat transfer to the batch blanket area is decisive for the possibility to reduce glass furnace size, and consequently lowering structural energy losses. Mathematical modeling studies support the development of new furnace designs and heating methods in industrial glass furnaces with targeted high energy efficiency and moderate / low NOx emissions. New technological elements, such as application of CFD models, innovative oxygen-firing glass furnace designs, improved refractory materials, and advanced process control systems support the development of new glass melting concepts and furnace designs.

INTRODUCTION

Glass industries are continuously searching for production cost reductions. Energy costs are an increasingly growing part of the total production costs. In Europe, since 2005, emission trading of greenhouse gas allowances has been in place. The costs for CO_2 emissions will grow in future, due to a limited availability (on purpose) of CO_2-allowances in the European Union.

In container glass melting, glass melting processes in Europe use about 45-50 % of all the (primary energy) energy [non-published LCA studies and references[1,2]], required in the production chain of that glass, including raw material supply. Benchmarking in the container glass industry shows that regenerative or oxygen-fired glass melting tank furnaces with capacities above 300 metric tons per day show lowest specific energy consumption (3.5 – 4.25 GJ/ton molten glass, depending on cullet% and quality demand). In most other sectors, oxygen-fired furnaces are often the most energy-efficient options, even including energy consumption of the oxygen generation processes. Even in the furnaces with highest energy efficiency, up to 55-60 % (regenerative fired container glass furnaces with batch preheating), average residence times of the melt in the glass furnaces are much longer than required for batch-wise melting processes as can for instance be performed in laboratory crucible tests.

Most industrial glass furnaces suffer from very wide Residence Time Distributions (RTD) of the glass in the melting tank, due to the poorly controlled glass melt flow patterns in conventional glass tank furnaces and because of the method of heat transfer to the glass forming raw material batch. The high levels of average residence times, require relatively large melt tank volumes and furnace sizes. Average residence time (hrs.), multiplied with pull rate (m^3 melt/hr) is the content of a melting tank. Although, approximately 20% energy savings are possible when comparing the average of the energy efficiency of EU glass furnaces to the best practice energy efficient furnace, see figure 1 for the container glass sector, the problem of the wide RTD and poorly controlled heat transfer and flow patterns is not solved by best practice today. Even the most efficient container glass furnace shows very wide RTD's and average residence times of 20 hours or more. Therefore new furnace designs are necessary. Such furnaces should split the melting process for the different functionalities: melting-in of batch (here most energy is required), dissolution of sand grains (strong convection and mixing is preferred, here temperatures should not be too high), removal of gases (bubbles, dissolved) from the melt (highest temperature level, shallow tank depth), secondary fining and conditioning (homogenization of composition and temperature) of the melt. Furthermore, for advanced glass furnace designs, the control of high intensity heat transfer, especially into the sections where highest energy supply levels are needed, should be improved, however without excessive evaporation from the melt of volatile glass species.

Although, modern glass production plants are equipped with Air Pollution Control systems (APC),[2] generally consisting of a scrubber and filter (to remove dust particulates), emissions of CO_2 and NOx are still of high concern. Primary measures are mostly preferred to reduce these emission levels of glass furnaces. NOx emission reduction methods have successfully been developed in the last 25 years. For instance in the Netherlands specific NOx emission levels dropped from 5.24 kg/ton molten glass down to 1.8 kg/ton melt from 1992 until 2009. This has been achieved by primary measures, including all oxygen-firing, only, and further reductions are expected.

Even with energy efficient regenerative glass furnaces, with high air preheat temperatures, specific NOx emission levels due to fossil fuel combustion can be kept at much lower levels (container glass furnaces < 1-1.2 kg NOx/ton molten glass around the year 2010) as was shown in the 1970-ties (> 4 kg/ton molten glass).[3] A combination of modified combustion chamber design, burner port geometry adaptations and optimized dimensions, type of burner, number of burners in use, burner settings (velocity of fuel injection, direction of fuel injection jets, creation of multiple jets from burner nozzles, positions of burners in burner ports), air number control (oxygen excess), and CO monitoring is a proven package of primary measures, being successful to reduce NOx emissions, often without the direct need for a DeNOx system. Emission levels in regenerative soda-lime-silica glass furnaces below 700-850 mg/Nm^3 are feasible for modern glass furnaces with high energy efficiency.[2,3]

ENERGY BENCHMARKING

Figure 1 shows the ranking of the energy efficiency of 168 container glass furnaces, starting from the lowest specific energy consumption (most energy efficient) up to the highest specific energy consumption. The energy values are normalized for the situation of 50% recycling cullet in the batch and re-calculating to total primary energy consumption per metric ton molten glass, taking into account the primary energy required for generating electric power (in case of electric boosting) and oxygen production.[1] From this figure 1, assuming it being a representative sample for worldwide container glass production, we can conclude that the average glass furnace consumes about 21% more energy than the best practice (most energy efficient) furnaces (average value for number 1-17) in the ranking of figure 1. However, glass quality demands may be different for the different furnaces, representing figure 1.

The theoretical energy demand, for heating the raw material batch and the endothermic fusion reactions for normal batch and bringing the fresh glass melt to temperatures of about 1350°C (throat temperature) is about 2.55-2.60 MJ/kg for a soda-lime-silica (SLS) glass melting process. For only cullet melting this would be 1.65 MJ/kg glass melt from this cullet. These are the lower limits, but even very well insulated glass furnaces will show some structural heat losses and fossil fuel firing is always associated with flue gas production. The flue gas heat contents can be partly recovered by batch preheating and combustion air preheating, but flue gas temperatures of at least 200-250°C will remain. Therefore the minimum achievable energy consumption level is at least 1.5 times the value determined by the conversion of raw materials into a hot glass melt and emitted batch gases.

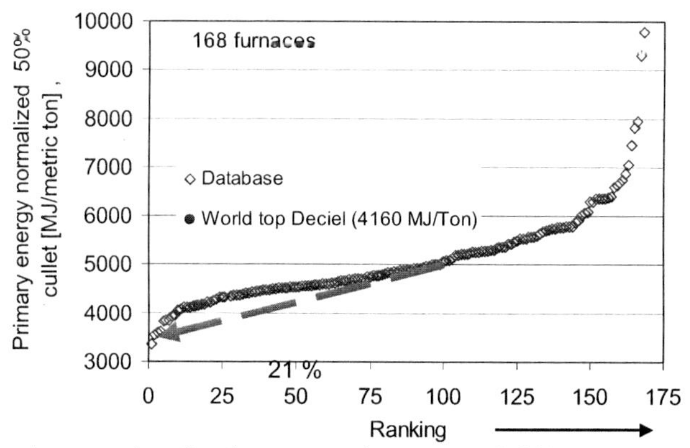

Average value of total set compared to average of 10 % most energy efficient: shows a potential of 21% energy savings on average

Figure 1. Example of ranking of energy efficiency from most efficient to lowest efficient container glass furnace of a set of 168 furnaces (excluding forehearths). Each data point refers to one existing container glass furnace. Specific energy data are normalized to 50% cullet and primary energy consumption electricity & oxygen production taken into account.

For normal batch this would be about 3.8 MJ/kg molten glass and for 100% cullet, about 2.5 MJ/ton molten glass. Today the most efficient container glass furnaces show energy losses / energy distributions similar to figure 2, in the case of charging room temperature (non-preheated) batch into the furnace. In all cases today, average residence times (typically 20-60 hours) of the glass in industrial melting tanks are much more than minimum melting times required for laboratory melts (typically 3-5 hours).

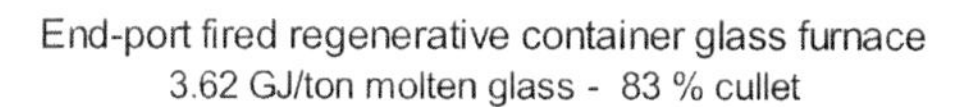

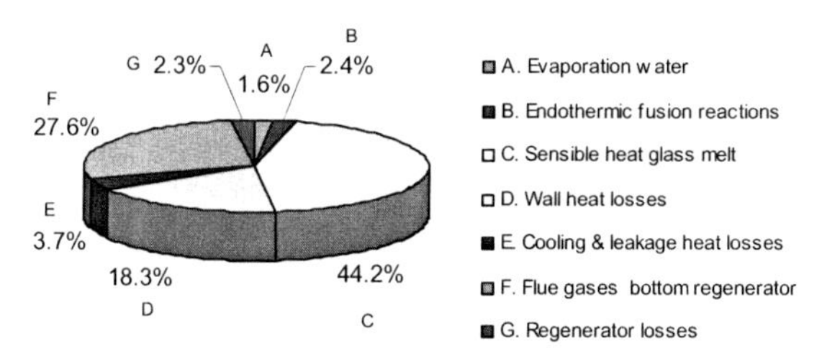

Figure 2. Energy consumption and distribution of energy (losses) in an energy efficient container glass furnace (top 10% in the benchmark pool).

NOx EMISSIONS

In this paper, only the effect of primary measures will be shown. In the 1980's and 1990's, many glass furnace combustion systems have been modified by changing the types of burners and adjusting the nozzle sizes to reduce fuel injection velocities. Mixing between fuel (natural gas and/or fuel oil) and combustion air can be retarded by lower gas velocities and fuel jets not directing sharply into the combustion air flow.[3] Furthermore, combustion control systems have been developed to limit the excess of combustion air or oxygen, preventing oxygen rich conditions in the flames, that usually lead to high NOx conversion rates (so called thermal NOx).

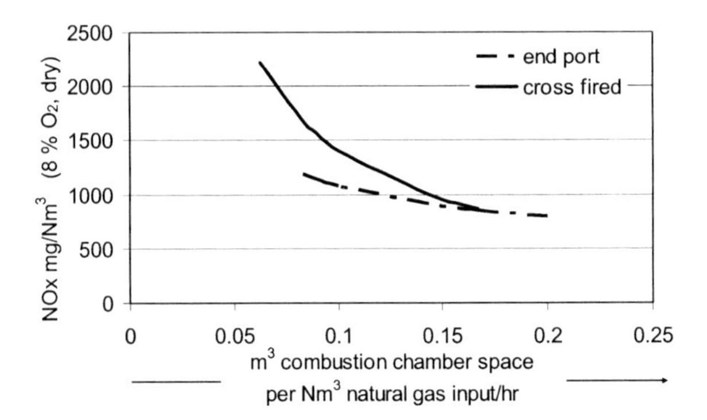

Figure 3. NOx emission levels in container glass industry for regenerative furnaces, depending on specific combustion space volume, curves are derived as best fit from many data.

Today, new furnace designs are considered to be important for further decreasing the NOx formation levels in fossil fuel fired glass furnaces, especially for regenerative furnace types. NOx benchmark studies, correlating furnace design parameters with specific NOx emissions and Computational Fluid Dynamic modeling of combustion processes in glass furnaces show that modifications in combustion chamber, burner port designs and number of burners are essential for a combination of high heat transfer and low NOx formation levels.

Table I summarizes some results from an inventory from approximately 40 end-port fired furnaces. This table shows clearly that an increase in combustion chamber volume, tighter control of oxygen excess and slower mixing by burner angle adjustments can result in strong reductions of NOx emissions of regeneratively fired glass furnaces. Further optimizations seem to be possible by optimizing burner port design and burner types.

Modern regenerative glass furnaces show a relatively high crown, spacy regenerators and wide/high burner ports with a small slope and a limited number of burners per port. The burner position, burner angles and nozzle size are important elements for flame shaping, for flame length, heat radiation to batch blanket & melt and NOx formation levels. Figure 3 shows results of measured specific NOx emissions in container glass sector for regenerative furnaces, depending on the relative size of the combustion chamber.

Table I. Effect of changes in air excess (oxygen residue in flue gas), burner angle (underport) and relative volume of combustion space on NOx emissions values normalized to base case (100%) with high NOx level.

End-port fired regenerative glass furnaces	"Low"-NOx glass furnaces	"Mid" NOx-group glass furnaces	"High"-NOx glass furnaces Base-case
NOx emission level range	500 – 800 mg/Nm^3 8% O_2, dry	~1200 -1600 mg/Nm^3 8% O_2, dry	~2000 mg/Nm^3 8% O_2, dry
Typical O_2 content in top regenerator	64 %	80 %	100 %
Vertical burner angle compared to base case	50 %	85 %	100 %
Relative volume combustion chamber compared to volume flow fuel	125 %	110 %	100 %

ENERGY TRANSFER TO MELT AND BATCH BLANKET

One of the most important drawbacks of continuous melting tank furnaces, as used today in almost all glass sectors, is the strong recirculation ("return flow") flow of glass melt from the hot-spot zone of the melting tank along the glass melt surface, back to the batch blanket tip, (see figure 4). In the hot-spot zone of the melting tank, with high heat transfer from the combustion chamber to this zone, the glass melt receives its quality. Here bubbles are released from the melt, viscosity is very low and homogenization is effective. This high temperature melt is suitable for making high quality glass products. However, most of the well (re)fined glass melt from this section of the tank is flowing upstream to the batch blanket area instead of being transported to the conditioning zones of the melting process and after that to the forming processes. The "return flow" can be 5 to 8 times stronger than the net pull. The heat input from the combustion space directly above the batch blanket to this blanket is generally not sufficient to heat a batch effectively in order to keep the surface area of the melt covered by the blanket as small as possible. Thus, the return flow is essential to limit the surface area covered with non-molten batch. This return flow of "hot-spot" glass melt is providing most of the heat required to heat and melt (fuse) the batch materials. Without the return flow, batch blankets could extend over the whole furnace length. These very massive return flows, necessary for batch heating in conventional furnaces however lead to very wide residence time distributions for the glass in the melting tank. Some volumes of melt will not re-circulate (often this gives the minimum residence time), but some parts may re-circulate up to 10 times, giving very large residence time values. Figure 4 schematically shows this situation. Fig. 5 shows determined residence time distributions for 3 industrial furnaces. A fundamental aspect associated with this behavior is that this strong re-circulation flow determines the necessary volume of a melting tank for a certain production capacity while still keeping a certain level of minimum residence time for glass melt quality. In the case that the part of the heat supply by this re-circulation flow can be replaced by another heating source for melting the batch blanket, the re-circulation can be retarded and the required tank volume can be decreased.

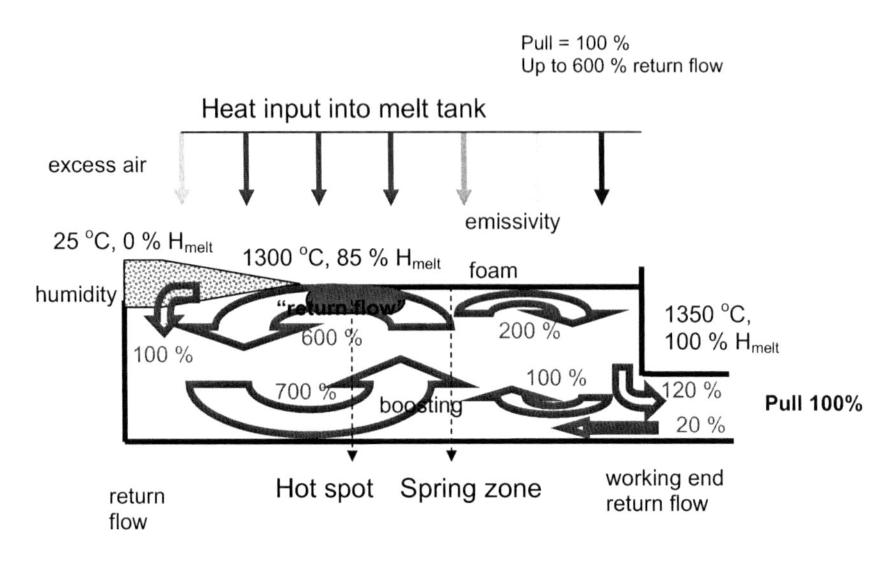

Figure 4. Scheme of main flow patterns in glass melting tank for continuous glass production. The "return flow" from spring zone/hot spot to batch blanket can be more than 6 times the pull. Batch absorbs within about 1 hour, 80-90% of the total net heat input in the melting process.

Modeling calculations for a typical container glass furnace showed that a reduction of the well-insulated container glass furnace volume by 30% will result in about 5% less energy consumption. For very well-insulated furnaces, which operate already at high specific pulls, potential energy savings by an even more compact melter is rather limited. But, smaller furnaces require lower capital investments. In the theoretical extreme case of infinite large regenerators and a perfectly insulated furnace the energy consumption is about 3.0 MJ/kg molten glass, this is the theoretical limit for normal batch (assuming no batch preheating). This limit however is not of practical value, since perfect insulation is not possible and infinite large regenerators are not feasible.

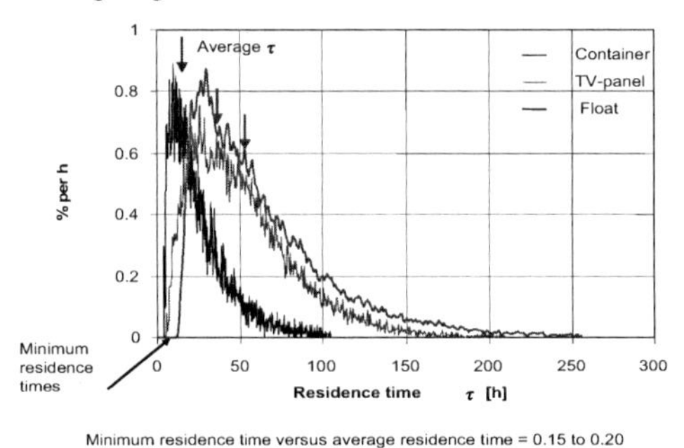

Figure 5. Typical residence time distributions for the melt in different glass melting tanks for float glass, TV panel glass and soda-lime-silica container glass production.

Possible approaches to tackle the problem of obtaining high heat fluxes to the batch blanket without requiring the strong return flow from the hot spot area are:

- Applying so-called batch electric boosting systems with electrode configurations just underneath the batch blanket.
- Applying a separate section for batch melting using submerged combustion technology,[4] and bringing the just melted batch to a second stage for further complete melting & fining, but without recirculation (return flow of melt) between the fining and submerged combustion section.
- Use very thin batch blankets in a shallow section of the melting tank.
- Apply batch preheating, and reduce the amount of heat transfer necessary for batch heating & melting in the furnace itself.
- More effective heating of batch from the top by improved coverage of flames above the batch blanket (e.g. using crown burners).

Most of the above mentioned approaches however show drawbacks such as use of expensive electricity, carry-over of loose batch by high convection through or above the batch, large surface areas (thin batch blanket) or extra investments for pre-heaters.

PROCESS STEPS IN INDUSTRIAL GLASS MELTING

Generally in industrial glass melting, starting with raw materials and cullet, different process steps can be distinguished. Preferably these process steps should take place in different sections of the melting tank and in the correct sequence. Return flow from one section to a previous section is not beneficial for glass melting and energy efficiency performance. Overlap of different process steps should be avoided since different process steps require different conditions and operational settings for individually optimal performance. A comparison of the optimum conditions for batch heating, sand grain dissolution, clay or feldspar grain or alumina grain dissolution and removal of bubbles from the melt show significant differences. In the following sections the different process steps will be shortly described and from a theoretical point of view, optimum process conditions will be indicated and qualitatively discussed.

BATCH HEATING/MELTING-IN OF BATCH

Modeling studies[5] and observations in industrial glass melting tank furnaces show that the batch blanket extends from 25 to 75% of the furnace length dependent on the batch charging system, the pull and efficiency of heat transfer to the batch to be heated. The apparent density of loose batch is lower than the resulting melt phase. The batch is floating on the glass melt surface, but is partly immersed in the melt (like icebergs in water). The batch can form a blanket over the width of the tank or it forms small piles or individual islands, floating on the glass melt surface. Sometimes, the batch (especially when using a screw batch charger system) forms a worm shape "caterpillar" irregularly moving on top of the glass melt. From entering the furnace till reaching the batch blanket tip, it takes typically about 40-60 minutes for the batch material. Within that time, the batch is heated up to about 1250°C. About 80-90% of the total energy that has to be transferred to the batch/melt has taken place in that short time period. Further heating for fining and sand dissolution requires 10-20% extra energy, but these processes take much more time. The heat conduction into the batch blanket interior determines the rate of batch melting, a thinner batch blanket or a batch island heated from all sides will enhance heating and melting of the batch. The use of batches with higher heat conductivity (cullet rich batches or pelletized batch[6,7]) offers the possibility of limiting coverage of the melt by non-molten batch. The heat conductivity of batch is particularly low for temperatures below 800°C and for powder batches. Heating from room temperature to 800°C takes much more time than the second stage up, to

1250°C. Thus, batch preheating will be very effective in reducing the time needed to heat the batch blanket, since the batch typically enters the furnace at already 250-350°C, instead of room temperature, and the initially slow heating process of the cold batch is partly circumvented when using batch preheating. Thus, the use of preheated batch, very cullet rich batches or pelletized batch will reduce the batch blanket coverage on top of the glass melt for a certain pull rate and enlarges the free non-covered glass melt surfaces. The heating of the batch takes place from 1). above by the flames/crown; 2). at the batch blanket tip by the re-circulating melt coming from the hot spot, and 3). from underneath by the slowly moving melt flowing below the batch. This melt might be extra heated by electric (batch) boosting.

It should be recognized that the heating process of the batch blanket is probably the most essential process step that determines the size (and level of required re-circulation of hot glass melt flows) of the glass furnace. To be able to make more intensive melters with significantly smaller tank volumes as the currently known industrial glass furnace types, the way of heating the charged batch in the furnace is decisive.

DISSOLUTION OF SAND GRAINS

During batch heating, and batch melting (in batch blanket area) by the chemical reactions between different batch components, a large part of the material (including sand) will be transformed into a liquid (melt) phase. This melt phase will flow from the batch blanket area into the re-circulating melt. However, some batch components that do not show a sharp melting point, nor have a high reaction rate, may release their non-molten particles to the glass melt. The dissolution process of these particles in the molten silicate phase is rather slow and determined by the maximum solubility of batch components in the glass melt and the diffusion rate of the dissolving component in the melt. Both properties depend on glass composition and temperature. Dissolution of sand or alumina rich grains further depends on their grain sizes. It is important to avoid large particle sizes in these raw materials and wide grain size distributions. The finer parts would dissolve rather quickly and it becomes increasingly difficult for the larger grains to dissolve in a melt that reaches saturation when most fine material (e.g. fine sand / SiO_2) is already dissolved. During sand dissolution, the local glass melt just surrounding the grains will experience an increased acidity level (SiO_2 rich melts are rather acid) and the solubility of CO_2 in this melt becomes low: dissolved CO_2 (originally coming from the carbonates) will form small bubbles at the interface of the sand grains. It is very important that all batch grains are completely dissolved before the fining (bubble removal processes) start. There should be no formation of new fine bubbles (seeds) during the fining process and especially not at the end of the fining process. Small seeds in the melt at the end of fining will give glass quality problems. Such seeds can not be effectively removed anymore.

Thus, sand dissolution should be finished before the fining process starts, above the so-called fining onset temperature. Typically reactions and dissolution of sand or alumina rich grains takes place in the batch blanket or in the melt underneath the batch blanket and before reaching the hot spot zone. It is recommended for soda-lime-silica glass production to avoid temperatures >1350°C in this zone (below fining onset) and to select raw materials with the proper grain size distribution (preferably < 0.20 mm sizes). The dissolution process will be supported by velocity gradients in the melt, stirring or bubbling of the melt, or strong free convection will aid the dissolution processes for sand grains, feldspars or clay particles.

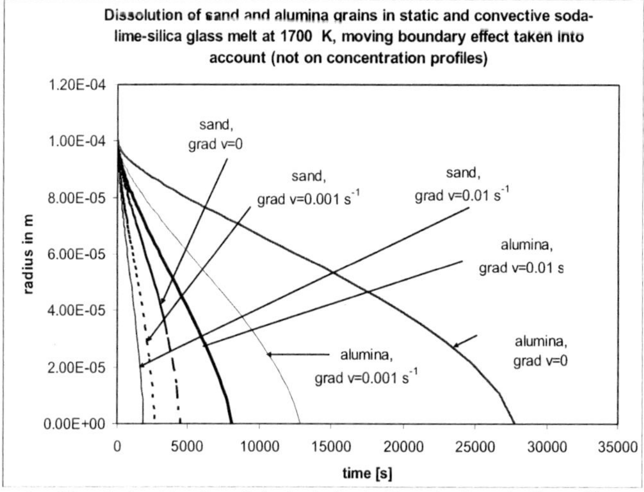

Figure 6a. Dissolution kinetics of single (spherical) grains in soda-lime-silica melts, at 1427 °C (1700 K) depending on velocity gradients in melt according to grain dissolution modeling in viscous melt. Initial diameter 0.2 mm (R = 0.1 mm)

Thus, the dissolution rate depends on:

- Composition of the grains
- Size of the grain
- Convection flow in melt
- Temperature: solubility & diffusion
- Glass composition.

In figures 6a and 6b, the grain dissolution in soda-lime-silica melts is presented for sand grains and different levels of convection (the parameter grad v gives the average velocity gradient of the melt and is a measure for convection/stirring intensity). The residence time in the melting-in and dissolution zone should be sufficient to allow all grains to be dissolved before the melt enters the fining zone. This is extremely difficult in conventional melting tanks, without physical separation between the sections (zones) and no control of temperatures and flow conditions per zone.

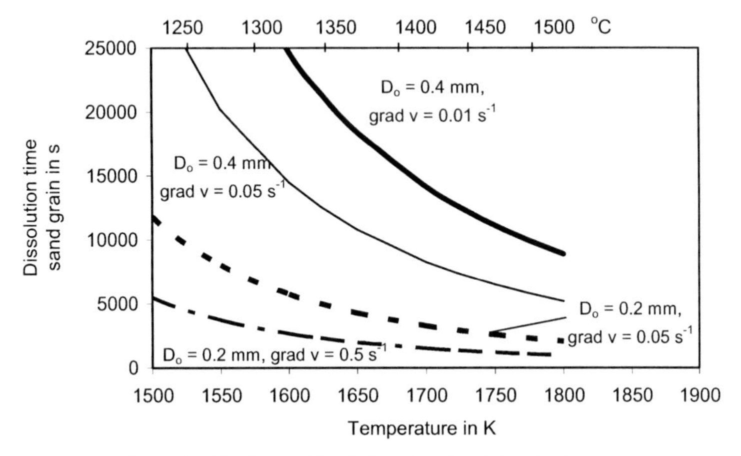

Figure 6b. Maximum dissolution time for 0.2 and 0.4 mm (D_o = initial diameter) sand grains depending on glass melt temperature

FINING PROCESSES

There are two mechanisms for removal of gas bubbles from molten glass. Therefore two stages are generally exploited, the primary fining and secondary fining process. There should be no overlap of these processes and secondary fining should be taking place downstream of the primary fining step. In the next sections both mechanisms will be described and the kinetics of primary fining will be demonstrated. Primary and secondary fining processes depend on temperature, oxidation state of the melt, contents of dissolved gases and the concentrations of added fining agent to the melt. The behavior of gas bubbles in such melts can be observed by laboratory tests.[8] With these tests, bubble growth rates by exchange of dissolved gases from the melt and release of fining gases, bubbles ascension in the melt and fining on-set temperature for certain glass melt composition can be determined.

PRIMARY FINING

The number density of fine bubbles (typically diameters in the range of 0.05 – 0.4 mm) in the freshly obtained glass melt from melting batch is in the order of 10^5 per kg.[9] In most glass products seed counts should be very low, depending on the type of glass product. Often, the occurrence of bubbles with sizes (D = diameter) above 0.2-0.3 mm in the glass, will lead to reject or lower quality applications (e.g. in the flat glass sector). Therefore, the removal of gas bubbles needs to be very effective. The primary fining process is based on the ascension of the bubbles in the viscous melt up to the glass melt surface. At the surface these bubbles form a foam layer (which is generally not desired) or they collapse. According to Stokes' law (valid for small bubbles), the bubble/seed ascension in a static melt is proportional to the square of the bubble diameter (D^2) and inversely proportional to the glass melt viscosity. Figure 7 shows the bubble removal time as a function of glass melt temperature and bubble size (D) for a float glass soda-lime-silica (SLS) melt, assuming bubbles initially at the bottom of a 1 meter deep tank. From this figure it is obvious that even at relatively high temperatures, bubbles smaller than D = 0.2 mm (200 μm) need a time in excess of the minimum residence times (3-8

hours), but even exceeding average residence times (24-60 hours). Thus, such bubbles will not be completely removed without changing these conditions.
Fining agents are added to the batch with the following characteristics:

- The fining agent will be dissolved in the glass melt, with low losses during melting-in.
- The fining agent is compatible with the glass properties, it does not negatively affect the glass.
- The fining agent is decomposing or evaporating at a temperature level above the so-called fining-onset temperature: the fining agent releases fining gases. The fining gases partly diffuse into the fine bubbles present in the melt.
- The oxidation state of the melt may effect the fining onset temperature (e.g. for sodium sulfate or antimony fining) and should be matched in order to obtain sufficient fining gas release at the optimum temperature level. The oxidation state of the glass can be influenced and controlled by the batch composition and should be tuned to obtain the required glass properties, transmission and glass color.

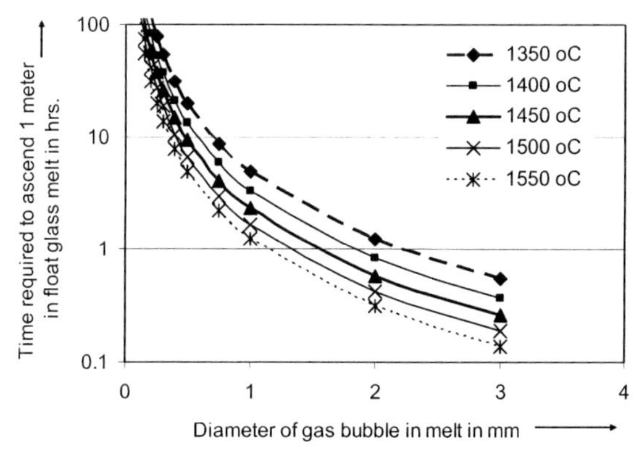

Figure 7. Ascension of bubbles, depending on temperature and bubble size (diameter D), in float glass (soda-lime-silica) melt, assuming Stokes' law

Fining agents can be sulfates (e.g. sodium sulfate), sodium chloride, arsenic oxides, antimony oxides or tin oxide. They deliver fining gases: SO_2 ($+O_2$), NaCl vapor or only oxygen. The fining gas evolution, starting at certain temperature level, with suitable glass melt viscosity, leads to bubble growth. Fining gas diffusion from the melt into a small bubble will always lead to bubble growth, since bubble internal pressure will not increase. To accommodate the increased gas species (moles of gas) the bubbles consequently have to grow. The figure 7 shows that larger bubbles will more easily ascent in the melt. Typically viscosity levels for a glass melt during fining are 5-20 Pas (at temperature levels of ±1500 – 1350°C in float glass melts). Bubbles in the bottom area of the tank and smaller than 0.5 mm diameter will not be removed within 10 hours from the glass melt. Thus, it is important to obtain bubble growth or to reduce the glass tank depth (fining shelf). Furthermore, the flow patterns in the melting tank should ensure that the glass melt flowing along the tank bottom underneath the batch blanket will experience upward forces in the hot spot area of the tank. This upward force in the spring zone, see figure 4 (e.g. by bubbling, buoyancy of extra heating of the melt by barrier boosting) additionally brings melt with bubbles to higher temperature sections and to the glass met surface apart from the Stokes flow driven ascension.[9,10]

Fining is effective when all seeds are growing to sizes above a diameter of about 1 mm. Figure 8 shows the experimentally derived fining onset temperature for a sulfate (re)fined float glass melt, depending on the level of sulfate in the melt.

GAS STRIPPING FROM GLASS MELT

During bubble growth by diffusion of fining gases (e.g. SO_2, O_2) from the molten glass into the seeds, the original gases in the bubbles will be diluted. For instance, initially a seed may contain almost pure CO_2 gas. The partial pressure of CO_2 is about 1 bar (or even more, dependent on the hydrostatic pressure). But, during the fining process, the CO_2 partial pressure in this seed/bubble will decrease, due to the dilution by fining gases. This will disturb the equilibrium between the partial pressure of CO_2 in the bubble and the partial pressure in equilibrium with the dissolved CO_2 in the melt. The disturbed equilibrium for CO_2 will result in diffusion of CO_2 from the melt into the growing bubbles. Thus, CO_2 is stripped from the melt and the content of dissolved CO_2 in the melt will decrease. This also takes place for other gases dissolved in the molten glass: H_2O, N_2, Ar etc. Thus, during primary fining, not only bubble growth and accelerated bubble ascension, but also gas out-stripping from the molten glass takes place. After an efficient primary fining process, the molten glass is lean in dissolved gases. Primary fining is one of the most critical steps in industrial melting of high quality glasses. All glass melt flow patterns in the melting tank should reach the fining onset temperature and the duration time of the melt above this temperature should be sufficient to achieve bubble growth above D = 1 mm and to bring all bubbles to the surface of the melt. It is extremely important to design a furnace and to optimize process settings to achieve optimum flow patterns in the melting tank, such that all the melt is exposed to temperatures above the fining onset temperature level.

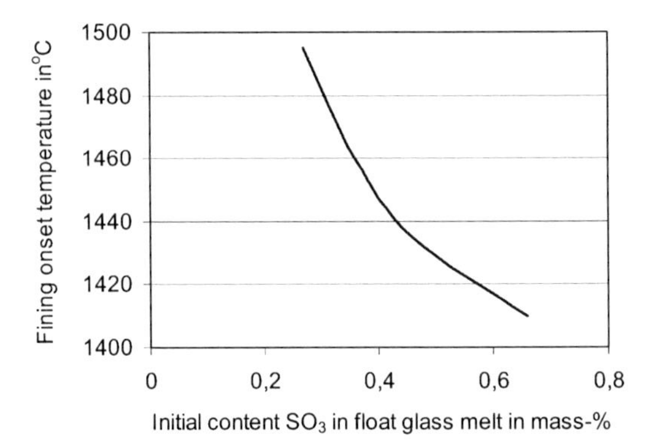

Figure 8. Experimentally determined fining onset temperature for typical float glass melt in air atmosphere, dependent on sulfate content in the fresh melt (no cokes added to batch).

Figure 9 shows an example of the temperature-time relation of a glass melt volume from charging end (doghouse) to the exit of the tank for the most critical flow pattern (lowest temperature course). The fining onset temperature depends on glass composition, water content in the melt (generally fining onset temperatures decrease for melts with increased water contents), fining agent content (see figure 8) and redox state of batch or addition of reducing agents (e.g. cokes) to the batch. For glass melting tanks with lower temperature levels, the addition of reducing agents in combination with sulfates in the batch will lower the fining onset temperature[10], but bubble growth takes place at the higher viscosity levels (20-25 Pas). For strongly oxidized molten soda-lime-silica glass and sulfate fining, the onset temperature may exceed 1440^{0}C (< 10-15 Pas). Thus, for these furnaces it is very important to achieve temperature levels above this value and all glass melt flow patterns need to pass through the zone with T > fining onset temperature. This zone is generally close to the surface of the melt. The glass melt should be typically at least 30-60 minutes above the fining onset temperature to enable sufficient fining.

Important process conditions and design features for optimum primary fining:

- All glass melt flow patterns should reach the fining onset temperature, which depends on glass melt composition and content of fining agent and oxidation state. Use of CFD (Computational Fluid Dynamics) modeling (simulation of glass melting process and calculation of heat transfer, flows and temperatures) to identify the critical glass melt flow patterns with lowest temperatures is important to design a high performance tank furnace and to choose optimum operational furnace conditions.
- Avoid downward directed mixing of the melt in the fining zone (improve space utilization[11]), because this would hinder the steady ascension of bubbles towards the glass melt surface: the flow in the primary fining zone should show no strong vertical mixing tendency.
- Optimize the content of fining agent and oxidation state of the batch to control the fining onset temperature, but avoiding excessive gas formation leading to foaming. Glass melt sulfur and redox sensors[12,13] can be used, especially important when applying post-consumer cullet with varying oxidation state and colors.
- Apply well-designed barrier boost systems,[14] acting over the whole tank width. The glass melt flowing from the charging area (doghouse) to the throat or waist should experience, over the whole width of the tank, upward buoyancy forces by the heat supplied by the barrier boost system (generally rows of vertical electrodes).
- Avoid too strong forced bubbling or bubbling pipes too far apart from each other, to avoid channeling of the glass melt along the bottom, between the bubblers.
- Use shallow tank sections for the primary fining zone[15,16] to reduce the height that bubbles have to overcome, but also to increase the local tank bottom temperature where small seeds may not experience bubble growth if the tank depth is too large.
- Apply a separated fining section, without back-flow of melt into the melting-in-zone: however this would require major modifications of heat transfer modes into the batch blanket.
- Design the space and dimensions of the fining section to ensure sufficient residence time of the all the melt flow patterns (figure 9), above the fining onset temperature (typically 30-60 minutes).
- Use very high quality refractory materials for the melting tank downstream the fining section or spring zone. Blistering from refractory surfaces, downstream the fining process, will lead to bubble defects that cannot be digested anymore by the cooling melt.

For a melting process for approximately 300 tons molten glass per day, the volume of a fining section (without strong mixing there!), should be about 7.5 -12 m^3 assuming temperatures above the fining onset level: for a fining shelf of 0.4 meter height, for instance a width of this shelf of 7 or 8 meter and length of about 2-2.5 meters.

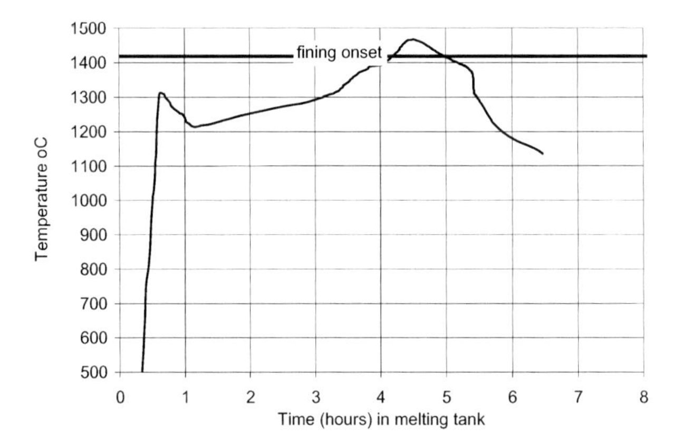

Figure 9. Example of the temperature – time curve of the most critical glass melt flow pattern in an industrial float glass melting tank according CFD modeling.[5] The melt is about 45 minutes above the fining onset temperature in this case.

SECONDARY FINING

After primary fining, the glass melt flows are slowly cooling in the downstream part of the melting tank, waist, throat and refiner or working ends. Small bubbles containing gases that chemically dissolve in the melt may shrink due to a diffusion of gas species from the bubbles into the melt. Bubbles with gases that show increasing solubility at decreasing temperature (e.g. oxygen, sulfur oxides, CO_2) may even be completely re-absorbed. Especially in glass melts with very low concentrations of dissolved gas components after primary fining (by stripping of gases from the melt at high temperatures during bubble growth and removal), this stage of the fining process is effective in removing residual small bubbles during controlled cooling. It will have no or hardly any effect on residual bubbles, that contain gases with very low solubility, such as nitrogen (N_2) rich bubbles.

However, laboratory studies at TNO (not published) showed that this secondary fining process generally takes place in a very narrow temperature range. Above this temperature, gas solubility levels in the melt are too low and at temperatures below this range, the diffusion process of gases into the (more viscous) melt will strongly slow down. It is very important to control the cooling of the melt after primary fining and to achieve sufficient time for this process step at the optimum secondary fining temperature range, for SLS glass often around 1300-1350°C. The glass melt oxidation state and presence of polyvalent ions will have an effect on the re-absorption of SO_2-rich bubbles.

HOMOGENIZATION

During melting and fining, convection flows: glass melt velocity gradients, will stretch the cords in the melt and the thickness of the inclusions will decrease. The chemical homogenization by the combination of stretching and diffusion (diffusion to eliminate concentration differences, taking place over a thinner cord) is necessary to obtain a glass product with uniform composition and homogeneous optical and mechanical properties and homogeneous thermal expansion.
In the melting tank, bubbling can aid the homogenization process, but too strong bubbling may negatively affect the primary fining efficiency. During stirring or bubbling lapping in of bubbles should be avoided. Stirring can be applied in the waist or in forehearths/feeders of glass furnaces.

However, stirring is only effective when special designed stirrer spaces are constructed. All glass melt flowing through the stirring chamber should be homogenized to eliminate cords coming from refractory corrosion, surface glass (depleted with volatile glass melt species), caused by non-homogeneous batch or pollution of batch (e.g. by glass ceramics in cullet).

POSSIBLE FURNACE CONCEPTS

There are two main approaches to innovate glass melting processes:

1. The evolution approach: based on existing continuous glass melt tank furnaces.

Changes include the modifications in combustion chamber design to increase the residence times of the combustion gases, to improve the surface of flames covering the glass melt surface area, and to decrease mixing rate between oxygen or air and injected fuel jets. These changes are mainly important to minimize formation of so-called thermal NOx and are considered as effective primary measures in combination with a). the application of modern burner designs with adjustable orifices and burner angles and b). the use of advanced control systems for air / oxygen excess based on the on-line measurement of oxygen in the exhausting combustion gases. The next improvement is the application of in-situ and on-line monitoring systems for CO in the exhaust, entering the regenerators.[17] Concerning the melting tank, changes in boosting systems[14] can be very effective to optimize energy efficiency and glass quality, e.g. using dedicated electrode configurations for barrier boosting. The application of a shallow fining shelf is successfully applied in the LoNOx® melter of SORG.[15,16] Important for glass quality is the selection and use of high quality refractory especially for sections of the tank downstream the fining (spring) zone.

A next step in optimum operation of glass furnaces is the use of model-based predictive control. With modern furnace control systems,[17,18] temperatures and glass melt flows can be accurately controlled in order to avoid excessive temperatures (which lead to increased energy consumption) and to achieve the optimum temperature ranges in the different areas of the melting tank, dedicated for sand grain dissolution (below fining onset), primary fining (at least 30-60 minutes above the fining onset temperature), secondary fining and conditioning. The combination of these evolutionary developed furnaces based on the conventional single melting tank approaches with batch & cullet preheating can give low NOx emissions, high energy efficiency (up to 65% of the energy for heating and melting the batch in case of container glass production and about 50% for float glass) and thus reduced CO_2 release.

2. The revolution approach

Segmented melter designs have been reported since the last few decades, also rapid melter devices have been patented. Barton[19] presented a comprehensive overview, already about 20 years ago. However, these types of glass melting systems have hardly been applied in glass industries even up to now. The advantages of compact or rapid melters compared to the voluminous conventional melting tanks with long average residence times and hold-up, are considered to be not sufficient to compensate for the increased complexity of these innovative designs, unknown risks, more expensive refractory materials or expected shorter furnace lifetimes. The most essential problem to overcome is the intensive direct supply of energy/heat to the batch blanket without the necessity of a strong return flow bringing hot molten glass to the batch blanket tip. Apparently within a very short time (less than 5 % of the average residence time), the batch materials have to receive more than 85 % of the theoretical heat demand for melting-in in the first step of glass melting in conventional melting processes. Intensifying heat transfer to the batch, without strong recirculation flows from the fining sections is essential for compact melters. Submerged combustion melter for first stage: Intensive heat transfer can be achieved by the so called submerged combustion melter. This may be the base for the first step in the glass melting process with average residence times < 1-2 hours. Main problems here are the high heat losses through water cooled walls and the risk of segregation and batch carry-over. Downstream the submerged combustion melter, complete dissolution of sand grains should take place, for soda-lime-silica glass, taking typically about 1-2 hours (depending on convection flows and sand grain size distribution) for temperatures around 1300-1350°C. The next stage, with probably a shallow fining shelf for primary fining, needs 1-2 hours as well and secondary fining an additional 1-2 hours. The total average melting process would need in that case about 6 to 8 hours. This is at least a factor 3 less than for conventional tank furnaces.

HIGH PULL OXYGEN-FIRED FURNACES

Since the 1990's all oxygen firing is successfully applied for all types of glasses. However, the furnace designs have never been modified and optimized for the potentially increased heat transfer intensity from oxygen-combustion compared to air firing. In almost all cases the conventional unit type melter is used as basic design. The only difference is that instead of many small fuel (or natural gas) – preheated air burners, a few oxygen-gas or oxygen–oil burners are placed in burner blocks mounted in the combustion chamber sidewalls/breast walls. One development is the CGM technology (top burners) used to intensify the heating of batch.[20] This technology is considered as providing a high energy flux to speed up melting or enable melting process for batches with a low level of fluxing agents. An increased coverage of the batch blanket and spring zone area of the tank by wide radiant flames could improve overall heat transfer and allows the operation of melting tanks with lower return flow intensities. In case of high convection flames, however, carry-over may be strongly enhanced, which will increase particulate emissions and fouling of the flue gas system. Therefore it is of utmost importance to optimize combustion systems without increasing batch losses or evaporation losses by high combustion gas velocities just above the molten glass or batch.

CONCLUSIONS

It can be concluded that the future glass furnace concepts that answer to the needs of energy efficiency, flexibility, low melting volume and low emissions, especially NOx and CO_2 will be based on:

- Improved control of temperatures and flows of the melt using advanced control systems based on sophisticated simulation models (e.g. CFD models).
- Dedicated sections, if possible separated from each other, for batch heating & melting-in of batch, sand grain dissolution, primary fining, secondary fining and finally temperature homogeneity conditioning, see the scheme of figure 10.
- New combustion technologies for regenerative and oxygen-fired glass furnaces, improving heat flux distribution and flame shaping.
- Intensification of direct heating of the batch blanket.
- Use of improved refractory materials downstream the fining (spring) zone.
- Application of dedicated primary fining sections, using one of the following options:
    - Fining shelf, with typically glass level of 0.3 – 0.5 m.
    - Barrier electric boosting system over complete width of the tank.
    - Physical fining aids in case of segmented melters with specific fining section (e.g. helium bubbling[21] or centrifugal fining).[22]
    - Thin film melting and fining (but this method requires relatively large areas).

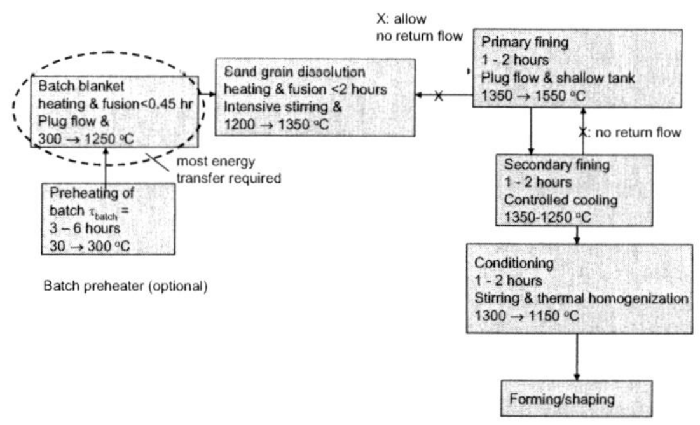

Figure 10. Scheme of required process steps & conditions for high quality glass melting, minimizing residence times (example float glass)

Furthermore, the energy contents of flue gases comprising 25-40% of the energy, that is generally supplied to the fossil fuel fired glass furnace, is an important energy loss when not recovered. Currently batch preheating systems are in development and are tested to allow preheating of raw materials with less than 50% cullet.[23] A more sophisticated technology is the pelletizing of the normal raw material fraction of the batch. The pelletizing process requires water or aqueous solution for binding of the powder raw materials into pellets (agglomeration of all raw materials in 3-15 mm sized granules). For the drying process of pellets and preheating (200-400°C), flue gases can be used. The use of preheated pellets will lead to improved batch homogeneity (no segregation of pelletized batch) and significant increases in energy efficiency and melting capacity: up to 15%. CFD modeling tools[5] support the design of more complex glass furnace concepts and will help to understand the flow patterns and glass melting performance of new melting and heating methods depending on the process settings. In the next decade: 2010-2020, these models are the basis for creating a next generation glass melter, probably based on dedicated zones for the different process steps to achieve a more compact glass melt system with lower CO_2 footprint and lower NOx emissions and more stable melting conditions.

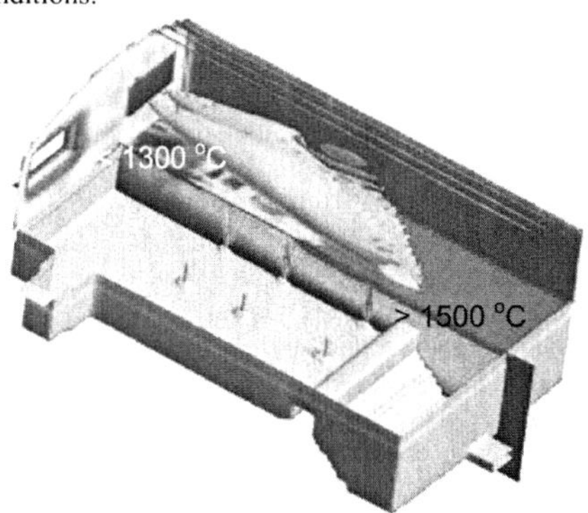

Figure 11. Example of flame contours & temperature fields determined for glass melting furnace according CFD simulation modeling: Furnace with vertical electrodes, a weir and deep refiner.

REFERENCES

[1]Beerkens, R.G.C.; Limpt, van H.A.C.; Jacobs, G.: Energy efficiency benchmarking of glass furnaces. *Glass Sci. Technol.* **77** (2004) no. 2, pp. 47-57.
[2]Glass BREF Reference Document on Best Available Techniques in the Glass Manufacturing Industry, October 2010, *European IPPC Bureau*, Sevilla.
[3]Beerkens, R.; Van Limpt, H.: Reduction of NOx emissions in regenerative fossil fuel fired glass furnaces: a review of literature and experimental data. *Glass Technology: Eur. J. Glass Sci. Technol. A*, December 2008, **49** (6), pp. 279-288.
[4]Rue, D.; Kunc, W.; Aronchik, G.: Operation of a Pilot-Scale Submerged Combustion Melter. *Proceedings 68th Conference on Glass Problems*. Ed. Charles Drummond, III. 16.-17. October 2007, Columbus, OH. pp. 125-135.
[5]Beerkens, R.G.C.: Modeling of the Melting Process in Industrial Glass Furnaces in *Schott Series on Glass and Glass Ceramics, Science, Technology & Applications*, editors: D. Krause & H. Loch. ISBN 3-540-43204-3, Springer (2002) Chapter 2. Melting and Fining. pp. 17-73.
[6]Daniels, M.: Einschmelzverhalten von Glasgemengen. *Glastech. Ber.* **46** (1973), no. 3, pp. 40-46.
[7]Costa, P.: Untersuchung des Einschmelzverhaltens von pelletiertem Gemenge zur Glasherstellung. *Glastech. Ber.* **50** (1977), no. 1, pp. 10-18.
[8]Rongen, M; Beerkens, R., Faber, A.J.: Observation of chemical and physical fining processes in glass melts. *81. Glastechnische Tagung Deutsche Glastechnische Gesellschaft*, Aachen 4.-6. June 2007.
[9]Mulfinger, H.O.: Gasanalytische Verfolgung des Läutervorganges im Tiegel und in der Schmelzwanne. *Glastech. Ber.* **49** (1976), no. 10, pp. 232-245.
[10]Collignon, J.; M. Rongen, M.; Beerkens, R: Gas release during melting and fining of sulphur containing glasses. *Glass Technol.: Eur. J. Glass Sci. Technol. A*, (2010), **51**, no. 3, pp.123-129.
[11]Němec, L.; Cincibusová, P.: Glass melting and its innovation potentials: the role of glass flow in the bubble-removal process. *Ceramics – Silikáty* (2008) **52**, no. 4. pp. 240-249.
[12]Laimböck, P.R.; Beerkens, R.G.C.: Oxygen Sensor for Float Production Lines. *The Glass Researcher, in Am. Ceram. Soc. Bull.* 85, no. 5 (2006) pp. 33-36.
[13]Bauer, J.: Measuring the Sulfur Content of Industrial Glass Melts Using Square- Wave Voltammetry. *Proceedings 61st Conference on Glass Problems*, October 17-18, 2000, OSU Columbus OH, Editor Charles H. Drummond III, Am. Ceram. Soc. (2001) pp. 205-219.
[14]Stormont, R.: Electric melting technologies for energy efficiency and environmental protection. *Glass Worldwide* (2008) no.18.
[15]Ehrig, R.; Wiegand, J.; Neubauer, E.: Five years of operational experience with the SORG LONOx® Melter. *Glass Sci. Technol. (Glastech. Ber.)* **68** (1995), no. 2, pp. 73-78.
[16]Nebel, R.: Application of Fining Shelf to Furnace Melting Technology. *Proceedings 61st Conference on Glass Problems*, October 17-18, 2000, OSU Columbus OH, Editor Charles H. Drummond III, Am. Ceram. Soc. (2001) pp. 21-26.
[17]Dang, D.D.; Bjorøy, O.; Kaspersen, P.; Measurement of CO and O_2 in Gases at High Temperatures. *Proceedings of GlassTrend seminar: Advanced Sensors & Control in High Temperature Processes*, 4-6 October 2010, Maastricht, The Netherlands.
[18]Huisman, L.; Reijers, J.; Brugman, R.: Advanced and Supervisory Process Control of Glass Furnaces. *Glass Machinery Plants & Accessories* (2009) no.6. pp. 72-75.
[19]Barton, J. L.: Innovation in glass melting. *Glass Technol.* (1993) **34**, no. 5 pp. 170-177.
[20]Simpson, N.; Marshall, D.; Barrow, T.: Glass Furnace Life Extension Using Convective Glass Melting. *Proceedings 64th Conference on Glass Problems,October 28-29*, 2003 University Urbana-Champaign. Editor Waltraud M. Kriven, Am. Ceram Soc. (2004) ISSN 0196-6219, pp.129-140
[21]Beerkens, R.G.C.: Analysis of advanced and fast fining processes for glass melts, *In Proceedings of the Advances in Fusion and Processing of Glass III*, July 27-30, 2003 Rochester NY USA, The American Ceramic Society, Westerville. *Ceramic Transactions* **141** (2004) pp. 3-34.

[22]Němec, L.; Tonarová, V.: Glass melting and its innovation potentials: bubble removal under the effect of the centrifugal force. *Ceramics-Silikáty* (2008), 52 no. 4. pp. 225-239.
[23]Beerkens, R.G.C.: Energy Saving Options for Glass Furnaces and Recovery of Heat from Their Flue Gases and Experiences with Batch and Cullet Pre-Heaters Applied in the Glass Industry. *Proceedings 69th Conference on Glass Problems*, November 4 – 5, 2008, editor Charles H. Drummond III Am. Ceram. Soc., (2009), pp. 143 -162.

*Corresponding author: ruud.beerkens@tno.nl

THERMAL VERSUS CHEMICAL CONSTRAINTS FOR THE EFFICIENCY OF INDUSTRIAL GLASS MELTING FURNACES

Reinhard Conradt
RWTH Aachen University
Aachen, Germany

ABSTRACT

From a formal point of view, a glass furnace is a heat exchanger comprising a hot stream passing through the combustion space and a cold stream passing through the tank. While the hot stream may be treated as a one-phase flow, the cold stream also involves phase transformations and latent heat storage. Such a set-up is subject to several constraints: a combined area-volume constraint, and a time constraint for both heat transfer and phase transformations. This rather abstract concept is most helpful in assessing the efficiency limits of glass melting. From a practical point of view, it is essential to know if the efficiency of an individual furnace is limited by heat transfer, or rather by the batch-to-melt conversion rate. Consequently, measures aiming at enhancing the operation efficiency must match the individual situation. This principle is demonstrated for campaigns with industrial furnaces. In many cases, the use of low-enthalpy batches and fast conversion batches is an efficient way to improve glass melting.

INTRODUCTION

In view of increasingly stringent rules for CO_2 emission, and of rocketing energy costs, the reduction of the energy demand of glass melting remains one of the key challenges for the entire glass industry. Countless evolutionary steps of optimization led to a situation where the performance of traditional furnaces seems to approach its physical limits. This is clearly reflected by benchmark surveys performed during the past years.[1] While a number of technologists work on truly revolutionary concepts of glass melting, the main efforts during the coming decade will most certainly focus on exhausting the limits of the conventional process. It is chiefly the long lifetime of traditional furnaces which lends support to this expectation. In spite of the considerable progress made in the FEM modeling of furnace behavior, it is still difficult to identify (a) which exactly are the physical limits of a traditional furnace and – even more important – (b) which processes constitute these limits. The performance of a furnace may be constrained by thermal processes, i.e., by processes related to heat transfer, or by the heat demand and the kinetics of chemical turnover, i.e., by the intrinsic rates of batch melting, quartz dissolution, and bubble release. In the following text, these processes shall be summarized by two categories termed thermal constraints and chemical constraints, respectively. It is true, the performance of a real furnace will reflect a blend of these influences. But we may expect that there is always one principal constraint governing the overall performance, and this may differ from furnace to furnace. A successful optimization campaign should therefore aim at identifying and targeting the principal constraint of an operation.

In the first section of the present paper, the nature of the above constraints will be pointed out in terms of a zero-dimensional analysis of what may be termed an "ideal furnace."[2] In the second section, a concept is presented by which the performance of real furnaces may be evaluated by a retrospective analysis of process data recorded in daily production routine. Finally, it is outlined how the use of fast conversion batches may help to push the chemical constraints far beyond the present limits.

CONCEPT OF AN IDEAL FURNACE

This section deals with an idealized concept of an air-gas fired continuous furnace with a Siemens type regenerator in comparison to an oxy-fuel furnace. In order to clearly elaborate some fundamental aspects, a few simplifications are adopted: Electrical boosting is not taken into consideration, the volume flow of batch gases entering the combustion space is neglected, and the wall losses are put to zero. These details may be easily added to the picture at a later stage. The furnace itself serves a dual function: It is a heat exchanger attached to a chemical reactor. Let us first look at the function of the heat exchanger. It transfers heat from a hot stream H passing through the combustion space to a cold stream L passing through the tank (see figure 1). The hot stream may be considered as a fully reacted offgas entering at adiabatic flame temperature T_{ad} and leaving at offgas temperature T_{off}. It carries a heat capacity flow $m'_H \cdot c_H$ given in units of kW/K; m′ denotes a mass flow and c a mass related heat capacity. The cold stream enters at ambient temperature $T_0 = 25$ °C and leaves at pull temperature T_{ex}; its heat capacity flow is $m'_L \cdot c_L$. Any temperature differences ΔT refer to T_0, thus $\Delta T = T - T_0$. The ambient pressure is kept at $P_0 = 1$ bar. Yielding to the habits of combustion engineers, any gas volumes are referred to a reference state of 0 °C, 1 atm; this does, however, not interfere with the choices made previously. The heat exchange area is A, and the effective heat exchange coefficient is α_{ht}. The power P_{sf} unleashed to the combustion space is given by $P_{sf} = m'_H \cdot c_H \cdot \Delta T_{ad}$, the power transferred to the tank is $P_{ht} = m_H \cdot c_H \cdot (T_{ad} - T_{off})$. The efficiency of heat transfer $\eta_{ht} = P_{ht}/P_{sf}$ is thus given by the power drawn from the furnace is $P_{ex} = m'_L \cdot c_L \cdot \Delta T_{ex}$. In the simplified set-up of figure 1, it is identical with P_{ht}.

$$\eta_{ht} = 1 - \Delta T_{off}/\Delta T_{ad} \tag{1}$$

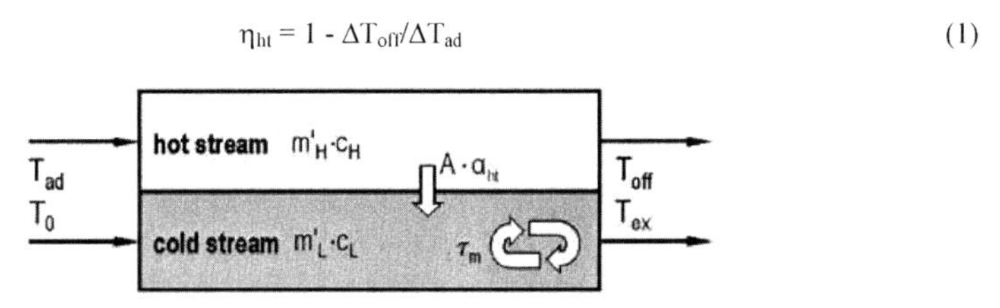

Figure 1. Glass furnace presented in its function as heat exchanger with contact area A between hot stream (H) and cold stream (L); $T_0 = 25$ °C; T_{ad} = adiabatic flame temperature; T_{off} = offgas temperature; T_{ex} = pull temperature; α_{ht} = heat transfer coefficient; τ_m = mean dwell time; $m'_H \cdot c_H$ and $m'_L \cdot c_L$ are the heat capacity flows of H and L, respectively.

A 1st law presentation of the furnace in its function as a heat exchanger is given in figure 2. This time, the regenerator is also taken into consideration, and the wall losses are no longer neglected. Figure 2 shows that the power P_{sf} set free in the combustion space is composed by the power P_{in} delivered by the input of fuel and by the power P_{re} recovered from the hot offgas. The power P_{ex} drawn from the furnace is given by the pull p in t/h and the exploited heat H_{ex} in kWh/t

$$P_{ex} = H_{ex} \cdot p \tag{2}$$

where the unit of 1 t refers to the amount of workable glass melt. As explained previously in detail,[3] H_{ex} is given by

$$H_{ex} = (1 - y_C) \cdot \Delta H^\circ_{chem} + \Delta H(T_{ex}) \tag{3}$$

where y_C is the cullet fraction in the batch in t of cullet per t of glass, ΔH°_{chem} is the standard heat of formation of glass and batch gases from the batch (25°C, 1 bar), and $\Delta H(T_{ex})$ is the heat physically stored in the workable melt at pull temperature T_{ex} (relative to T_0). The heat physically stored in the batch gases is not considered to be a part of H_{ex}, but of the offgas. With H_{ex} known for a real batch[3], the heat capacity flow of the cold stream may now be quantified by

$$m'_L \cdot c_L = H_{ex} \cdot p / \Delta T_{ex}. \tag{4}$$

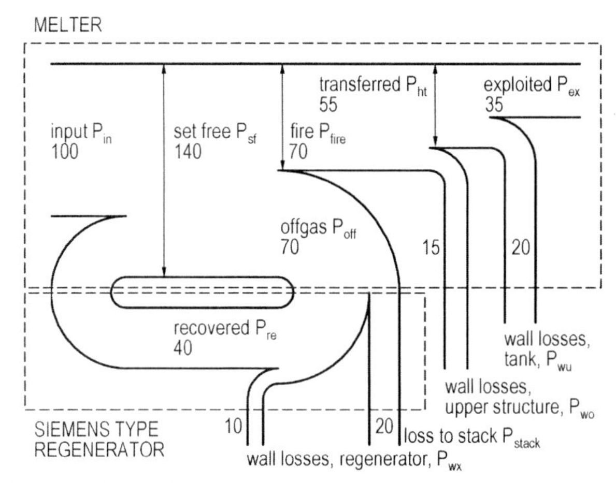

Figure 2. Heat balance of a glass furnace with Siemens type regenerator; P_{in} = power input; numbers denote the amounts of the other power terms (typical values) in % of P_{in}

Typical numerical examples for a soda lime glass batch containing 50% cullet are: H_{ex} = 550 kWH/t, ΔT_{ex} = 1325 K; thus $m'_L \cdot c_L/p$ = 0.42 kWh/(t·K). As for the hot stream, P_{in} is given by the volume flow of fuel V'_{fuel} in m^3/h and the net calorific value of the fuel H_{NCV} in kWh/m^3,

$$P_{in} = H_{in} \cdot p = V'_{fuel} \cdot H_{NCV} \tag{5}$$

Equation (5) allows one to relate the flow of fuel to the overall specific heat demand of glass melting H_{in} in kWh/t. For CH_4(g) as an idealization of a real mineral gas, H_{NCV} = 10 kWh/m^3. For the further analysis, we make use of the following near-stoichiometric approximations of the heat capacities of air, an air-fuel offgas, and an oxy-fuel offgas, c_{air}, c_{off}, $c_{off\text{-}oxy}$, (data calculated after tabulated thermodynamic data, see, e.g., ref.[4]), and the corresponding volume ratios relative to the fuel, n_{air}, n_{off}, $n_{off\text{-}oxy}$. The data are compiled in Table I. The auxiliary quantities z_j are useful to avoid lengthy formulae. The heat capacity flows of the hot stream are thus given by

$$m'_H \cdot c_H = P_{sf} \cdot z_{off} \tag{6}$$

for the air-fuel operation with regenerator. For an idealized regenerator, P_{sf} may be approximated by $1.5 \cdot P_{in}$; this cannot be shown in detail within the scope of this paper. For oxy-fuel operation without heat recovery, we obtain

$$m'_H \cdot c_H = P_{in} \cdot z_{off\text{-}oxy} \tag{7}$$

The typical constraints under which a heat exchanger works are[5]

- a potential imbalance of the heat capacity flows $m'_H \cdot c_H$ and $m'_L \cdot c_L$;
- a limited heat transfer coefficient α_{ht}
- a limited heat exchange area A
- a limited volume of the combustion space constraining the dwell time of the hot stream
- a potential pressure drop

Among these, the imbalance of the heat capacity flows seems to be the most severe constraint of traditional glass furnaces. In order to demonstrate this, let us vary H_{in} from a value as high as 2500 kWh/t to the (unphysical low) threshold of H_{ex} = 550 kWh/t and compare $m'_H \cdot c_H/p$ to $m'_L \cdot c_L/p$. The result is shown in figure 3. In fact, both air-fuel and oxy-fuel operation suffer from a gross imbalance of the heat capacity flows of hot and cold stream. Figure 3 also suggests that a mixed air-fuel oxy-fuel operation with heat regeneration is more than a temporary compromise to brush up the performance of a furnace. In fact, mixed combustion concepts are a valid approach to ease the flow imbalance problem. The traditional furnace with Siemens type regenerator suffers from another flow imbalance anyway. This is the imbalance between offgas and combustion air. Thus, the efficiency of heat recovery $\eta_{re} = P_{re}/P_{off}$ is limited to $\eta_{re} < z_{air}/z_{off} \approx 80$ % even in an infinitely large regenerator with infinitely fast heat exchange ($\alpha_{ht} \cdot A \rightarrow \infty$). To summarize: Flow imbalance seems to be the principal thermal constraint in conventional glass furnaces.

Table I. Properties of flows through the combustion space; c_j = heat capacity, n_j = volume ratio relative to the fuel; off = air-fuel offgas, off-oxy = oxy-fuel offgas; $z_j = c_j \cdot n_j / H_{NCV}$ with H_{NCV} = net calorific value of the fuel.

j =	c_j in $\frac{Wh}{m^3 \cdot K}$	n_j in $\frac{m^3}{m^3}$	z_j in $\frac{1}{10^3\,K}$
off	0.49	10.6	0.52
off-oxy	0.65	9.6	0.20
air	0.43	3.1	0.41

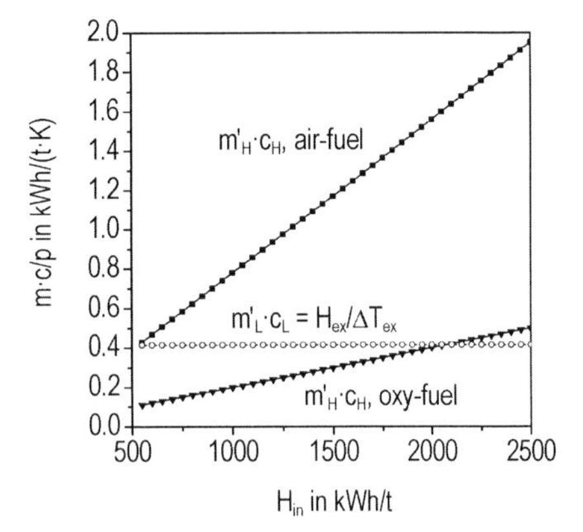

Figure 3. Heat capacity flows per pull as a function of heat input H_{in} required to melt 1 t glass

ANALYSIS OF INDUSTRIAL FURNACES

Daily production routine yields a host of most valuable data elucidating the performance of glass furnaces. Practically all glass producers collect and record data on

- the batch and glass composition
- the production rate
- the input of fuel and air
- temperatures at various positions of the furnace, etc.

Such a database allows to retrospectively construct heat balances of the type as shown in figure 2, and to trace them as a function of the production rate p in t/h. Figure 4 shows the results of such an analysis performed on a small oxy-fuel fired specialty glass furnace. As a general observation, the power input P_{in} linearly varies with the production rate. No theoretical justification is given within the scope of this paper. Anyway, the observation is very well substantiated on an empirical basis already. Here, it may be useful to add a short comment on the numerical meaning of the squared regression coefficient r^2. In fact, r^2 is an indicator of the probability W_N telling that an observed linear correlation is non-accidental. The mathematical expression for $W_N(r^2)$ is quite complicated (see, e.g.,[6]) and shall not be presented here. It is, however, worth realizing that, with an increasing number N of observables, r^2 becomes increasingly distinctive. The linear correlation $P_{in} \propto p$ shown

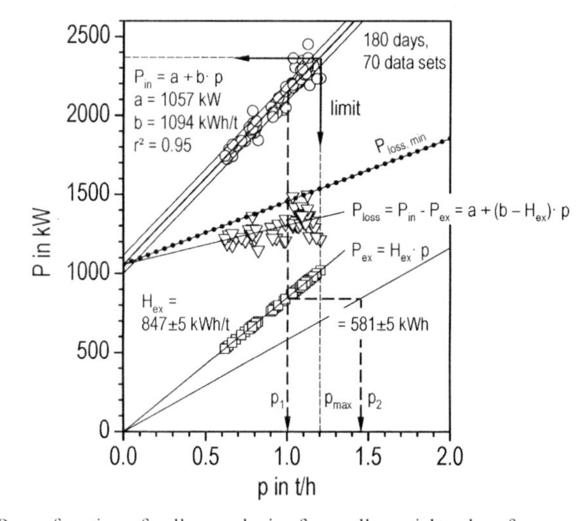

Figure 4. Power P as a function of pull p; analysis of a small specialty glass furnace; observed data and regression lines; P_{in} = power input (standard deviation is ±57 kW); P_{ex} = power drawn from the furnace; P_{loss} = cumulative losses; H_{ex} = exploited heat; $P_{loss,min}$ = threshold until which the increase of P_{loss} with p is governed by flow imbalance only.

in figure 4 is based on 70 observables. This corresponds to $W_N > 99.95$ %. In practical terms, a linear correlation is beyond any doubt. Thus, for the power input P_{in}, the relation holds.

$$P_{in} = a + b \cdot p \tag{8}$$

The power P_{ex} drawn from the furnace is given by

$$P_{ex} = H_{ex} \cdot p \; ; \tag{9}$$

For $p \rightarrow 0$, P_{ex} vanishes. The difference $P_{in} - P_{ex}$ summarizes the cumulative losses P_{loss} given by

$$P_{loss} = P_{in} - P_{ex} = a + (b - H_{ex}) \cdot p > 0. \tag{10}$$

In figure 4, the actual process data for P_{in}, P_{ex}, and P_{loss} are shown along with the calculated straight lines. At a pull of $p_{max} = 1.2$ t/h, the furnace obviously reaches its limit of performance as reflected by an increasing scatter of the data. At this point, $P_{in} = 2370$ kW and $H_{in} = 1975$ kWh/t are reached, and the furnace is able to sustain an exploited power of $P_{ex} = 1020$ kW. Batch A used during this operation is characterized by a very high chemical heat demand $\Delta H^{\circ}_{chem} = 257$ kWh/t yielding H_{ex} as high as 847 kWh/t. The evaluation allows one to predict the pull increase attainable when a low-enthalpy batch is used instead. The suggested alternative batch B uses calcined raw materials (of course leading to the very same glass composition) and has $\Delta H^{\circ}_{chem} = 56$ kWh/t only, yielding $H_{ex} = 581$ kWh/t. Let us depict $p_1 = 1$ t/h as a reference pull for batch A. Under these conditions, the furnace requires $P_{in} = 2150$ kW input and easily accommodates an exploited power $P_{ex} = 850$ kW. With batch B, the pull can now be safely raised to $p_2 = 1.45$ t/h. The only thing which needs to be clarified ahead is to show that the conversion rate of the alternative batch does not lag behind the original one. This can be checked by a simple batch-free time lab test. The industrial campaign following the above assessment verified the expectations.

If ΔH°_{chem} and hence H_{ex} are considered as variables, then the above furnace certainly operated under a chemical constraint. In order to explore the nature of the constraint under the boundary condition of a constant H_{ex} let us express P_{loss} in terms of the quantities shown in figure 2:

$$P_{loss} = P_{wu} + P_{wo} + P_{wx} + P_{stack}. \tag{11}$$

The wall loss terms are essentially independent of the production rate p and may be summarized by a cumulative wall loss P_w. Inserting (11) into (10) yields

$$P_{stack} = (a - P_w) + (b - H_{ex}) \cdot p \tag{12}$$

Consequently, the slope of the term P_{loss} in figure 4 is stipulated by the power P_{stack} carried away by the offgas. The slope of P_{stack} given by $b - H_{ex}$ may be taken as an indication of the nature of the principal constraint under which a furnace works. The conclusions are summarized in Table II. These conclusions must be read with caution. It is admitted that some of the potential constraints listed below eq. (7), such as the combustion space volume and the pressure drop, are not taken into consideration. Nevertheless, Table II allows one to distinguish between three different cases: In case I, the performance of the furnace is definitely constrained by the chemical reactions. Any measures aiming at improving heat transfer would yield disappointing results. However, case I is a case not frequently encountered in industry. In case III, by contrast, the performance of the furnace is definitely constrained by a poor heat transfer coefficient or a too small heat exchange area. Such a type of furnace should never be used to run a campaign aiming at testing the effect of different raw materials because, most likely, no difference at all would be observed. In order to distinguish between cases II and III, we have to estimate the threshold until the thermal performance of the furnace is governed by the flow imbalance alone. The information is contained in the coefficient b_{imb}. The procedure to determine b_{imb} is again demonstrated for the furnace analyzed in figure 4. Let us recall that the power carried by the hot and cold stream is given by eq. (4) and (7), respectively.

Table II. Evaluation of the nature of constraints as concluded from an analysis like the one shown in figure 4.

case	characteristic threshold of the slope of P_{stack}	meaning of the threshold; principal constraint involved
A	$b - H_{ex} = 0$	P_{stack} remains independent of the pull p; it does not increase with increasing power input P_{in}; the heat transferred to the tank just follows in proportion to the pull; no thermal constraint involved at all; the principal constraint is a chemical constraint *)
B	$0 < b - H_{ex} \leq b_{imb}$	P_{stack} increases when P_{in} increases; in other words, an increasing portion of P_{in} cannot be transferred to the tank; this is, however, due to the flow imbalance only; the heat transfer coefficient α_{ht} does not pose any constraint; with prevailing flow balance, the principal constraint is a chemical constraint
C	$b - H_{ex} > b_{imb}$	P_{stack} increases when P_{in} increases; like in case B, an increasing portion of P_{in} cannot be transferred to the tank; this is due to a constraint related to α_{ht}; the principal constraint is a thermal constraint

*) even cases with $b - H_{ex} < 0$ have been observed and documented

A heat exchanger not constrained by a finite heat transfer coefficient α_{ht} – this is the situation for case II – reaches a maximum of transferred power $P_{ht,\ max}$ (see, e.g.[7]). It is limited by the maximum temperature difference and the flow balance ratio $\xi = m'_H \cdot c_H / m'_L \cdot c_H$. In the case of a glass furnace, the maximum temperature difference is ΔT_{ad}. Among the different arrangements of a heat exchanger (co-, counter-, or cross-flow), $P_{ht,\ max}$ reaches a value of at least

$$P_{ht,\ max} = z_{off,\ oxy} \cdot \Delta T_{ad} \cdot (1 + \xi)^{-1} \cdot P_{in} \tag{13}$$

When P_{ht} assumes a maximum, the cumulative losses reach a minimum,

$$P_{loss,\ min} = P_{in} - P_{ht,\ max} \tag{14}$$

Let us adopt $\Delta T_{ad} = 4000$ K for the oxy-fuel operation; $z_{off,\ oxy}$ is taken from Table I. Then the slope of $P_{loss,\ min}$ in a P vs. p plot, and according to eq. (12), the slope of P_{stack}, is given by

$$\text{slope} = b - H_{ex} = (1 - z_{off,\ oxy} \cdot \Delta T_{ad} \cdot (1 + \xi)^{-1}) \cdot b \approx 0.32 \cdot b. \tag{15}$$

Resolving eq. (15) for b yields the threshold value $b_{imb} \approx 0.47 \cdot H_{ex}$. The dotted line in figure 4 marks this threshold. So, the particular furnace analyzed in figure 4 operates under conditions of case II. Apart from the above calculations, the slope criterion may always be used for a quick qualitative assessment of furnace operation. A small slope of P_{loss} always points to a case II situation. This situation is frequently encountered in industry. In a case II situation, however, it is always commendable to look into the options to ease the chemical constraints.

OPTIONS TO EASE THE CHEMICAL CONSTRAINTS

It is the conviction of the author; the batch-to-melt conversion is one of the yet unexploited options which may help to significantly push the efficiency of glass melting beyond its present limits. As shown in the previous section, this may not apply to all, but to a large number of existing glass furnaces. This assessment goes together well with some conclusions published in a recent roadmap.[8] In the drafting process of this roadmap, the potential future actions in the field were given the acronym "tailored batches." Let us briefly summarize the different types of chemical constraints possibly encountered in glass melting.

These are, with respect to the batch:

- the amount of latent heat required to melt the batch
- the reaction rate of the batch reactions
- the temperature level required to promote the batch reactions
- the rate of quartz dissolution
- the rate of bubble release from the melt

- with respect to the furnace:
- the dwell time characteristics of the melt in the tank
- the space utilization of the reactor volume.

The individual points are not necessarily independent from each other. A brief look to the mass flow balance through the tank of a furnace (figure 5; designed after some typical figures presented by Beerkens[9]) reveals details on some well-known features. By far the most voluminous loop of mass flow is related to batch melting and quartz dissolution. In traditional melting this is required for two reasons. First, the batch blanket is an area drawing a power almost equal to the entire quantity P_{ex}. So the heat cannot be supplied from the combustion space above the batch blanket alone, but essentially relies on the counter-flow of very hot melt flowing from the hot spot to underneath the batch blanket. Second, the large extent of the loop corresponds to comparatively high velocities and velocity gradients, which helps to promote the quartz dissolution process. It goes without saying that any significant changes in the course of batch melting reactions would allow to substantially redesign the pattern shown in figure 5. The option to design low-enthalpy batches has been discussed in the previous section. Another way is the design of glass compositions with a reduced amount of constitutional silica. This is the amount of crystalline silica found in the heterogeneous phase equilibrium corresponding to the glass composition. The amount of constitutional silica is equivalent to the amount of sand in a batch which cannot be digested by silicate formation reactions, but has to be dissolved by slow diffusion processes. Even surprisingly small changes of glass composition – insignificant for the working end – may considerably lower the amount of constitutional silica and yield remarkable effects.

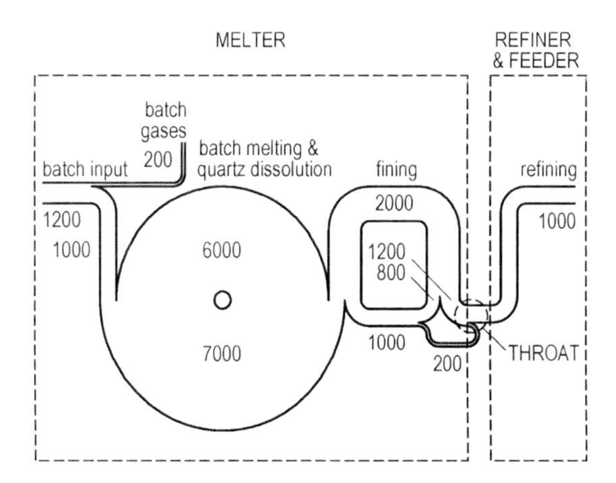

Figure 5. Mass balance of a glass melting tank; numbers denote the amounts of matter involved in units of kg per 1000 kg of produced glass, after ref.[9]

Let us have a look at the reaction temperatures: A glance at some fundamental silicate formation reactions compiled in Table 3 reveals a surprising fact: Their equilibrium temperatures are located well below 300 °C. The challenge consists in overcoming the kinetic barriers by:

- generating a low-T fluid phase which helps to enhance the atomic mobility in the system this may be achieved by making use of low-T eutectics, or by employing hydrothermal conditions like in the lime silica brick process
- increasing the surface-to-volume ratio of the reactants and
- minimizing diffusion path lengths by selectively bringing into contact suitable combinations of reactants

Table III. Equilibrium temperatures T_{eq} and standard heats of reaction of some fundamental silicate formation reactions, calculated from tabulated data.[4]

Reaction	T_{eq} in °C	$\Delta H°_{chem}$ in kWh/t
$Na_2CO_3 + SiO_2 \rightarrow Na_2SiO_3 + CO_2$	287	185
$Na_2CO_3 + 2SiO_2 \rightarrow Na_2Si_2O_5 + CO_2$	256	127
$2Na_2CO_3 + CaCO_3 + 3SiO_2 \rightarrow 2Na_2O \cdot CaO \cdot 3SiO_2 + 3CO_2$	281	191
$Na_2CO_3 + 3CaCO_3 + 6SiO_2 \rightarrow Na_2O \cdot 3CaO \cdot 6SiO_2 + 4CO_2$	152	128

Much of the effort in this area is not present in freely accessible literature (see ref.[10-11] as earlier examples). It has been known for a long time[12-13] that the turnover in a conventional soda lime silicate batch with grain sizes of typically 250 µm does not reach any considerable rates until the onset of physical melting of the soda ash. In fact, the appearance of a salt-like low-viscosity flux in the batch lifts the kinetic constraints of the thermodynamically already favored silicate formation reactions. This is demonstrated by the left graph of figure 6 presenting the release of CO_2 from a batch of soda ash, limestone, and quartz sand, normalized to the overall amount of CO_2 initially contained in the batch.

The right graph of figure 6 illustrates the scientific background using the model system KCl. When approaching the melting point of KCl at 772°C, the atomic mobility of the cation increases by several orders of magnitude. It is worth noting that almost the same level of ionic mobility can be reached in an aqueous or hydrothermal system at temperatures below 300°C. In industry, the principle is exploited in the production of lime silica bricks, however, not yet in the glass industry. In any case, the reaction rate between a highly mobile cation and a granular reaction partner is expected to follow the pattern

$$\text{reaction turnover} = \frac{\text{contact area}}{\text{distance}} \times \text{atomic mobility}$$

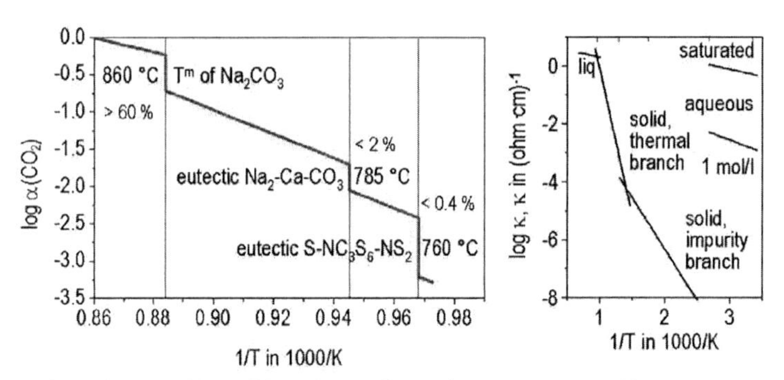

Figure 6. Left graph: Logarithm of the release of CO_2 from a soda ash + limestone + sand mixture, relative to the overall content of CO_2, as a function of reciprocal temperature; temperatures of points where a liquid phase is generated are given along with the observed amounts of CO_2 release in %; data taken from ref.:[11] Right graph: Temperature dependence of the atomic mobility in the system KCl as reflected by electrical conductivity; left branches: system solid – liquid; right branches: aqueous system at 1 bar.

Without being specific, let us suggest that the glass industry should proceed along the same path of development as the food industry did previously. The development of baking mixtures and instant coffee powders followed the very same philosophy, i.e., to speed up thermodynamically favored but kinetically constrained reactions, and to enhance early homogenization of a system.

Figure 7 shows the results of a thermo-optical analysis of the local micro-kinetics of batch melting under a heating rate of 10 K/min. The left-hand series (a) shows the reaction between the conventional glass grade raw materials sand, soda ash and limestone. The arrangement is chosen in such a way that the soda ash may "choose" its preferred reaction partner. This turns out to be the sand grain. Obviously, it is the larger difference of the Gibbs energy of the sodium silicate formation in comparison to the formation of a mixed salt Na-Ca-CO_3 which drives the process. A previous study[14] had shown that under extremely high heating rates, the inverse reaction order is observed. Such high heating rates are, however, not reached in a batch heap – except for the outermost layers. The sequence also shows that recrystallization immediately follows the primary fusion reaction. The decomposition of limestone is known to take place at about 900°C. This cannot be seen in the hot-stage microscope, except probably from the slight shrinkage of the grains. Beyond 900°C, it takes quite some while until the sodium silicate melt is able to absorb the calcia. The right-hand series (b) shows the micro-kinetics between soda ash, sand and cullet. In contrast to its neutral role in the heat balance, see eq. (3), cullet is by no means a neutral species kinetically. It vigorously reacts with the sodium silicate, which is primarily formed like in sequence (a). (Readers having imagination may even recognize a small elephant sucking in the cullet melt by its trunk). Another series not shown here was performed on the following arrangement of grains: cullet / sand / soda ash / cullet / sand, so as to again make the soda ash "decide" on its preferred partner. In this test, the primary melt was formed between soda ash and cullet – which is not a preferred reaction path as it delays quartz dissolution.[15] Finally, some very promising results were obtained with conventional glass batches prepared by an "instant coffee" strategy. After a vigorous onset of melting, the turnover of all reaction partners was completed at temperatures well below 1000°C.

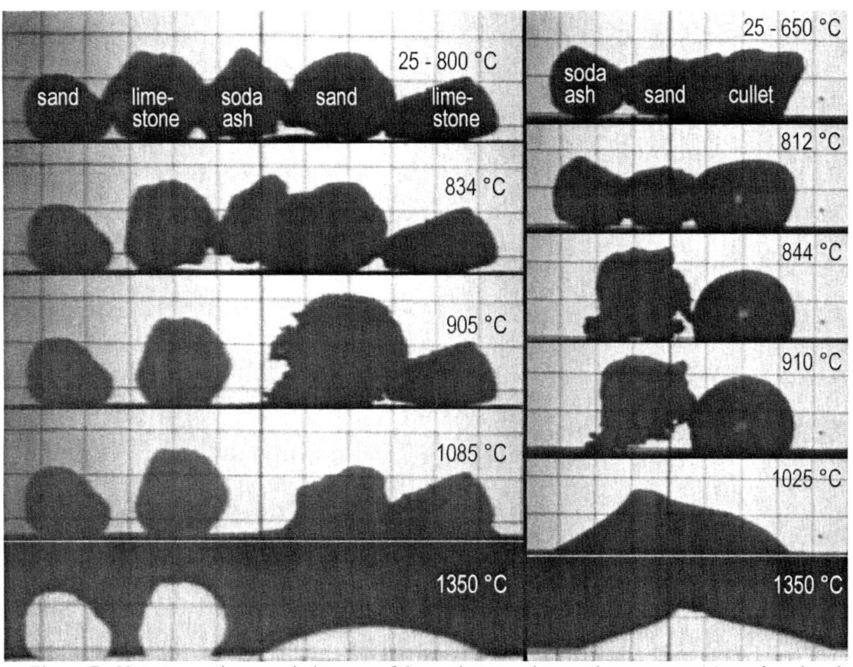

Figure 7. Hot-stage microscopic images of the grain-to-grain reactions among (a) sand, soda ash, limestone and (b) sand, soda ash, cullet, heated up at 10 K/min; the grid has a width of 500 μm

CONCLUSION

The physical limits of the performance of conventional glass furnaces are determined by constraints due to the heat transfer conditions or by constraints related to the heat uptake and the kinetics of chemical reactions. It was shown that the heat capacity flow imbalance of hot and cold stream, i.e., of the mass flows through the combustion space and through the tank, respectively, is probably the principal and most severe thermal constraint of conventional furnaces. Using the example of a small oxy-fuel furnace, it was demonstrated how a retrospective analysis of furnace operation data can be used to unveil the nature of the principal constraint under which an individual furnace works. Three categories of performance were derived: (I) the distinct absence of any thermal constraint, i.e., an operation limited by chemical constraints alone, (II) an operation under a flow imbalance as the exclusive thermal constraint, and (III) an operation under a constrained heat transfer coefficient. Finally, it was outlined how the rate of batch-to-melt conversion can be manipulated by its micro-kinetics, with an "instant coffee" type of batch as the target of future development.

REFERENCES

[1]R. Beerkens, Energy balances of glass furnaces: Parameters determining energy consumption of glass melting processes, in: W. H. Drummond III, ed., 67th Conference on Glass Problems, Ceramic Engineering and Science Proceeding 28 (2008).
[2]The concept was inspired by an oral comment by H. J. Barklage.
[3]R. Conradt, Thermodynamics of Glass Melting, 385-412, in: *Fiberglass and Glass Technology*, F.T. Wallenberger, P.A. Bingham, eds., Springer Verlag, Berlin (2010).
[4]O. Knacke, O. Kubaschewski, K. Hesselmann, eds., *Thermochemical Properties of Inorganic Substances*, Vol. I + II, 2nd Edition, Springer Verlag, Berlin (1991).
[5]A. Bejan, *Entropy Generation Minimization – The Method of Thermodynamic Optimization of Finite-Size Systems and Finite-Time Processes*, CRC Press, Boca Raton (1996).
[6]J.R. Taylor, *Fehleranalyse – Eine Einfuehrung in die Untersuchung von Unsicherheiten in physikalischen Messungen*, VCH Verlagsgesellschaft mbH, Weinheim (1988).
[7]A. Schack, *Der industrielle Waermeuebergang*, Verlag Stahleisen mbH, Duesseldorf (1969).
[8]K. Bange, M. Weissenberger-Eibl, eds., *Making glass better. An ICG roadmap with a 25 year glass R&D horizon*, International Commission on Glass (ICG), Madrid (2010).
[9]R. Beerkens, see the contribution in this present volume.
[10]W. T. Barrett, P. M. Brown, Eutectic mixture as a flux for glass melts, U.S. Patent 4,341,566 (1982).
[11]R. Conradt, PRILOX® Softbatch – Comparative study of a conventional and a pre-conditioned ready-to-use glass batch. XVI. ICF Technical Exchange Conference, Karlovy Vary, Czech Republic (2004).
[12]C. Kroeger, G. Ziegler, Reaction rates of glass batch melting. III, Reaction rates in the quaternary system Na_2O-CaO-SiO_2-CO_2, Glastech. Ber. 28, 199-212 (1954). (in German)
[13]C. Kroeger, F. Marwan, Reaction rates of glass batch melting. VI, Effects of additions on the reaction of soda – limestone – quartz batch. Glastech. Ber. 29, 277-289 (1956). (in German)
[14]L. Riedel, The wetting of limestone and quartz by molten soda ash. Glastech. Ber. 35, 53-56 (1962). (in German)
[15]W.H. Manring, A.R. Conroy, Influence of cullet in the melting of soda-lime glasses. Glass Ind. 49 no. 4, 199-203; no. 5, 269-270 (1968).

*Corresponding author: conradt@ghi.rwth-aachen.de

FUTURE OF GLASS MELTING THROUGH THE IN-FLIGHT MELTING TECHNIQUE

S. Inoue,[1] T. Watanabe,[2] T. Yano,[2] O. Sakamoto,[3] K. Satoh,[4] S. Kawachi,[5] T. Iseda[5]
[1]National Institute for Materials Science, Tsukuba, Japan; [2]Tokyo Institute of Technology, Tokyo, Japan; [3]Asahi Glass Co., Ltd., Yokohama, Japan; [4]Toyo Glass Co., Ltd., Kawasaki, Japan; [5]New Glass Forum, Tokyo, Japan

ABSTRACT

The principle of the in-flight glass melting technique is reviewed and the future of the industrial glass melting is discussed based on the achievements of the on-going project on the development of the in-flight glass melting technology. In the in-flight melting, a granular batch is charged directly into a burner flame. The heat transfer from heating source to a glass batch is enhanced very much and the glass melting energy can be saved by more than 50% of the current glass tank furnaces. The innovative glass melting technique will trigger the revolution in glass production to push the glass industry into a new era. The technique can also reduce the melting furnace sizes, which will contribute to the reduction of furnace construction cost. The standard soda-lime-silica glass composition can be melted using only an oxy-gas firing burner. The oxy-gas burner can be combined with an arc plasma torch to compose the hybrid heating in the new technique. The hybrid heating source can generate higher temperature than the burner heating and is applicable to the melting of high liquidus temperature glass forming systems. The feature will enable the production of the new functional glasses which are difficult to be fabricated by the currently popular Siemens type melting furnaces.

INTRODUCTION

The consumption of the energy generated by firing fossil fuels increases the CO_2 gas emission in the atmosphere, which causes the greenhouse effect to raise the global temperature. In the glass industry the mixtures of raw materials are melted in tank furnaces heated by firing of fossil fuels to produce sheet glasses, bottles, tablewares etc. Though the inhibition of the greenhouse effect can be controlled by using energy sources other than fossil fuels, the energy savings directly reduce CO_2 gas discharged into the atmosphere and is a direct solution to alleviating the greenhouse effect.

In glass production, the Siemens type tank furnaces have been used for more than one and a half centuries under the improvement of thermal efficiency by modifying the furnace design and the refractory. The raw materials are melted and held at 1500°C~1600°C for a prolonged time in the high temperature melting zone of the furnace to remove bubbles and to make the melts uniform. The residence times in the melting zone depend on the required qualities of the glass products. The residence time is estimated to be about 7 days for LCD (liquid Crystal Display) panel glass and to be about 1.5 days for bottle glass. The energy for glass production is roughly allocated into 20% for melting, 60% for heat loss caused by exhaust gas and 20% for radiation loss from the furnace wall. The energy taken away by exhaust gas is rather large and the Siemens type furnaces are equipped with regenerators to recover exhaust heat. Many ideas have been applied to improve efficiency on the Siemens type furnaces and the efficiency values have been increased step by step. Though the Siemens type furnaces are excellent glass melting furnaces, the efficiency can be improved drastically by removing the regenerators and supplying the energy necessary only for the glass melting. Such a process is expected to give 60% energy reduction at a maximum. Recently the glass industry has employed the non-Siemens type direct fired furnaces not equipped with regenerators. The report published by the U.S. Department of Energy (DOE) in 2002 points that the direct firing melting potentially improves the efficiency by about 40% in contrast with 15-20% on the regenerative Siemens type furnace.[1] The direct heating non-Siemens type glass melter is expected to show a good performance in glass production.

In Japan, the project on the development of the innovative glass melting process based on the in-flight glass melting technique was proposed by the National Institute for Materials Science, Tokyo Institute of Technology, Asahi Glass Co., Ltd., Toyo Glass Co., Ltd. and New Glass Forum in 2007 and was accepted as the National Project "Energy Innovation Program/ Development for Innovative Glass-Melting Process Technology" by New Energy and Industrial Technology Development Organization (NEDO). The project has been executed from FY2008 to FY2012. The project organization is shown in Figure 1 with the lists of the techniques in which the project members are developing. In this paper, the principle of the in-flight melting technique and the project intermediate progress are reviewed and the future of glass melting is discussed.

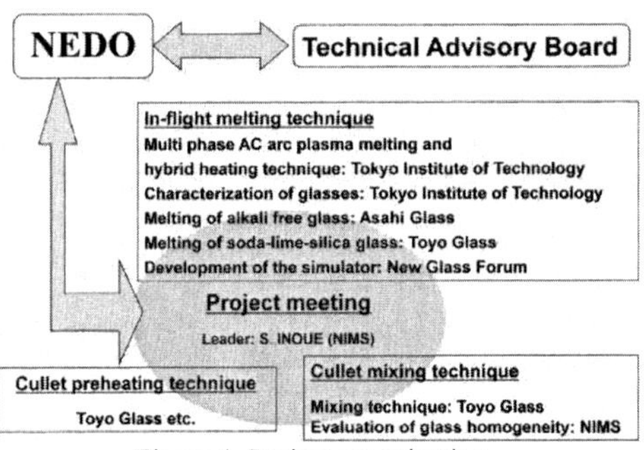

Figure 1. Project organization.

NON-SIEMENS TYPE GLASS MELTER

Some non-Siemens type furnaces have been developed in the United States. From 1984 to 1990, the Advanced Glass Melter (AGM) was developed to drastically improve the heating efficiency by introducing a glass batch directly into an oxygen gas or air stream to be melted within the burner flame.[2] The glass batch melting was greatly enhanced, leading to an energy savings of 24%. The high density plasma glass melter was developed in the time between 2000 and 2010.[3] The high density DC-arc-plasma was employed for the rapid melting of the glass batch possessing high liquidus temperature. The melter was tested to produce E-glass fiber marbles and the energy saving was reported to be about 50% at a maximum under the production scale of 3 t/day. The Submerged Combustion Melter (SCM) has been studied since 2001.[4] The SCM is designed to melt glass batch by the oxy-fuel burner firing in the melt. The firing in the glass melt intensifies heat transfer from flame to melt remarkably. The energy saving is expected to be more than 20%. These three methods are examples for non-Siemens type glass melters which have some principal relations with the in-flight melting technique. The development history of the non-Siemens type furnaces indicate that the non-Siemens type melter will be effective for the drastic energy savings in glass melting.

IN-FLIGHT GLASS MELTING TECHNIQUE[5]

The principle of the in-flight glass melting technique is illustrated in Figure 2 in contrast with the conventional Siemens type melting technique. In the in-flight melting, the glass batch is charged into an oxygen gas stream to be introduced directly into an oxy-fuel burner flame similar to the AGM

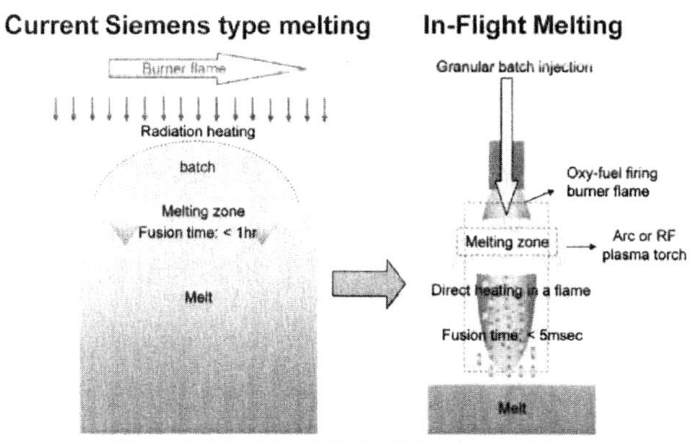

Figure 2. Principle of the in-flight glass melting.

melter. The burner flame including glass batch in the in-flight melting corresponds to the burner flame radiation melting zone in the Siemens type melting. The glass batch for the in-flight melting is granulated as 100~300μm size particles by spray drying. The particles of granular glass batch are melted while passing the burner flame for about 10msec and converted into glass melt droplets. The distance between the flame end and the glass melt surface is longer than the AGM case. Employing granular glass batch accelerates the reactions among the raw materials in granules and makes it possible to melt within a short flight time in a flame. Moreover the granular batch heating principally avoids the segregation of silica sand during melting. This is another effect enhancing the melting of glass batches and is an advantage which the AGM technique cannot produce. The oxy-fuel burner can be combined with a multi phase AC-arc type or RF type plasma torch. The heating system is selected from two types, an oxy-fuel burner type or an oxy-fuel burner-plasma torch hybrid type, according to the glass compositions. The composition possessing high liquidus temperature requires high melting temperature and the hybrid type heater is selected. The balance of the powers between a gas burner and a plasma torch affects the melting temperature and the energy cost. As the electricity price is rather high in Japan, the portion of the plasma torch is taken as small as possible.

INTERMEDIATE PROGRESS OF THE PROJECT

The project has developed three different test furnaces, which are the 1t/day oxy-gas firing furnace (furnace-1), the 200kg/day oxy-gas firing-multi phase AC arc plasma hybrid furnace (furnace-2) and the 300kg/day oxy-gas firing-RF plasma hybrid furnace (furnace-3). In the project, two kinds of glass compositions were employed for the melting tests. One is a standard commercial soda-lime-silica glass composition and the other is an alkali free borosilicate glass composition similar to the Corning 7059 glass. Alkali free glasses require melting temperature higher than standard commercial silicate glasses. Furnace-1 is usually used for the melting of standard silicate composition and furnace -2 and -3 are used for the melting of both compositions.

The granular glass batches were prepared by a spray dryer. The conditions and the block diagrams of the spray drying process are summarized in Figure 3. The suspension slurry prepared by mixing the fine powdered raw materials with water was sprayed through a disk type or a nozzle type atomizer into a heated air cyclone. In practical use, the spray dryer is designed to be heated by the exhaust gas from the furnace. A few hundred micrometer diameter granules were prepared for the soda-lime silica glass and about 100μm diameter granules were for the alkali free borosilicate glass.

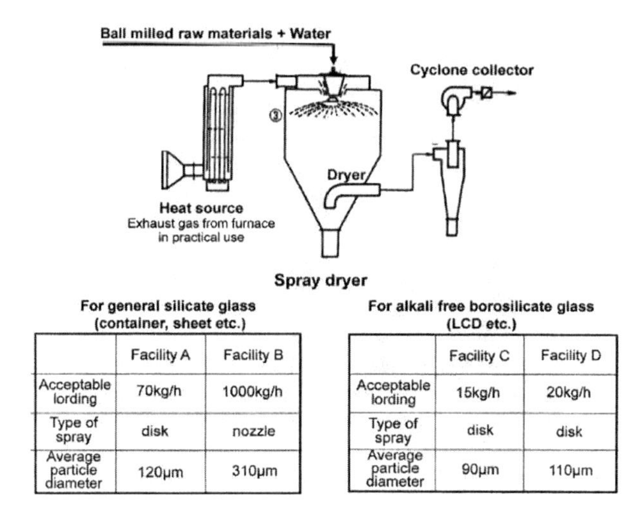

	Facility A	Facility B
Acceptable lording	70kg/h	1000kg/h
Type of spray	disk	nozzle
Average particle diameter	120μm	310μm

	Facility C	Facility D
Acceptable lording	15kg/h	20kg/h
Type of spray	disk	disk
Average particle diameter	90μm	110μm

Figure 3. Operation conditions and schematic illustration of the spray dryer.

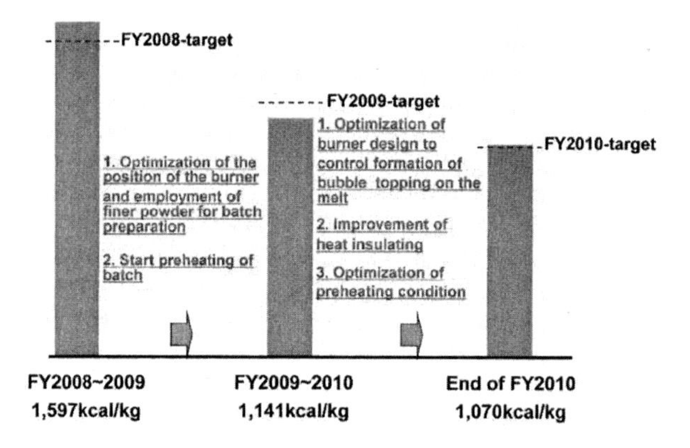

Figure 4. Achievements of the on-going project.

Figure 4 shows the history of the project accomplishment on the melting soda-lime-silica glass by the 1t/day oxy-gas firing test furnace (the furnace-1). The heat necessary to produce 1 kg glass was reduced from 1,597kcal/kg-glass to 1,070kcal/kg-glass which attained the project target. From FY2008 to FY2009, the burner position above the glass melt was optimized and the batch preheating was started. The most effective point was the burner position. From FY2009 to FY2010, the modification of the burner design and the burner position was found very important to suppress the formation of foam layer on the glass melt. Further optimization of the burner is necessary to attain the final target (900 kcal/kg-glass) at the end of FY2012. The key point of the optimization of the burner is the broadening and elongation of the flame which increases heating zone volume and flight time.

The various energy efficiency values are plotted against the scale of the furnace production in Figure 5. The values are estimated or collected from the literatures.[6,7] The final target value of 900kcal/kg-glass is expected on the production scale of 1t/day. In general, the glass production efficiency decreases with the increase of the furnace scale. The project final target value is about 20%

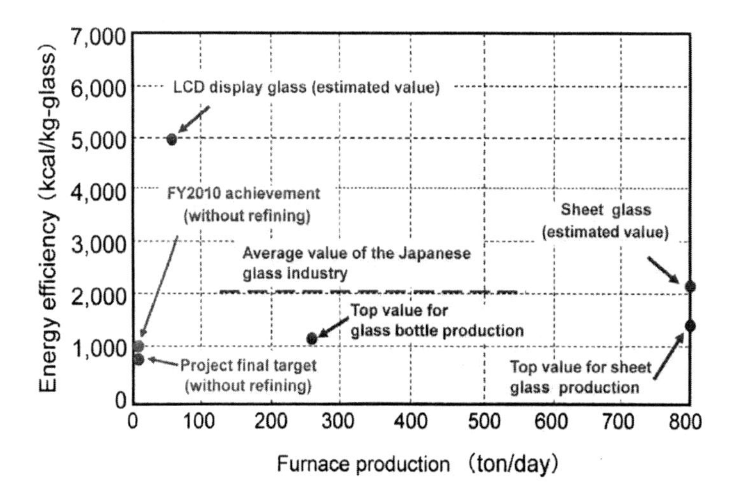

Figure 5. Comparison of the energy efficiency among the various glass melting processes.

of LCD glasses, 45% of the average value Japanese glass industry and 43% of sheet glass production. The final target value of the project (900kcal/kg-glass) is low enough at the 1t/day scale production.

THE FUTURE OF GLASS MELTING THROUGH THE IN-FLIGHT MELTING TECHNIQUE

Figure 6 shows an image for the expected drastic change from the Siemens type tank furnace to the in-flight melting tank furnace at 100t/day scale production. In the figure, the bubble eliminator is added to the in-flight melter, which is necessary to produce bubble free glasses like LCD panel glasses. The volume of the Siemens type melter can be decreased drastically by about 97% on the in-flight melter and by about 94% on the in-flight melting tank furnace. The decrease of the furnace volume extremely shortens the residence time in the melter. In the Figure 6 case, the residence time is expected to decrease from 5days to several hours. Moreover, the big volume reduction of tank furnaces will also cut construction costs of furnaces drastically.

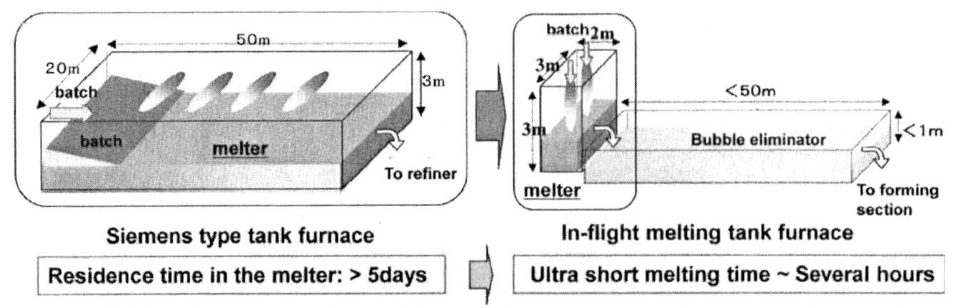

Figure 6. Schematic illustration for the change of the tank furnace on the in-flight glass melting.

For the project, the simulator for the in-flight melting tank furnace has been developed to estimate the energy balance at the 100t/day scale production based on the experimental results obtained from the 1t/day scale test melting. Figure 7 shows the in-flight melting tank furnace model for the simulation. The melting scale is 100t/day and the surface area of the glass melt in the melter is 7.4m^2. The fuel is methane gas and the combustion rate is 15Nm^3/hr. The refiner, bubble eliminator, is added to the in-flight melter. All the heat necessary for the spray drying and the cullet preheating is supplied

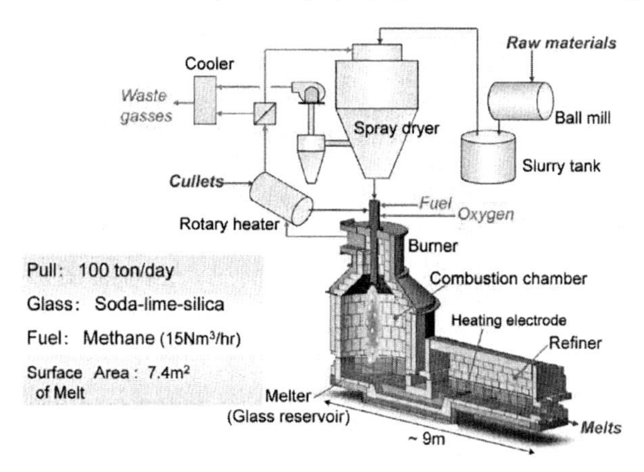

Figure 7. Simulation furnace model for the in-flight glass melting on 100t/day scale.

from the furnace exhaust. The grinding of the raw materials and the batch granulation system are not included in the simulation but the energy necessary for the spray drying, the grinding of the raw materials and the cullet preheating are taken into account in the calculation of the efficiency. Figure 8 shows the heat balance calculated from the above simulation. 70% of the input energy is used for the glass melting and the efficiency of the glass melting is expected to be 902kcal/kg-glass. The big energy savings is estimated on the in-flight glass melting tank furnace.

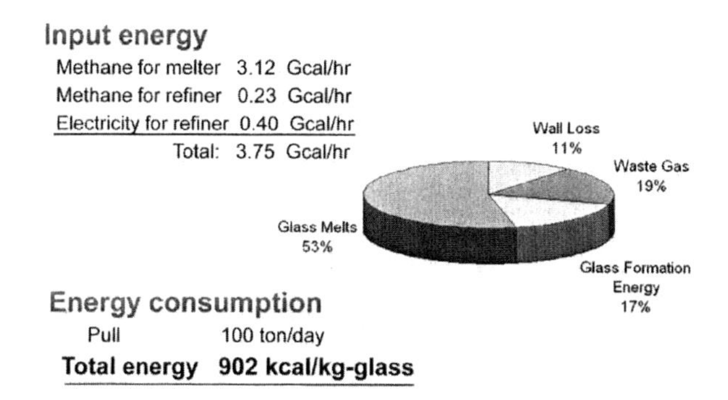

Figure 8. Heat balance in the 100t/day in-flight glass melting simulation furnace.

In the in-flight melting process, the plasma heating is used to accelerate the melting of the granular glass batches. The power balance between the oxy-gas firing and the plasma arc should be adjusted to get sufficient energy efficiency taking into consideration the energy costs. The high power plasma glass melting can melt the glass batches possessing high liquidus temperature which are difficult to melt in the Siemens type furnaces. Therefore the in-flight glass melting operated at the height of the plasma heating will be able to produce the new functional glasses containing the high melting point compounds.

After completing the project, the in-flight melting technique is expected to be applied at first to the small scale glass production like car lamp lenses, colored glasses etc. Extension to the production of LCD panel glasses, small bottles and bulbs etc. will follow around 2017. The larger scale glass production like bottle, filament fiber, solar panel cover glass etc will start around 2023. Application to huge scale glass production of sheet glass for buildings and cars is expected to start around 2027 depending on the preceding developments where the benefits of energy saving will be fully enjoyed.

CONCLUSION

The in-flight glass melting technique contributes very much to the reduction of the energy consumption in glass production and the sizes of the furnaces. The various merits other than the drastic energy saving are summarized below:

1) Merits of the size reduction of the furnace
   - a) Reduction of the construction costs and time of the furnaces.
   - b) Reduction of running costs derived from the energy saving.
   - c) Quick exchange of the glass composition.
   - d) Easy adjustment of the melting scale to the production plans by one furnace for one production line system.
   - e) Reduction of the costs to scrap the furnaces.

2) Merits of the employment of various heating sources
   - a) Melting of the glasses possessing high melting temperature.
   - b) Production of the high melting temperature glasses with reasonable costs.

The employment of the in-flight glass melting will step up the glass industry to the new stages, which will also contribute to the improvement of global circumstances.

ACKNOWLEDGEMENT

The financial support provided by Strategic Development of Energy Conservation Technology Project of NEDO (New Energy and Industrial Technology Development Organization) is gratefully acknowledged

REFERENCES

[1]Energy & Environmental Profile of the Glass Industry – April 2002, prepared by ENERGETICS Incorporated, Columbia, Maryland for U.S. Department of Energy Office of Industrial Technologies, pp.53 (2002).

[2]R. Tiwary, D. Stickler and J. Woodroffe, J. Am. Ceram. Soc., **71** (9) 748-53 (1988).

[3]R. Gonterman and M. Weinstein, "High Intensity Plasma Glass Melter – GO13093 Project" presented at Glass Problems Conf., GMIC Workshop, Oct. 26, 2005.

[4]M. Greenman, Am. Ceram. Soc. Bull., **83**(4) 14 (2004).

[5]Y. Yao, T. Watanabe, T. Yano, T. Iseda, O. Sakamoto, M. Iwamoto and S. Inoue, Sci. Technol. Adv. Mater., **9** 02513 (2008).

[6]G. Jacobs and H. V. Limpt, Glass, **81**(4) 108 (2004).

[7]Glass Technology Handbook (in Japanese), Asakura Publishing Co., LTD., pp.313 (1999).

*Corresponding author: inoue.satoru@nims.go.jp

APPLICATION OF THE IN-FLIGHT MELTING TECHNOLOGY TO AN ALKALINE FREE BOROSILICATE GLASS

O. Sakamoto, C. Tanaka, S. Miyazaki, N. Shinohara, S. Ohkawa
Asahi Glass Co., Ltd.
Yokohama, Japan

ABSTRACT

Application of the in-flight melting (IFM) technology to an alkaline free borosilicate glass was investigated. Granules for the glass composition of $50SiO_2$-$15B_2O_3$-$10Al_2O_3$-25BaO (wt %) were prepared by the spray drying method and fed into the radio-frequency (RF) induction thermal plasma or oxy/gas combustion flame. Although the granules were successfully melted and converted into the vitreous state, volatilization of B_2O_3 occurred severely. The results suggested that the vitrification degree was in a tradeoff relation with the residual amount of B_2O_3. RF plasma was more suitable than oxy/gas combustion for IFM of borosilicate glass in order to avoid the tradeoff relation. The use of granules with high mechanical strength, made from very fine raw materials, was found to be effective to achieve little volatilization of B_2O_3 and high vitrification degree simultaneously.

INTRODUCTION

We have been developing an innovative glass melting technology which could reduce the melting energy to less than half compared to the conventional system. In this technology, premixed and granulated raw materials are fed into the high temperature zone produced by means of oxygen combustion burner and/or plasma, then converted into glass while rapidly passing through it. This technology is called "in-flight melting (IFM)" because the glass formation proceeds during flying. Since the melting is completed within a very short time during flying, IFM can be expected to reduce significantly both the residential time in the furnace and the furnace size, contributing to the energy savings. The applicability of IFM has been investigated in an alkaline free borosilicate glass, as well as a soda lime glass, using RF plasma and oxy/gas burner.

In this study, granules for the glass composition of $50SiO_2$-$15B_2O_3$-$10Al_2O_3$-25BaO (wt %) were prepared by spray drying method. They were fed into the hot zone, which was generated by the radio-frequency (RF) induction thermal plasma or oxy/gas combustion flame, and were rapidly heated. The heated granules were collected after quenching, and the glass composition and the vitrification degree were examined for them. The effects of the particle size of the granules and the primary raw materials on the glass constituents and the vitrification degree were precisely investigated.

EXPERIMENTAL

An alkaline free borosilicate glass with the composition of $50SiO_2$-$15B_2O_3$-$10Al_2O_3$-25BaO (wt %) was used in this study. Three types of silica sand (SiO_2) with different particle sizes (average particle sizes are 0.5, 5.0 and 8.0 micrometers respectively for silica sand (A), (B) and (C), boric acid (H_3BO_3), alumina (Al_2O_3) and barium carbonate ($BaCO_3$) were used as the starting raw materials. Those raw materials were mixed and dispersed in water to make slurries. The slurries were then spray dried to form granules. Prepared granules were fed into the hot zone, which was generated by the radio-frequency (RF) induction thermal plasma or oxy/gas combustion flame, and were rapidly heated to convert into glass. Operational conditions of IFM were as follows: Feed rate of granules / 80 – 120 g/min, Power input for RF plasma/ 90 – 110kW, Power input for the oxy-LPG gas burner/ 40-60kW.

The in-flight melted glass particles were then quenched and collected in the stainless steel bowl. Size distributions of the starting granules and the melted glass particles were measured by using laser diffraction particle size analyzer (Microtrack, Nikkiso Co., Ltd. MT3200) in a dry state. Microstructures of the granules were observed by SEM (KEYENCE Corp., VE-9800), and the compositional distributions on the cross section of the granules were analyzed by EPMA (SHIMADZU Corp., EPMA-1610). The chemical composition of the obtained glass particles were analyzed for SiO_2, Al_2O_3 and BaO by the X-ray fluorescence analyzer (Rigaku Corp., ZSX Primus) using glass beads. And the content of B_2O_3 in the glass was determined by wet analysis (JIS R2015). The amount of crystalline SiO_2 (quartz) in the particles was measured by powder X-ray diffraction analysis (Rigaku Corp. Rotaflex RU-200B) in which ZnO was used as an internal standard. The vitrification degree was calculated from the ratio of residual silica/glass.

RESULTS AND DISCUSSION

Particle size of granules prepared by spray drying varies ranging from 30 to 250 micrometers and the average diameter was approximately 100 micrometers. Figure 1 shows the SEM photograph of the granules made from silica sand (C) as a starting material.

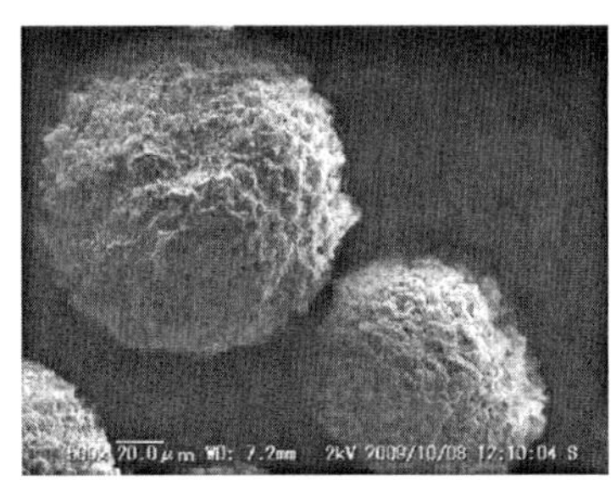

Figure 1. SEM photograph of granules made from silica sand (C).

Figure 2 shows the relation between the residual amount of B_2O_3 and the vitrification degree in in-flight melted glass. As shown, the vitrification degree and the residual amount of B_2O_3 were found in a tradeoff relation. The results clearly show that RF plasma was more suitable than oxy/gas combustion for IFM of the alkaline free borosilicate glass in order to increase the vitrification degree with suppressing the volatilization of B_2O_3.

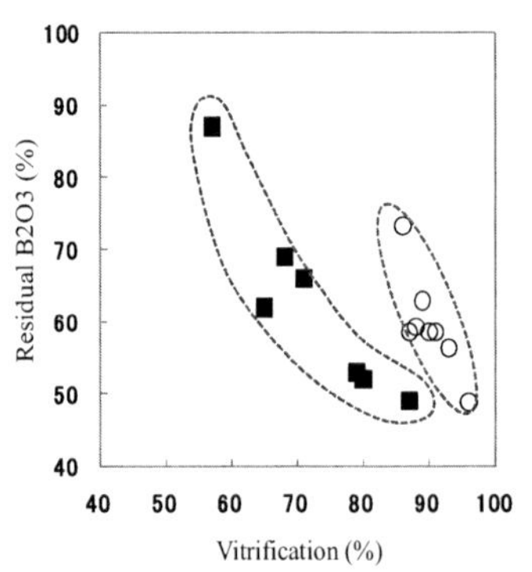

Figure 2. Residual amount of B_2O_3 and vitrification degree for the in-flight melted glass.
■: In-flight melted by oxy/gas combustion.
○: In-flight melted by RF plasma.

In order to examine the influence of the particle size of the in-flight melted glass on the glass composition, obtained glass particles were sieved into four classes and vitrification degree and residual amount of B_2O_3 were determined for the particles in each class. The results are summarized in Table I. It was revealed that the smaller the glass particle, the less the amount of B_2O_3 remained in the glass particle and the higher the degree of vitrification achieved.

Table I. Residual amount of B_2O_3 and vitrification degree for the in-flight melted glass particles categorized by sieving.

Category	Particle size of in-flight melted Glass	Residual B_2O_3 (%)	Vitrification degree (%)
0	As melted (before sieving)	69.5	86.1
1	Sieved 45-75μm	48.2	94.2
2	Sieved 75-105μm	63.2	92.2
3	Sieved 105-150μm	81.2	88.3
4	Sieved 150-250μm	84.7	85.7

Besides the small agglomerates existing in the original granules, fracture of the granules, which may occur during IFM process, was considered to be responsible for forming small fragments and, as a result, forming small glass particles. Increase of the surface area in such small agglomerates or fragments resulted in the vigorous volatilization of B_2O_3 during IFM process, causing less B_2O_3 to remain in the glass particles. Because the strength of granules was considered to be the significant characteristics governing the fracture behavior of granules, effect of the mechanical strength of the granules on the amount of volatilization of B_2O_3 during IFM process was investigated. The mechanical strength of granules can be estimated by applying the following Rumpf's equation (1);

$$\sigma = kF1(1-\varepsilon)/\pi D_p^2 \quad (1)$$

where σ is the mechanical strength of the granules, D_p is an average diameter of the primary raw materials, k is the coordination number, F1 is the interparticle force, and ε is the porosity of the bulk (granule). According to the above equation, the strength of granule is inversely proportional to the square of the diameter of raw material, provided that k, F1 and ε are almost the same.

Because silica sand is the major component in the granule, effect of the particle size of silica sand on the mechanical strength of the granules was investigated using three kinds of silica sands (A), (B) and (C) with different particle sizes. In this study, the mechanical strength of the granules was estimated from particle size distribution data under different dispersion conditions as described below. In the particle size analyzer which we used in this study, measurement is conducted just after blowing the compressed air to the particles in order to break soft agglomerates. Increase of the pressure of the compressed air blown to the granules, therefore, would result in the increase of small fragments or agglomerates caused by the fracture of granules, provided that the fracture strength of the granules is not so high. On the other hand, if the granules are sufficiently strong, the distribution curves measured under the different pressures of the compressed air would be identical.

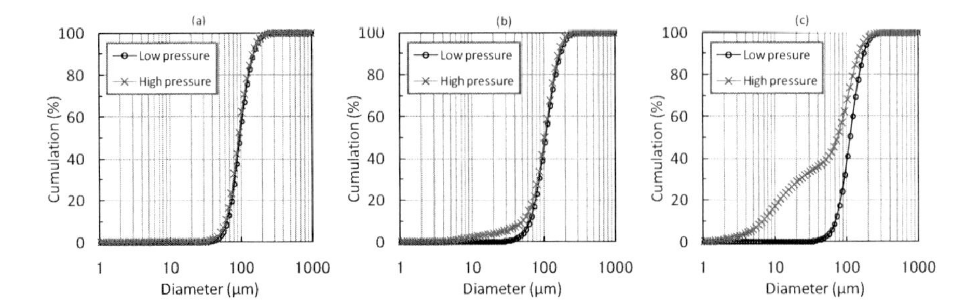

Figure 3. Particle size distribution of the granules made from silica sand (A), (B) and (C) measured in dispersion at low (5 psi) and high (50 psi) air pressure.
(a): prepared from silica sand (A), (b): silica sand (B), (c): silica sand (C)

Figure 3 (a), (b) and (c) show the particle size distributions measured under the different pressures of compressed air for the granules made from silica sand (A), (B) and (C) respectively. Circle and cross marks in each figure show the distribution measured at the air pressures of 5 psi and 50 psi respectively. As presented, the distribution curves in Figure 3 (a) were identical despite the pressures of the compressed air. Whereas the amount of small particles clearly increased for the granules containing silica sand (C) when the compressed air with the pressure of 50 psi was blown to the granules (Figure 3 (c)). The results show that the particle size of silica sand exerts an influence on the strength of the granules. As estimated from the above equation (1), strength of the granules increased as the particle size of silica sand decreased, resulting in the highest strength in the granules made from silica sand (A) which had the smallest particle size.

Figure 4 shows the relation between the vitrification degree and the residual amount of B_2O_3 in the in-flight melted glass particles, where the data for the granules made from three kinds of silica sand are plotted. Residual amount of B_2O_3 increased as the particle size of silica sand decreased, indicating that the use of granules with high strength resulted in the suppression of the volatilization of B_2O_3 caused by the prevention of the fragmentation during IFM.

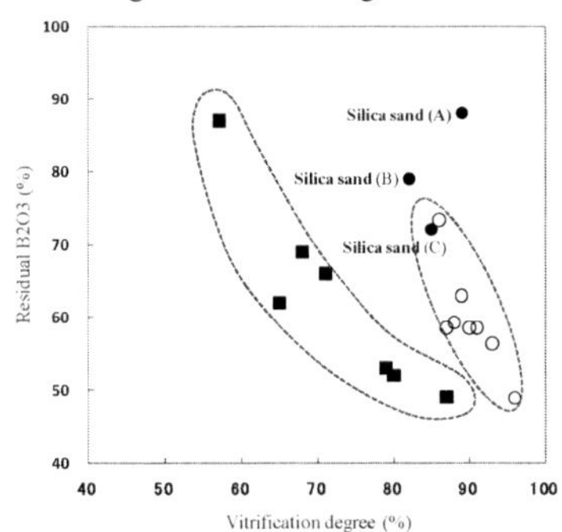

Figure 4. Residual amount of B_2O_3 and vitrification degree for the in-flight melted glass.
■: In-flight melted by oxy/gas combustion.
○,●: In-flight melted by RF plasma.

Figure 5 shows the results of EPMA analysis for Si and B components on the cross section of the granules made from three kinds of silica sand. This figure reveals that homogeneity for each component is better in the granules made from finer silica sand, that is, the granules made from the finest silica sand (A) shows the best homogeneity.

In the granulation process, boric acid dissolves in water of slurry and then may precipitate in the void spaces among the raw particles or on the surface of other raw materials during the drying process. Therefore, homogeneous distribution of boron in the granules made from silica sand (A) suggests the uniform distribution of the small interstitial void spaces between the uniformly arranged powder particles, which may have also contributed to the improved strength of the granule.

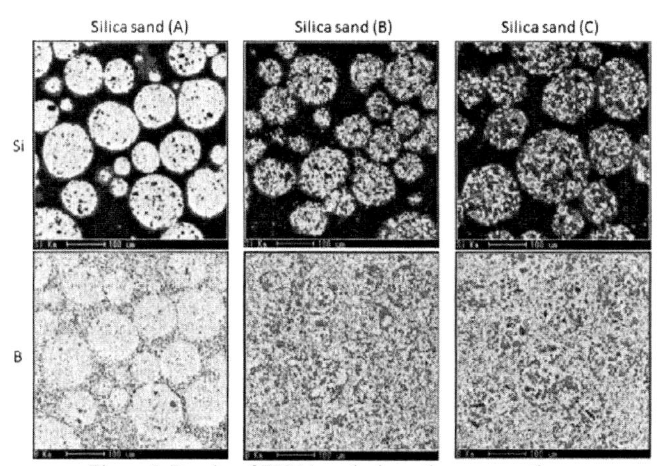

Figure 5. Results of EPMA analysis on the cross section of granules prepared from silica sand (A), (B) and (C).

CONCLUSIONS

In the IFM process for an alkaline free borosilicate glass, the vitrification degree and the residual amount of B_2O_3 were found in a tradeoff relation. In order to increase both the residual amount of B_2O_3 and the vitrification degree in in-flight melted glass particles, IFM by RF plasma rather than oxy/gas combustion flame and the use of granules with improved strength were significantly effective. The mechanical strength of granules was found to increase by using very fine primary raw materials. Uniform powder packing, which was responsible for the homogeneous distribution of the components, was also considered to be responsible for the increase in strength of the granules prepared from small silica sand.

ACKNOWLEDGMENT

The authors greatly acknowledge the New Energy and Industrial Technology Development Organization (NEDO) for financial support to this study.

*Corresponding author: osamu-sakamoto@agc.com

TEST RESULTS OF THE IN-FLIGHT GLASS MELTING USING ONE-TON/DAY LARGE SCALE EXPERIMENTAL MELTER

Masanori Iwamoto,[1] Keizoh Satoh,[1] Yasunori Ebihara,[1] Osama Sakamoto,[2] Chikao Tanaka,[2] H.Segawa[3]
[1]Toyo Glass Co., Ltd, Japan; [2]Asahi Glass Co., Ltd., Japan; [3]National Institute for Materials Science, Japan

ABSTRACT

It is explained that: 1). The granulated batch can be vitrified in the oxy-gas flame. 2). No seeds over 1mm were found in the melted glass. 3). The analyzed Na2O% is 15.4% while the analyzed batch of Na2O was 15.6%. 4). One pair of screw type stirrers gave good glass homogeneity. 5). The melted glass has low Fe3/Fe2 ratio. 6). The Rapidox measurement of the melted glass shows low oxygen activity. 7). Current specific heat with this experimental furnace is around 4.45GJ/ton-glass.

INTRODUCTION

We, the Glass industry have been known as one of the energy-consuming sectors. One of our major concerns is to minimize energy consumption. Various methods have been proposed and tried, i.e. the Submerged Combustion Glass Melting (SCM, Glass Institute and Gas Technology, 2005), the Advanced Glass Melter (AGM, Gas Research Institute 1984), Flame Impingement (CGM, BOC/Linde 1997).

Our In-Flight Melting has two stages. The first stage, from FY2005 to FY2007, we made a so-called feasibility study about In-Flight Melting and successfully melted the granular glass batch with an existing small facility. With this result, from FY2008, we have rolled out a new five-year project. We installed a newly designed 1ton/day In-Flight Glass Melting Experimental Furnace at the Toyo Glass Kawasaki Plant in 2008. Since that time we have been conducting In-Flight Melting trials.

PROJECT TEAMS AND THEIR ROLES

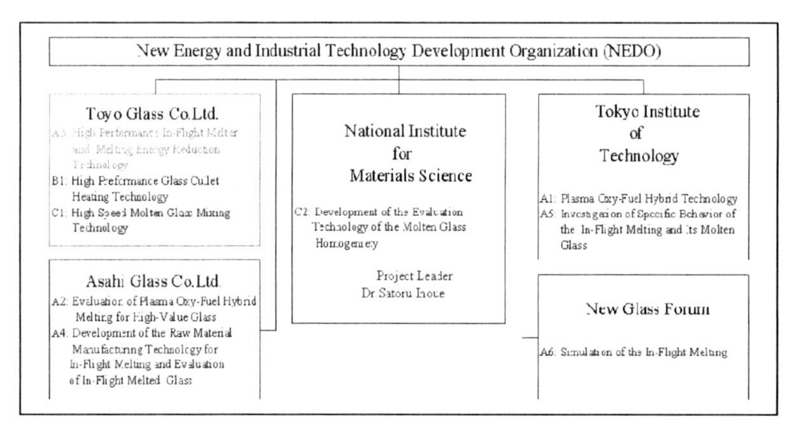

Figure 1. The project teams and their roles.

GENERAL VIEW OF THE IN-FLIGHT EXPERIMENTAL MELTER

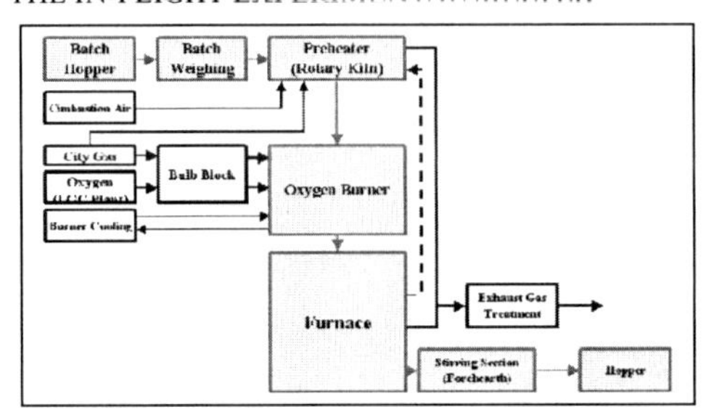

Figure 2. Flow diagram of the experimental melter

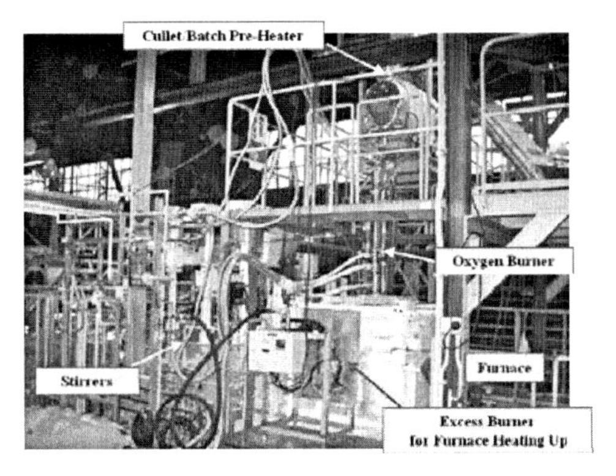

Photo 1. The experimental In-Flight Melter

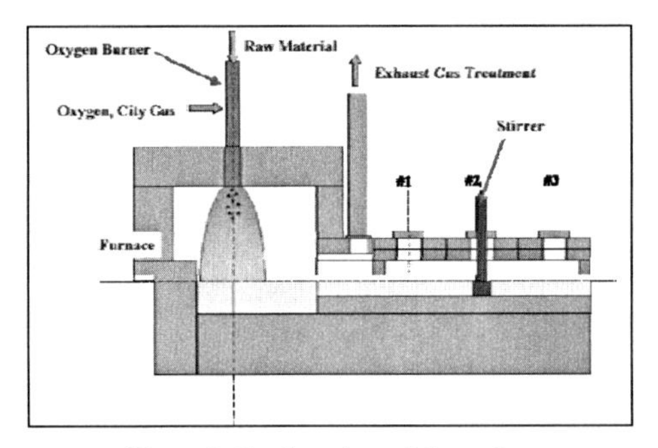

Figure 3. Section view of the melter

Fig. 2, Fig. 3 and Photo 1 show a schematic flow diagram, a melter schematic section view and the experimental IFM (In-Flight Melter).

A premixed fine granulated raw material will be conveyed into the batch preheater that then falls down into the water-cooled oxygen burner to melt in the flame. The melted and accumulated glass in the furnace chamber will flow into the stirring section. The Stirring section can be equipped with three pairs of stirrers but currently we are only using the #2 section.

The batch preheater is an indirect heating type. Currently, this preheater is heated by city gas. But it can be heated with exhaust gas to have over-all energy consumption as low as possible. An excess air burner is used for heating the furnace up. The batch hopper and batch weighing system and other equipment are not shown in Photo 1.

GRANULATED RAW MATERIALS AND PARTICLE DISTRIBUTION

Fig.4, Photo 2 and Table I show an example of the granulated batch particle distribution, a microscopic photo of the granulated batch particles and the analyzed batch composition.

As Table I shows, currently we are using soda-lime glass batch. Iron oxide is added to estimate the melting temperature and for checking the redox state of the melted glass. Using fine silica sand by the spray dry method to granulate the raw materials resulted in granulated particle size of around 125 micron.

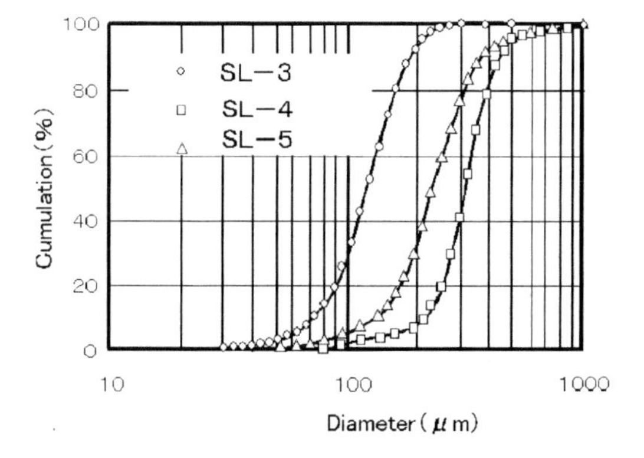

Figure 4. Granulated particle size distribution

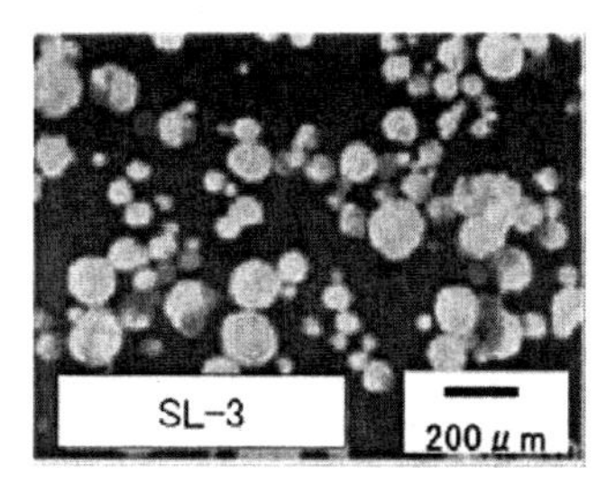

Photo 2. Granulated particle

Table I. Batch Composition.

	Mass%
SiO_2	71.4
Al_2O_3	2.0
CaO	10.1
MgO	-
Na_2O	15.6
K_2O	
Fe_2O_3	0.1
TiO_2	-
SO_3	0.5
Total	99.7

EXPERIMENT RESULTS

The Oxygen Gas Burner Flame

Fig.5, Photo 3a and Photo 3b show a simulated flame temperature profile, actual blue flame and the luminous flame with granulated raw materials respectively.

Since most of the container glass plants in Japan utilize city gas to minimize CO_2 amounts in the exhaust gas, this experiment has also used city gas. According to our simulation (Fig.5), the oxy-city gas flame has a substantial area that exceeds 1900°C. The actual flame (Photo 3a) is a so-called blue flame, but once the granulated raw materials fall into the flame it become a strong luminous flame (Photo 3b).

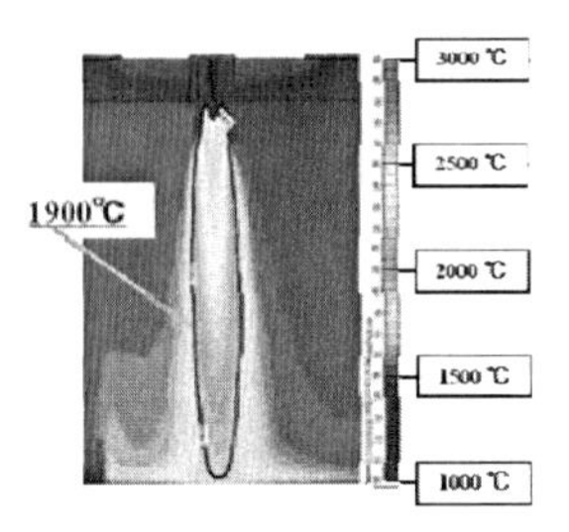

Figure 5. Oxy-gas flame temp. (simulated)

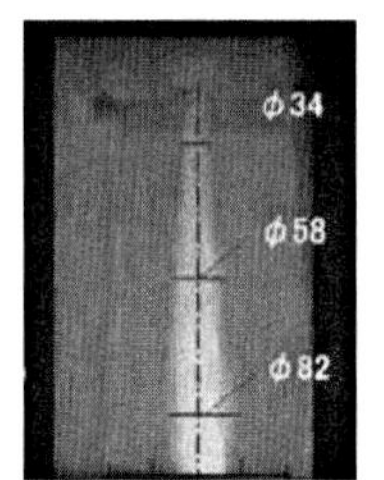

Photo 3a. Oxy-gas flame (no granulated batch, flame diameter in mm)

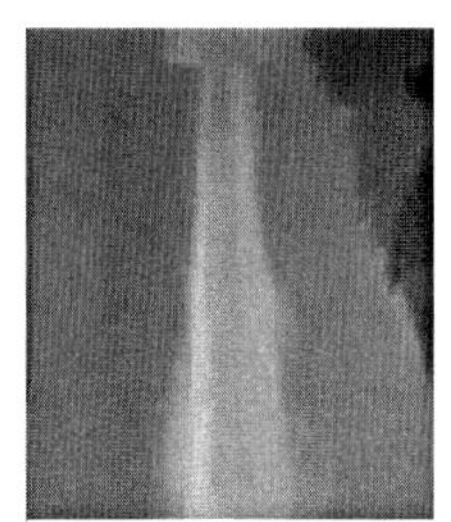

Photo 3b. Oxy-gas flame (with granulated batch)

Vitrification of Granular Raw Materials in the Flame

Fig.6 shows XRD SiO_2 peak of the flying granular raw material that has been sampled from the oxy-gas flame at various distances between the burner outlet and the molten glass surface. Fig.6 also shows that the SiO_2 peak of the samples has decreased along with the flame. The sample taken from the molten glass surface has shown no SiO_2 peak. From this result it can be said that the granular raw materials can be melted in the flame.

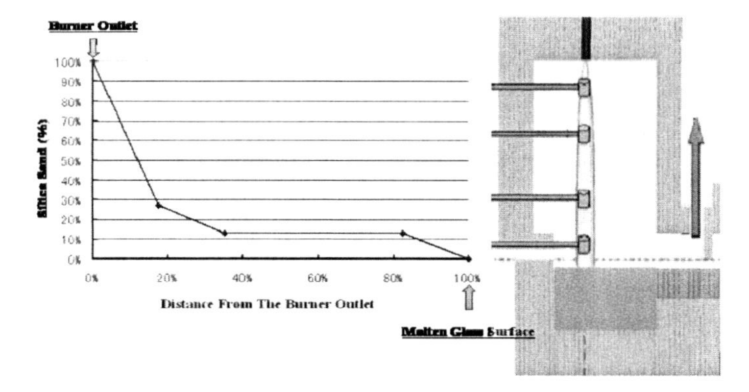

Figure 6. Vitrification of the granular batch along with the oxy-gas flame (Sampling position at the molten glass surface is not shown.)

Ladle Sample: Naked Eye Observation

Photo 4 shows the section view of the ladle samples that have been taken from the molten glass chamber during the In-Flight Melting. The surface of the molten glass chamber was covered by the foam layer. But, under the foam layer, we found the clear glass that had a seed count of 2.8 pcs/g by naked eye observation using a seed count observation box (Fig.7).

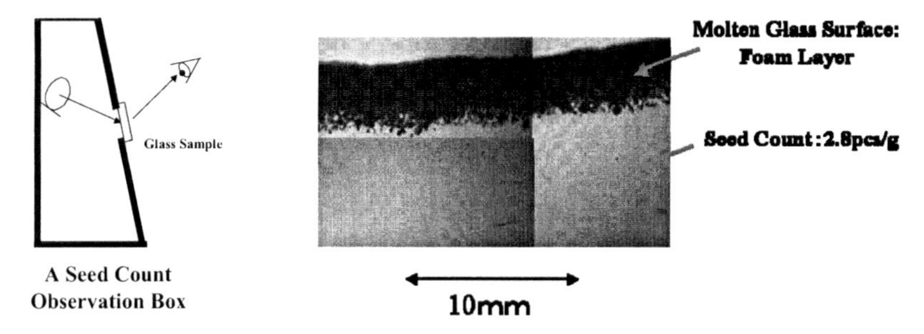

Figure 7. Seed Count Box

Photo 4. A section view of the ladle sample

Seed Counts (Microscope x10 Observation)

Fig. 8 shows the seed count results. Observation has been made with the Microscope(x10) showing the following results: 1. No seed over 1mmφ has been observed. 2. Seeds tend to increase with increasing feed rate and 3. Seeds tend to increase with decreasing melting energy. This suggests a need for further study for refining though the project itself is considered as rough melting.

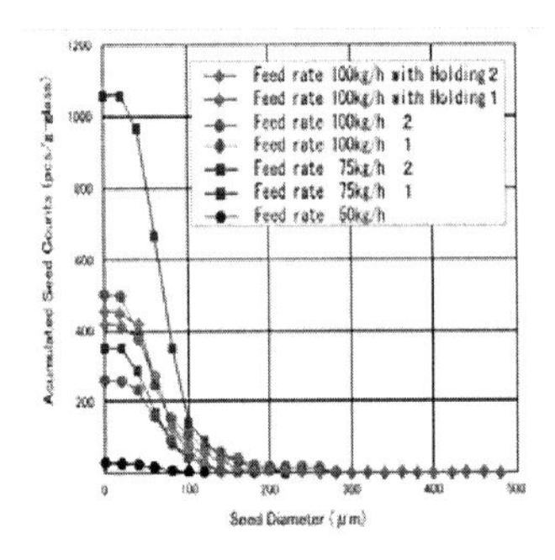

Figure 8. Seed Count using microscope(x10)

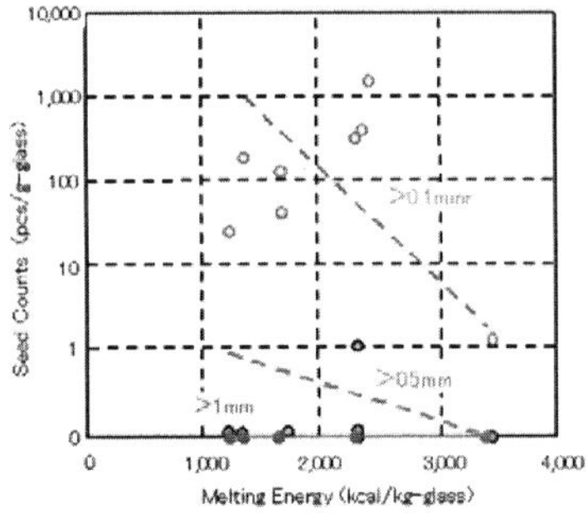

Figure 9. Seed count vs. melting energy

Homogenization/Effect of the Stirring

Photo 5a and Photo 5b show a striae observation result, before and after, the stirring. Fig.10 shows the sampling position. We currently use conventional screw type stirrers. With one pair of the stirrers, with 7.3 rpm upper direction rotation, and the temperature at 1,300℃, visible striae have almost disappeared. From this, we expect stirring by three pair will be more effective for good glass homogeneity.

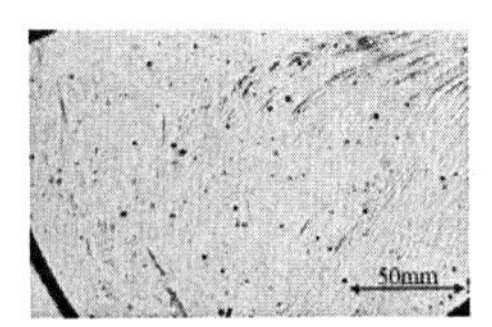

Photo 5a. Before stirring

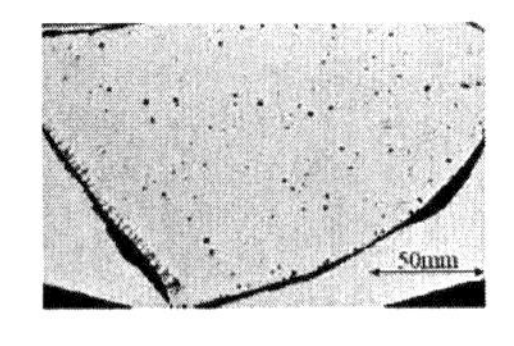

Photo 5b. After stirring

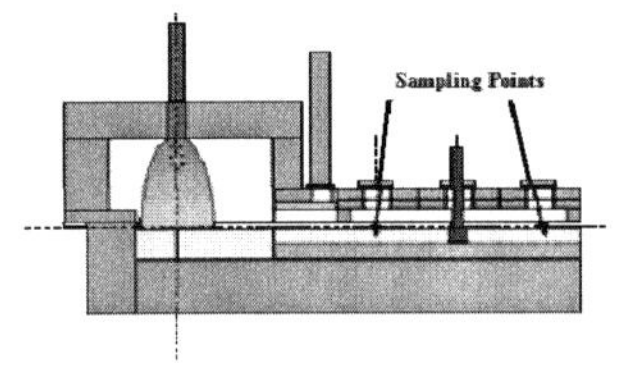

Figure 10. Sampling positions

Batch and Molten Glass Composition Comparison

Table II shows an example of molten glass samples composition (anal.) and batch composition (anal.). Based on the result, it can be said that for the Na_2O-CaO-SiO_2 system, there is no significant Na_2O loss during the high temperature In-Flight Melting. It is expected that we will be able to compensate this loss by adjusting the batch composition to get to the target glass composition.

Table II. Composition of the batch and melted glass

mass%

	Batch (anal.)	Molten Glass (anal.)
SiO_2	71.4	71.8
Al_2O_3	2.0	2.1
CaO	10.1	10.2
MgO	-	0.06
Na_2O	15.6	15.4
K_2O		0.04
Fe_2O_3	0.1	0.1
TiO_2	-	0.01
SO_3	0.5	0.3
Total	99.7	100

Molten Glass Redox State (Fe3+/Fe2+)

Fig.11 shows the comparison of the typical light absorption curves of the laboratory melted glass (Electric Furnace), glass sample of the conventional furnace and the In-Flight Melted glass sample. Fig.12 shows the Rapidox measured oxygen activity of those samples. From the oxygen activity and Fe3+/Fe2+ ratio comparison results, it can be said that the In-Flight Melted glass is melted within less of an oxygen atmosphere and high temperature.

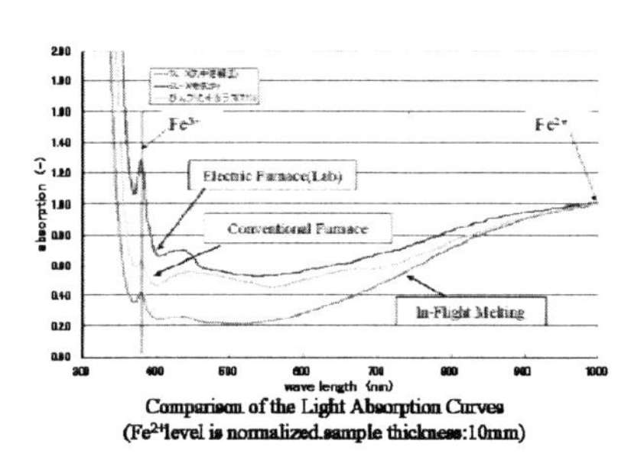

Figure 11. Fe3+/Fe2+

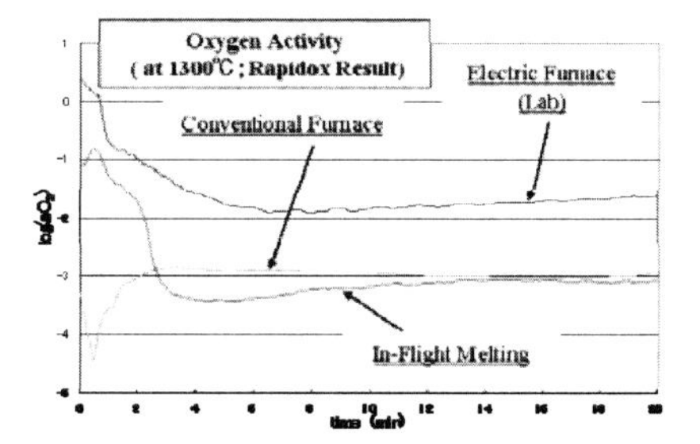

Figure 12. Oxygen activity Comparison (Rapidox)

Energy Consumption Comparison

Fig.13 shows the energy consumption level of the various size glass furnaces and the one ton experimental In-Flight Furnace that is 1000 kcal/kg-glass (4.18 GJ/ton-glass). This target does not include the refining process. An ultimate target level, with refining, is shown by the dotted line.

Our current level is 1070 kcal/kg-glass (4.47 GJ/ton-glass). But, we expect that we will be able to achieve our target level.

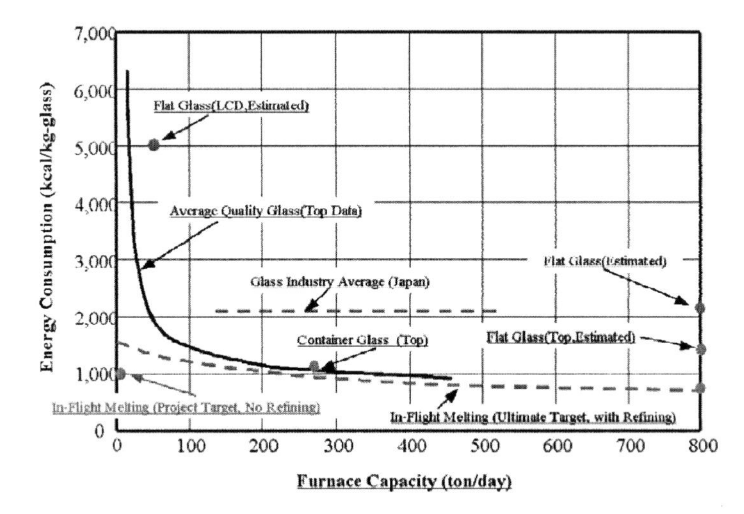

Figure 13. Energy consumption of the experimental in-flight melting vs. existing furnace

ACKNOWLEDGEMENT

This Project has been funded by the New Energy and Industrial Technology Development Organization (NEDO). We appreciate their continued support.

*Corresponding author: masanori_iwamoto@toyo-glass.co.jp

ENERGY EFFICIENCY SIMULATIONS USING FULLY COUPLED AND CONTROLLED REGENERATIVE FURNACE MODEL

Miroslav Trochta, Jiří Brada, Erik Muijsenberg
Glass Service, Inc.
Rokytnice 60
5501 Vsetin, Czech Republic

ABSTRACT

Full 3D CFD modeling of glass furnaces is increasingly becoming a standard tool for predictions of energy balance and glass quality. State of the art 3D furnace models usually include energy transfer inside the furnace and between the furnace and surroundings. Real regenerative furnaces are complex systems with complex interactions among regenerators, combustion chamber(s), glass melt and control systems. To achieve accuracy and realistic response of numerical models of glass furnaces, especially in optimization studies aimed at energy savings, the models should include all these interactions. GS Glass Furnace Model has been enhanced to have the 3D regenerator model integrated with the furnace model. Thus, changes inside the furnace influence behavior of regenerators and vice versa. Behavior of the control system is mimicked too – gas and air flow rates and other inputs are automatically adjusted to maintain specified temperatures at controlled thermocouple locations. An example case study, focused on energy savings, on an industrial size cross-fired regenerative float furnace, is presented. The study demonstrates how the model reacts on changes in furnace design and operating parameters. The cases are compared and differences in energy balance and glass quality are interpreted.

INTRODUCTION

Glass Furnace Simulations and Energy Savings

As was shown by extensive comparisons of energy efficiency of industrial glass melting furnaces, contemporary glass furnaces still have big potential for energy savings. To test new ideas, to get insight in the process, and to estimate impact of changes of furnace design and/or operation on energy savings, numerical (CFD, Computational Fluid Dynamics) simulation tools have been successfully used in many cases in the past.

Glass Service, Inc. has been developing a furnace simulation package, GS Glass Furnace Model (GS GFM) since 1990. The simulated phenomena include 3D glass flow, 3D combustion gas flow, heat transfer (convective, radiative, and conductive) heat transfer among glass melt, batch, combustion space, refractories, insulations and surrounding structures. GFM also contains special models that simulate batch melting, electric boosting, bubbling, stirring etc. Output glass quality is indicated by various quality indexes calculated from data obtained from particle tracing performed on the resulting fluid flow fields. The outputs also include various reports, including the global heat balance, as shown in Fig. 1.

To investigate various ideas of improving furnace efficiency, a case study is usually conducted: A base case, representing a "current" or "typical" operation of the furnace, is computed. As the base case can usually be compared with reality, this is also a useful for model validation. Once a good base case is ready, modifications (i.e., design and/or operating parameter changes) can be introduced and other cases are computed. The glass quality indicators, energy balance and emission indicators of the new cases are compared with those of the base case. The results can also be compared in much more detail to get insight into the process. Values of any calculated or derived physical variable can be easily obtained in any point of the furnace, statistics can be calculated, charts can be plotted etc.

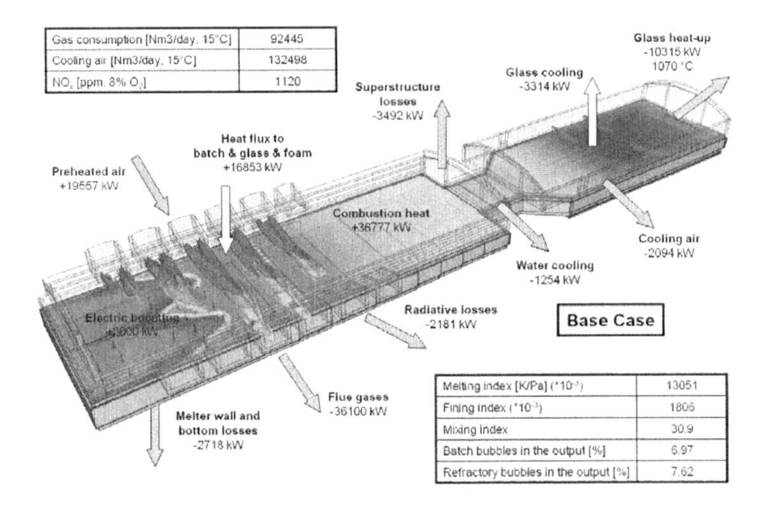

Figure 1. Sample result of a float glass furnace simulation – heat sources and losses, glass quality prediction.

METHODS AND PROCEDURES

Model 'Control'

When furnace design or operating parameters are modified and a new case is calculated, the user should make sure that the operating parameters are consistent, i.e., that an existing furnace would also use a similar set of operating parameters under the given conditions.

For example, if electric boosting power is increased on a real furnace, change of other operating parameters is automatically induced: The control system (or operator) would react on the increased energy input, probably by decreasing firing. The user should make sure that the simulation does the same, i.e., the model should also mimic this behavior of the furnace. To help the users in this task, GS GFM offers built-in PID controllers that regulate temperature in a specified model location (usually location of a real thermocouple) by manipulating flow rates of fuel, oxidizer and/or cooling media. Typical configuration of the controllers on a float glass furnace is shown in Fig. 2.

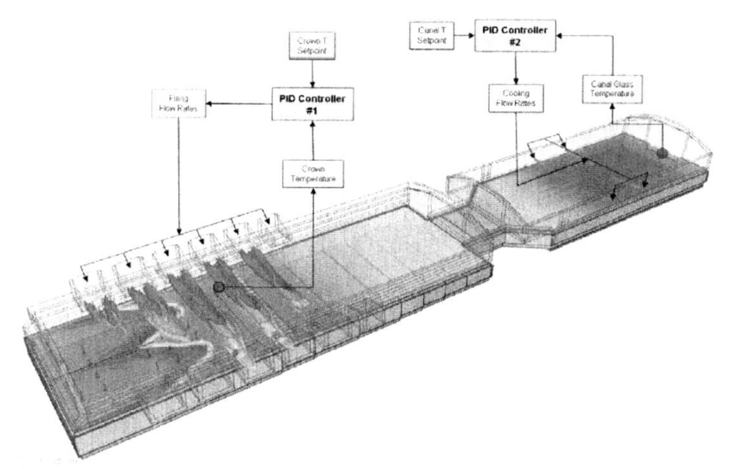

Figure 2. Typical control strategy of a model of a float glass furnace.

As soon as the controllers are properly set up and a realistic base case is calculated, simulation of new cases with different operating parameters and/or design changes can be performed very easily. As the controllers typically modify fuel flow rates, the fuel consumption of the new case is a result of such regulation and can be immediately compared with that of the base case.

Fig. 3 shows predicted changes of heat balance fuel consumption and glass quality when electric boosting is increased by 500 kW, compared to base case.

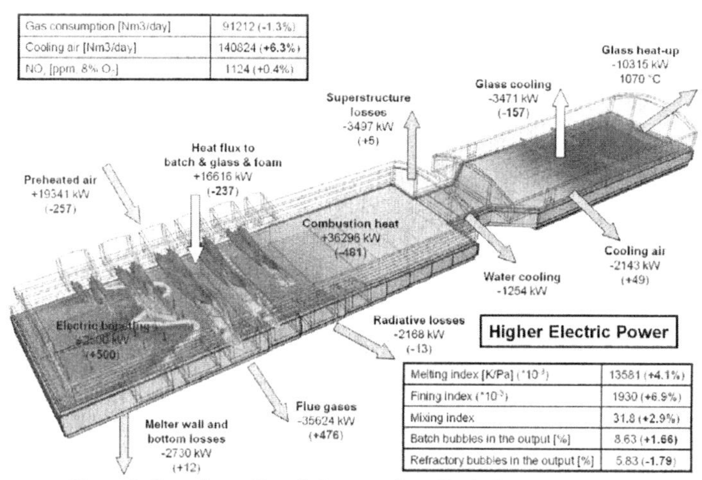

Figure 3. Overview of heat balance and quality indicators – electric boosting increased to 2500 kW (base case is 2000 kW).

PREHEATED AIR

Traditionally, combustion chambers of regenerative furnaces are simulated "port to port", i.e., regenerators are not included in the simulation. Preheated air temperatures are prescribed by boundary conditions. However, in case studies, it is not realistic to keep the preheated air temperature fixed. E.g., preheated air temperature varies with firing rate and flow distribution among the ports.

GS GFM offers a boundary condition that gives a more realistic result than fixed temperatures: Preheated air temperature is calculated from energy available at the exhaust port and user-defined regenerator efficiency, as shown in Fig. 4.

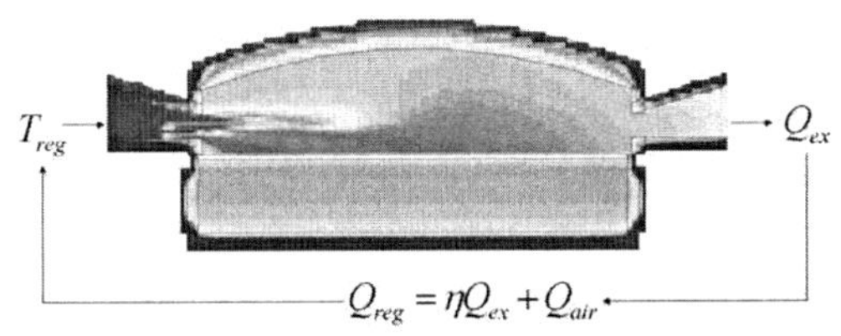

Figure 4. 'Heat Recovery' boundary condition.

This approach is more realistic than fixed temperature. However, in reality, the efficiency of heat recovery is also not constant – it varies with many parameters that influence properties of both air and flue gases, e.g. amount of batch gases (pull rate, cullet), combustion ratio, fuel type, air humidity etc. Probably the best solution is to have regenerators included in the simulation. As the checker geometry is complex, it would be very difficult, memory- and time-consuming to simulate the flow in the detailed checker geometry. GS GFM employs a special porous wall model to approximate the checkers and keep the computation efficient, while all important phenomena are included and all outputs of interest (e.g. flow patterns in the regenerators) are calculated. This approach also allows investigation of design and operation of the regenerators, change checker types etc. Before the porous wall approximation is used, detailed models of regenerators are calculated to find values of those parameters that are crucial for the approximation.

RESULTS AND DISCUSSION

Example Case Study – Base Case

To demonstrate function of the model, a sample case study was conducted. The base case is a model of a full industrial scale, 500 MTPD float furnace with 6 ports (5 ports firing), gas-fired, chimney block regenerators (separate chambers for each port), batch current, two rows of electrodes (1500 kW), one bubbler row, two coolers and a row of stirrers in the waist. The glass is a green float glass with 0.5% Fe_2O_3.

The following performance indicators were observed in each case: Fuel consumption (absolute or per kg of glass), heat recovery efficiency, NOx [g per kg of glass], melting quality (average of worst 0.1%), fining quality (average of worst 1%), mixing quality (average of worst 0.1%) and mean glass temperature.

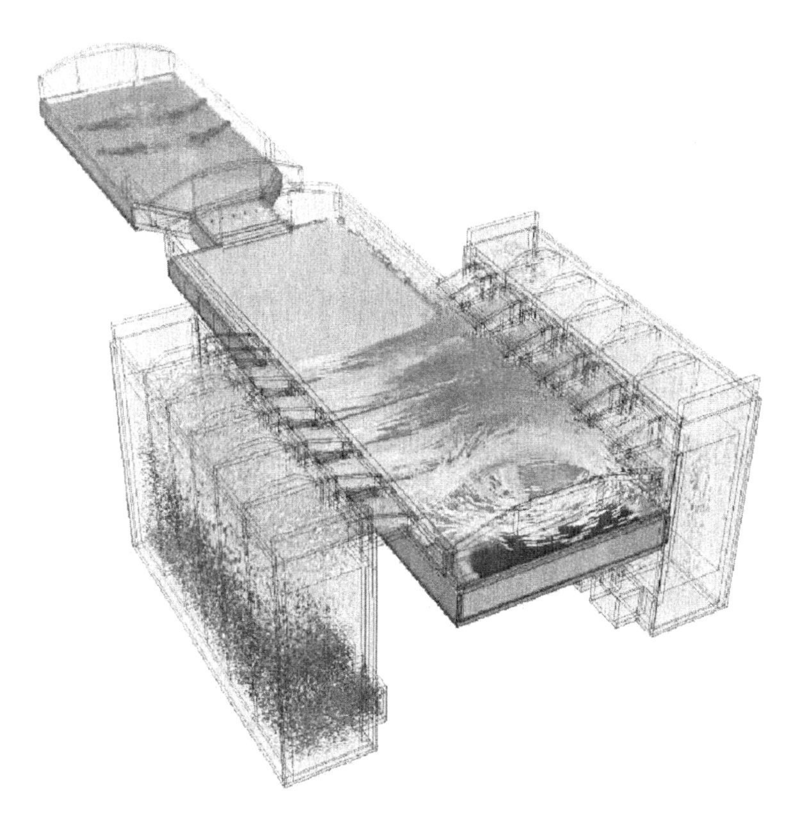

Figure 5. Float furnace model including regenerators

Case 1: Furnace Pull Decreased by 50 Metric Ton Per Day (MTPD)

In Case 1, furnace pull was decreased from 500 to 450 MTPD to demonstrate impact on energy consumption and glass quality. The following (Table I) shows the simulation result in absolute numbers and relative to the base case.

Values of changes which indicate any improvement in the furnace performance are indicated by bold letters, while values corresponding to any deterioration are indicated by italic.

Table I. Calculation results for Case 1

Parameter	Value	Change
Fuel consumption [kW]	34214	**-5.07%**
Fuel consumption [kJ/kg glass]	6569.2	*+5.48%*
Regenerator efficiency [%]	56.15	**+0.88**
NOx [g/kg glass]	4.610	*+10.85%*
Melting index	1.669	**+15.42%**
Fining index	1.369	**+57.14%**
Mixing index	3.534	**+9.26%**
Mean T [°C]	1289.4	**+6.89**

PID regulation decreased firing by 5%. Predicted glass quality is significantly better than in the base case, the mean glass temperature is almost 7°C higher. Thus, to retain the original glass quality, it would be possible to make a "second iteration", i.e. to setup another case, where the controlled temperature setpoint would be several °C lower, which would lead to a further decrease of firing and energy saving. In this way, the model can help find good operating parameters for a new glass or new pull rate.

Case 2: Cruciform Checkers

In Case 2, chimney block checkers were replaced by cruciform. The result showed what was expected: as fluids in cruciform checkers have partial horizontal freedom, distribution of the flow in the chamber is much more uniform, as shown in Fig. 6.

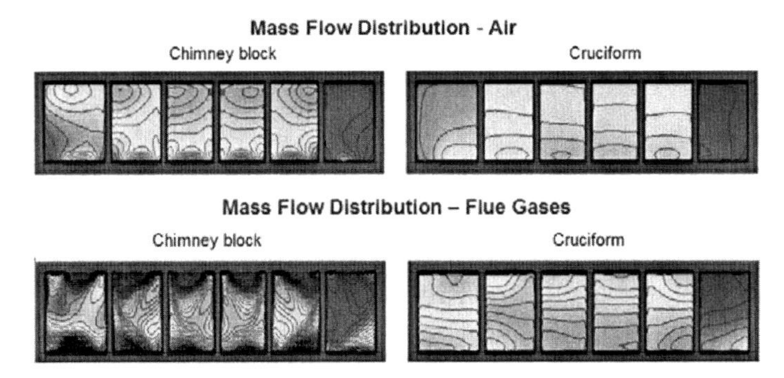

Figure 6. Comparison of flow distribution in cruciform and chimney block checkers.

The simulation predicted 4.74% better regenerator efficiency, which allows 7% fuel savings, still with slightly better glass quality.

Table II. Calculation results for Case 2

Parameter	Value	Change
Fuel consumption [kW]	33512	**-7.02%**
Regenerator efficiency [%]	60.02	**+4.74**
NOx [g/kg glass]	4.262	*+2.50%*
Melting index	1.546	**+6.93%**
Fining index	0.887	**+1.83%**
Mixing index	3.502	**+8.26%**
Mean T [°C]	1282.7	+0.16

Case 3: Cruciform Combined With Chimney Block

As Case 2 and also practical experience shows, cruciform checkers yield better heat recovery efficiency than chimney block, but there is also a difference in investment costs. To improve the flow distribution in the regenerator, the cruciform checkers do not have to cover all the height of the checkerwork – investigation of the flow pattern showed that the most critical part is the top of the checkerwork. Putting cruciform at the top allows the flue gases to have enough time to distribute horizontally. Air flow on the firing side is significantly slower and its distribution is more determined by distribution of the chimney effect, i.e., by checker temperatures.

In Case 3, the upper third of the checkerwork is cruciform, the rest is chimney block. The simulation predicted 4.4% fuel savings, with almost the same glass quality as the base case.

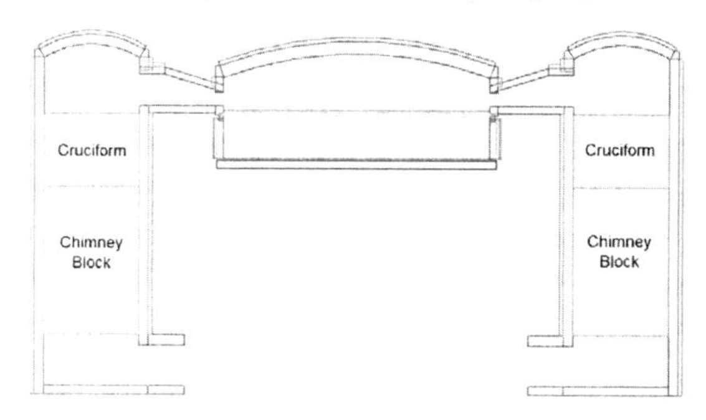

Figure 7: Regenerator setup and simulation result.

Table III. Calculation results for Case 3

Parameter	Value	Change
Fuel consumption [kW]	34456	-4.40%
Regenerator efficiency [%]	58.20	**+2.92**
NOx [g/kg glass]	4.199	*+0.97%*
Melting index	1.465	**+1.30%**
Fining index	0.865	*-0.75%*
Mixing index	3.479	**+7.57%**
Mean T [°C]	1282.6	+0.07

In practice, it would be possible to use the model to help find optimum checker layout, i.e., to find tradeoff between checker cost and fuel consumption.

Case 4: Flue Gases Injected in Air

One of the factors that limit efficiency of heat recovery is low absorptivity of air: While heat transfer from flue gases is mainly radiative, heat transfer from checkers to the air is dominantly convective. Thus, adding absorptive species to the air could significantly improve regenerator efficiency. The most easily available absorptive gas is the flue gas. As was already proposed by several authors, relatively small amounts of flue gas injected into air could significantly increase absorptivity.

In case 4, about 4% of flue gases are injected below the checkers. Although the injection is not very sophisticated and mixing is not perfect (further development would be useful here), fuel consumption was reduced by 1.45%.

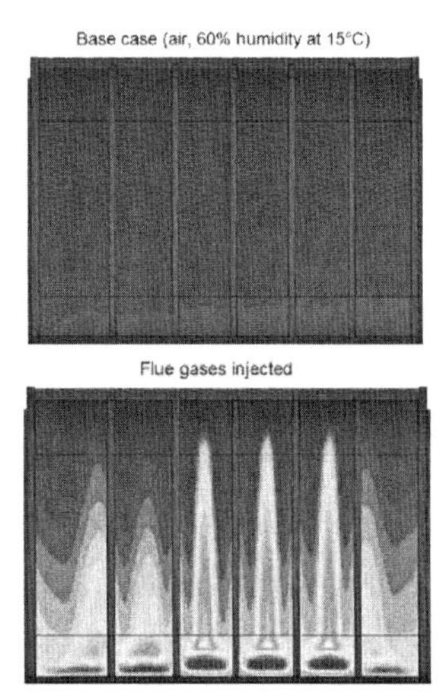

Figure 8. Absorption coefficient of air and air with flue gases injected.

Table IV. Calculation results for Case 4

Parameter	Value	Change
Fuel consumption [kW]	35518	**-1.45%**
Regenerator efficiency [%]	58.24	**+2.96**
NOx [g/kg glass]	4.132	**-0.64%**
Melting index	1.531	**+5.90%**
Fining index	0.858	*-1.49%*
Mixing index	3.369	**+4.17%**
Mean T [°C]	1283.2	**+0.64**

CONCLUSION

Two more steps were made to get modeling of regenerative furnaces closer to reality: The model itself mimics the furnace control which leads to more realistic behavior. Including regenerators in the model can give more accurate answers on energy balance.

GS Glass Furnace Model can assist in making decisions to save energy, reduce emissions and optimize furnace operation during its entire life:

- Optimize furnace design
- Optimize furnace operation and control strategy
- Investigate flexibility of furnace operation (color change, pull change, change of firing/boosting ratio, ...)
- Simulate furnace aging (e.g., what happens if the regenerators get plugged/corroded)
- Simulate changes during furnace lifetime (additional cooling, insulation addition/removal, ...)
- Simulate melting intensification and test new ideas of improving melting efficiency (bubbling, stirring, oxy boosting, regenerator improvements, ...)
- Optimize tradeoff between costs of investment and operation (e.g., profitability of electric boosting depends on investment costs, gas price, electricity price, CO2 penalties, future development of the prices, impact on glass quality, ...)

REFERENCES

[1]R. Beerkens, "Energy efficiency improvement potentials in glass manufacturing", *XXII International Congress on Glass*, Salvador, Bahia, Brazil, 2010.

[2]P. Schill et al., "Integrated Glass Furnace Model", *5th International Seminar on Mathematical Simulation In Glass Melting*, Horni Becva, Czech Republic, 1999.

[3]B.F. Launder and D.B. Spalding, *Computer Methods in Applied Mechanics and Engineering*, Volume 3, (1974), 269-289.

[4]J. Bauer, O.R. Hoffmann and S. Giese, *Glasstechnische Berichte*, Volume 67, (1994), 272-279.

*Corresponding author: bradaj@gsl.cz

STRATEGIC HIGH QUALITY QUARTZ SUPPLY FOR FUSION INTO SILICA GLASS

Carlos K. Suzuki, Murilo F. M. Santos, Eduardo Ono, Eric Fujiwara
State University of Campinas, UNICAMP-FEM, 13083-970 Campinas, SP, Brazil

Delson Torikai
Sao Paulo University, Escola Politecnica, Sao Paulo, SP, Brazil

Armando H. Shinohara
Federal University of Pernambuco, UFPE-DEMEC, Recife, PE, Brazil

ABSTRACT

An extensive in loco survey of main quartz producing regions in Brazil integrated with physicochemical characterization studies of quartz lascas and their fusion into silica glass bring the possibility of obtaining high quality quartz powder supplies for various industrial applications. The advantage is the relatively simple, cheap and environmentally friendly processing due to the high quality huge reserve of quartz ores of hydrothermal geological origin. The possibility of using highly transparent first grade quartz lascas in conventional fusion processes allows the fabrication of nearly bubble free silica glass. Due to the increasing demand of silica glass for processing solar cells and other products, this material is certainly a strategic supply for the present and future scenarios.

INTRODUCTION

The advent of semiconductor industries in the 1960's and their rapid expansion in subsequent years demanded a large amount of silica glass[1] produced by fusion of natural quartz lascas* from Brazil, the only raw material available for this purpose at that time. Nevertheless, Brazilian miners and exporters were not able to improve quality control in terms of impurities and transparency to fulfill the rapid evolution of raw material specifications for fusion into silica glass.[2-4] With the advent of Iota high quality pure quartz powder[5] processed by Unimin in the United States, the share of Brazilian quartz lascas in the international market has decreased to almost zero. In recent times, however, due to the short delivery of Iota Quartz, most of the silica glass manufacturers in the world have been looking for alternative sources of quartz powders. This scenario has become critical due to the increasing demand of China, especially to process solar photo-voltaic cells. In addition, the perspective of solar electricity becoming one of the major energy sources by 2050 will also increase the demand for such material.[6]

Many countries have quartzite and quartz sand typically of pegmatite geological origin, but it is unusual to find large reserves of hydrothermal origin quartz, which is much purer in composition with higher transparency than most pegmatite ores.[7] Quartz resources in Brazil surprisingly present a huge occurrence of both pegmatite and hydrothermal quartz. The main quartz locations in Brazil were delineated during the Second World War as a result of an intense search of piezoelectric grade quartz for strategic use in radio communications. Quartz ores are distributed along large areas called quartz belts (Fig. 1.a), such as the Central belt in the states of Goias and Tocantins, the Coastal belt in the State of Espirito Santo, and the Minas-Bahia belt, which extends for approximately 1000 km along the Diamantina mountains. Motivated by the scenario of high demand, we have conducted a large scale survey extending from the southern region to the north of the country in Amazon, State of Para.

The project was conducted by UNICAMP-The State University of Campinas with the financial support of JICA-Japan International Cooperation Agency. Approximately two hundred quartz mines were surveyed to make a kind of quality mapping of quartz mines in various states: Minas Gerais, Bahia, Goias, Para, Tocantins, Ceara, Rio Grande do Norte, Sao Paulo, and Rio Grande do Sul.[8]

Extraction sites in the mine in Oliveira dos Brejinhos, State of Bahia, and mine in Diamantina, State of Minas Gerais, are shown in Figs. 1.b and 1.c; respectively. Fig. 1.d shows a giant piezoelectric grade single crystal.

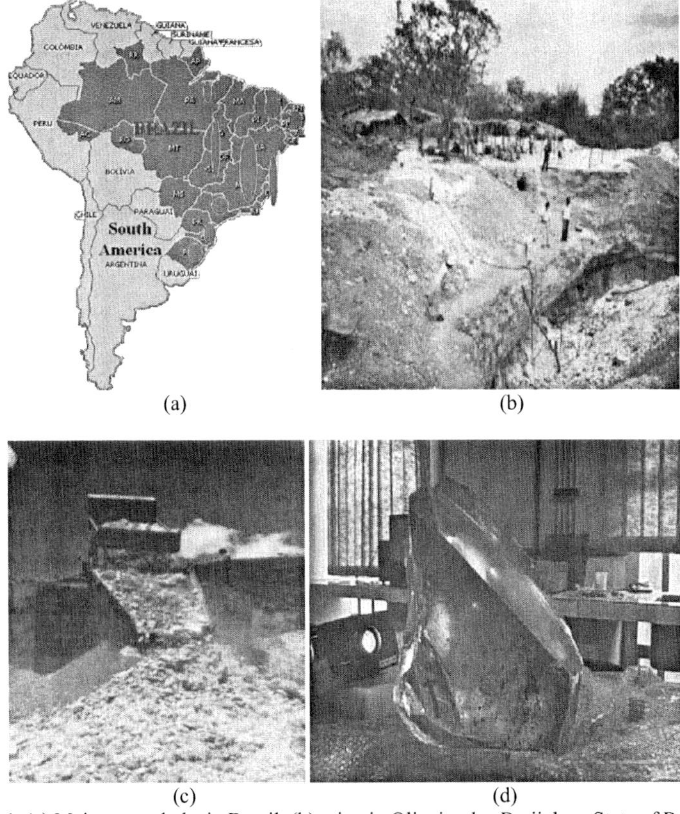

(a) (b)

(c) (d)

Figure 1. (a) Main quartz belts in Brazil; (b) mine in Oliveira dos Brejinhos, State of Bahia; (c) extraction in Diamantina, State of Minas Gerais; (d) giant piezoelectric grade single crystal quartz.

Graded Lascas and Fluid and Solid Inclusions

The degree of transparency of natural quartz varies according to the concentration of fluid inclusions and the number of small cracks. Depending on the type of lasca, it is possible to observe in the same lasca, parts of high transparency and different degrees of opacity. Therefore, for controlling transparency of quartz lascas, quartz ores are classified into graded lascas, denominated as first, mixed, second, third, and fourth (Fig. 2.a), and also to control eventual solid inclusions and large fluid inclusions by observation inside water (Fig. 2.b). Such a classification procedure can also be made by an automation system using a high resolution CCD camera.[9]

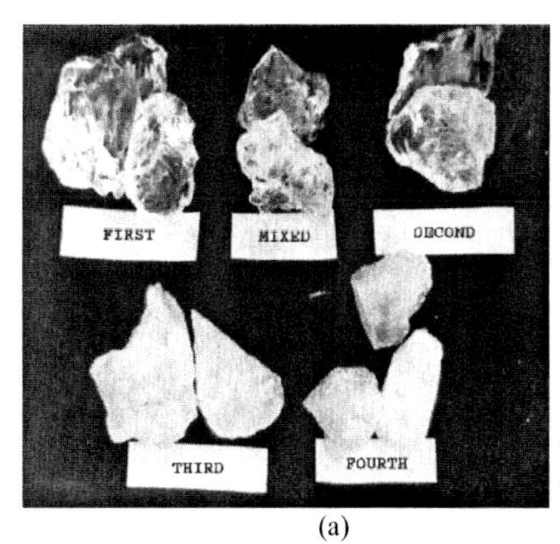

(a) (b)

Figure 2. (a) Graded lascas based on transparency degree. (b) Naked eye visual quality control of lascas in water tank. Inset shows optical microscopic image of solid and large fluid inclusions in lascas.

Silica Glass

Unique characteristics are found in silica glass, such as optical transparency and high resistance to corrosion, making it the only usable material for high temperature thermo-chemical treatment of semiconductors. However, physical properties of silica glass, such as viscosity, density, acoustic wave propagation, thermal conductivity, index of refraction, etc., depend on the fusion method and the metallic impurities contained in the raw material. In general, the silica glass manufacturers specify their products in various types according to the raw material, method of fabrication, and the type and concentration of impurities.[10]

Usual classification of silica glass based on natural quartz is:

(i) Type I, by fusion of natural or synthetic quartz powder in electric furnace in vacuum or in inert atmosphere at low pressure using plasma. Some of type I commercial products are Vitreosil, Infrasil, GE124, KI, KS4V, Puropsil A, Pursil, and T2030.

(ii) Type II, silica glass obtained by flame fusion, in general using hydrogen-oxygen flame. The OH content is ~120 ppm and metallic impurities are less than the initial raw-material by the effect of vaporization in high temperature flame. Some examples of type II products are Armesil T08, Heralux, Herasil, Homosil, KU-2, KV, OG Vitreosil, Optosil I, T-1030, and Ultrasil.

The present research aimed to obtain high quality and low cost quartz supplies for silica glass manufacture for various industrial applications, in particular, to fulfill the strong demand for processing solar photo-voltaic cells.

COMPARISON WITH IOTA QUARTZ AND DISCUSSION

Impurities content in natural quartz shows a good correlation with geological formation but it is particularly dependent on the region and the mine itself. It is possible to find a very high purity material even though in-natura. Table I shows the result of impurities analysis of lascas (first, mixed, second and third) from the same mine and their comparison with quartz powders Iota Standard and Iota 4. Surprisingly, they are comparable, even though practically without any purification processing.

Table I. Impurity contents in natural Brazilian quartz lascas compared to IOTA quartz (ppm).

Lascas	Fe	Al	Na	K	Li	Ti	Ca	Cu
First	0.2	9.7	0.3	0.2	1.2	0.5↓	0.1↓	0.02↓
Mixed	0.4	10.1	0.5	0.9	1.0	0.5↓	0.1↓	0.02↓
Second	0.4	8.3	0.4	0.2	0.9	0.5↓	0.2	0.02↓
Third	0.4	11.0	0.3	0.2	1.1	0.5↓	0.4	0.02↓
IOTA Standard	0.2	16.2	0.9	0.6	0.9	1.3	0.5	0.05↓
IOTA 4	0.3	8.0	0.9	0.35	0.15	1.4	0.6	0.05↓

Understanding the incorporation of different types of impurities in SiO_2 structure is of fundamental interest for the effective purification of natural quartz lascas. Particularly, it is important to understand how Al impurity behaves in quartz as it forms Al-related centers in conjunction with other types of impurities, e.g., Li^+, H^+, Na^+. Al impurity can enter the SiO_2 lattice as interstitial or substitutional. Interstitial Al can usually be removed by leaching, but substitutional one is quite stable by chemical leaching and remains even after fusion into silica glass. It is then essential and strategic to know a priori such characteristics in the quartz ores before spending time and energy for extraction and processing. The form of Al incorporation can be estimated by the correlation of Al and Li concentrations, which shows the tendency to form Al-Li in a larger or smaller scale. For example, in the material of mine 1 (Table II), the ratio Al/Li is about 1. It increases to 1.4 and 6.1 for mines 2 and 3, respectively. γ-ray irradiation causes the dissociation of Al-Li centers and formation of Al-hole and Al-OH centers[11-12] inducing the darkening effect by color centers related to the form of Al incorporation in the lattice. Higher the concentration of substitutional Al, more intense the effect of darkening, that can be quantified by the darkening factor (D), which corresponds to the quantity of Al-hole centers defined as:[8]

$$D = \alpha_\gamma - \alpha_0, \quad (1)$$

where, α_0 and α_γ are the absorption coefficients in 470 nm, before and after the irradiation, respectively.

Based on this method, it is possible to predict a priori that lascas from mine 1 are not suitable for purification treatment, as the main part of Al-content is substitutional.

Table II. Al incorporation in natural quartz as a function of Al/Li ratio in ppm and the darkening effect by γ-ray irradiation as adapted from Iwasaki et al.[8]

Natural quartz	Al	Li	Al/Li	Na	OH	Darkening degree by irradiation (cm^{-1})	Form of Al aggregation
Mine 1	282	284	1.0	1	32	5.9	Mostly substitutional
Mine 2	34	24	1.4	3	42	1.8	Part substitutional and part interstitial
Mine 3	55	9	6.1	0.3	67	0.4	Mostly interstitial

As aforementioned, the degree of transparency of lascas depends on the content of cracks and fluid inclusions. Parts of the fluid inclusions are eliminated by powdering, but another portion remaining in the grains can generate bubbles in fused materials. High viscosity of silica glass does not allow the elimination of bubbles, particularly in the case of flame fusion in air (Verneuil method). Therefore it is necessary to control good transparency of quartz ores in order to manufacture transparent silica glass. Strict selection of first grade lascas allows the manufacture of nearly bubble free silica glass by the Verneuil method (Fig. 3).

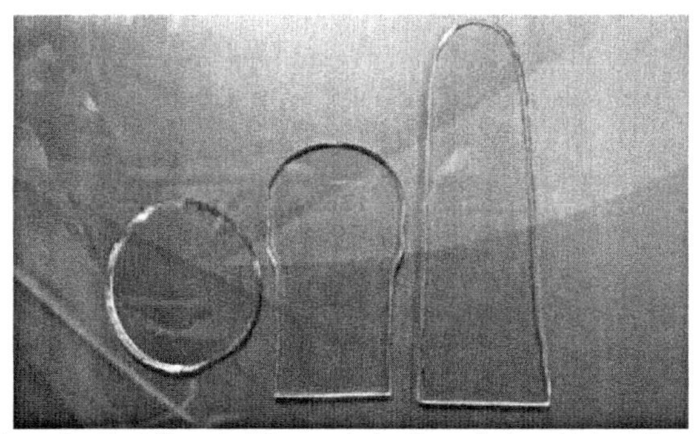

Figure 3. Silica glass of optical fiber cladding grade by flame fusion obtained from first grade lascas.

CONCLUSIONS

Characterization studies of Brazilian quartz extracted from various regions and their powdering and fusion into silica glass revealed technological and economic advantages of using quartz lascas for fusion into silica glass to fulfill the increasing demand for solar cells processing, optical fiber industries in terms of cladding and dummy rods, and other applications. Significant benefit to the environment can result by obtaining high purity quartz powder with a smaller quantity of energy, chemicals and infrastructure in comparison with the usual quartz ores; quartz sand or pegmatite origin quartz containing much higher concentrations of impurities.

ACKNOWLEDGMENTS

The authors would like to acknowledge the financial support of JICA, METI/AIST/ITIT, FINEP, CNPq, FAPESP and CAPES. They also wish to acknowledge Drs. Hideo Iwasaki, Fumiko Iwasaki, and Hiroshi Shimizu; Beneficiamento de Minerios Rio Claro Ltd., Sun Quartz CF, and Optron Micromecanica Optica Ltd.

FOOTNOTES

*lascas are small natural quartz fragments of 20-200g classified according to their transparency degree.[13-14]

REFERENCES

[1]R. Doering and Y. Nishi, *Handbook of Semiconductor Manufacturing Technology*. CRC Press. ISBN 1574446754, (2007).

[2]C. K. Suzuki, F. Iwasaki, and H. Iwasaki, Al impurity purification in quartz grown by the hydrothermal method, *Jpn. J. Appl. Phys.*, **28-1**, 68-72 (1989).

[3]D. C. A. Hummel, H. Iwasaki, F. Iwasaki, and C. K. Suzuki, Characterization of natural quartz lascas for high technology industries, part II: study of density and optical micrography measurements, *Ceramica*, **32**, 281-284 (1986).

[4]L. C. Diniz Filho, Quartzo Cristal, Sumario Mineral, DNPM/RN-National Department of Mineral Production, (2008).

[5]Iota High Purity Quartz, http://www.iotaquartz.com/world.html, (2011).

[6]German Advisory Council on Global Change, Fraunhofer Institut Solare Energiesysteme, (2003).

[7]J. Göetze, Chemistry, texture and physical properties of quartz – Geological interpretation and technical application, *Mineralogical Magazine*, **74-3**, p.645-671, (2009).

[8]F. Iwasaki and H. Iwasaki, Impurity species in synthetic and Brazilian natural quartz, *Jpn. J. Appl. Phys.*, **32**, 893-901 (1993).
[9]F. D. de Paula, R. Kakitani, M. F. M. dos Santos, E. Fujiwara, E. Ono, and C. K. Suzuki, Improvement of the lascas selection methodology applied to the manufacturing of solar photo-voltaic cells, In: IX Brazil-Japan International Workshop: Society, Energy and Environment, (2011).
[10]M. L. F. Nascimento and E. D. Zanotto, Diffusion process in vitreous silica revisited, *Phys. Chem. Glasses: Eur. J. Glass Sci. Technol. B*, **48-4**, 201-217 (2007).
[11]H. G. Lipson and A. Kahn, Infrared characterization of aluminum and hydrogen defect centers in irradiated quartz, *J. Appl. Phys.*, **58**, 963-66 (1985).
[12]L. B. F. de Souza, P. L. Guzzo, and H. J. Khoury, Correlating the TL response of γ-irradiated natural quartz to aluminum and hydroxyl point defects, *J. Lumin.*, **130**, 1551-56 (2010).
[13]H. Iwasaki, F. Iwasaki, V. A. R. Oliveira, D. C. A. Hummel, M. A. Pasquali, P. L. Guzzo, N. Watanabe, and C. K. Suzuki, Impurity content characterization of Brazilian quartz lascas, *Jpn. J. Appl. Phys. - Part 1*, **30**-7, 1489-1495 (1991).
[14]V. A. R. Oliveira, H. Iwasaki, and C. K. Suzuki, Study of impurities in Brazilian quartz lascas from various regions, *Ceramica,* **32-202**, 181-183 (1986).

*Corresponding author: suzuki@fem.unicamp.br

Processing and Applications of Glass

MEDICAL INTERACTIONS WITH GLASS PACKAGING SURFACES

R. G. Iacocca
Senior Research Advisor, Analytical Sciences R&D
Lilly Research Laboratories
Eli Lilly and Company
Indianapolis, IN 46205

ABSTRACT

Recent product recalls in the pharmaceutical industry over the detection of visible glass flakes (known as delamination) in parenteral (injectable) products has focused a great deal of attention on glass/product interactions. Because the presence of flakes is often detected after the product has been released to the market, delamination events can result in product recalls and product shortages. If one examines the glass science literature, one will learn that the generation of flakes from glass surfaces exposed to water-containing environments (known as weathering) is not a new phenomenon, and in fact, the first mention of glass/water interaction was reported in 1770. This paper will look at the mechanism behind the weathering of glass, and describe specific tests that can predict glass/product interactions for glass pharmaceutical containers.

INTRODUCTION

Because many pharmaceutical liquids are injected directly into the human body, thereby bypassing many of the filters that exist in the gastrointestinal, the existence of glass flakes is completely undesirable, posing possible risks to patient safety, and severely impacting final product quality. Glass delamination has been described in the pharmaceutical literature.[1,2] Currently, the definition of delamination relies on the visual detection of glass flakes (also known as lamellae), which can present some interesting challenges. Differences in human visual acuity will manifest in varying degrees of detection for a product that is delaminating glass. Second, the actual appearance of glass flakes is the final step in the delamination process. The actual loss of corrosion resistance of the glass surface occurs much sooner than the manifestation of visual glass flakes.[3]

Decades ago, Dimbleby recognized the challenges of relying solely on the visual detection of glass flakes as an indicator of glass suitability with a given pharmaceutical liquid,[4-8] and developed multiple alternate analytical tests for the assessment of glass durability with pharmaceutical liquids. This work was further progressed by Iacocca et al.[3] who showed that the so called "precursor" tests can produce leading results of glass incompatibility, compared with the weeks or even months required for the manifestation of visible glass flakes.

Glass reactions with water are well known in the glass science literature. Lavoisier was the first to observe that boiling water produced changes in the appearance of the glass container used to hold the water.[9] For glass exposed to ambient moisture or naturally occurring sources of water, the loss of chemical durability on the glass surface is referred to as weathering, and is commonly encountered with glass antiquities (and recognized as detrimental).[10-13] Figure 1 shows a low-magnification optical micrograph of a piece of glass that has been weathered.

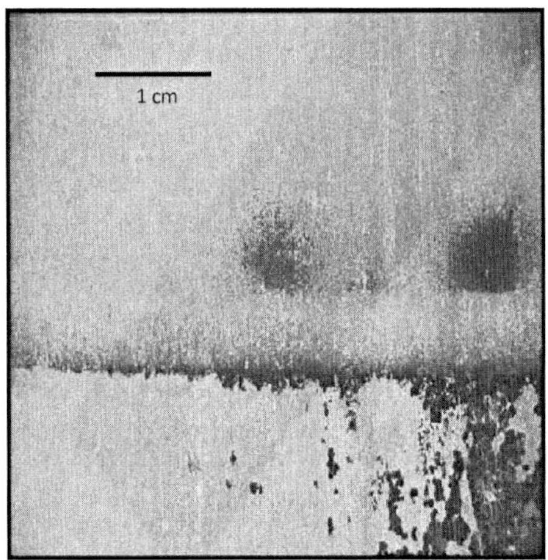

Figure 1. Optical micrograph of weathered glass showing the appearance of visible glass flakes on the exposed surface.

Figure 2 is a scanning electron micrograph of the surface of a glass antiquity that has undergone weathering. Later in this paper, it will be shown that a glass vial that has undergone delamination strongly resembles this cracked appearance as well.

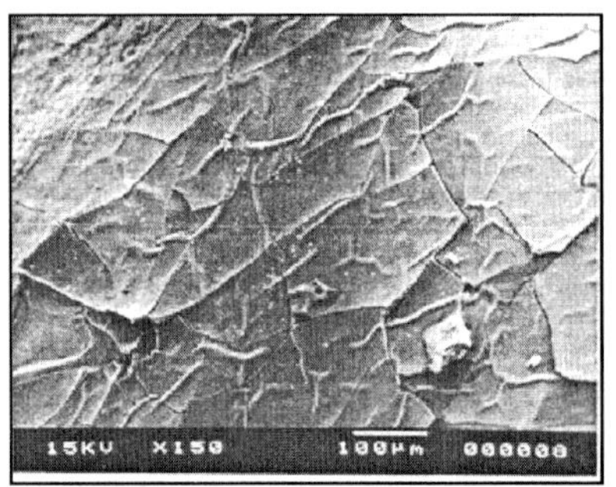

Figure 2. Scanning electron micrograph of a weathered glass antiquity.[13]

The mechanism behind the formation of these flakes in weathered glass is fairly well understood. White has put forth six basic mechanisms by which glass and ceramics undergo chemical degradation; however, his explanation does not specifically refer to the generation of flakes.[14] In weathering, the silicate network is disrupted and dissolved by the water that is present. This reaction is favored under basic conditions.[15] The dissolution of the silica in the glass leaves behind a glass (or a gel) that has a chemical composition that is different from the bulk glass. Upon drying, this surface layer, which contains less silica than the bulk composition, will exhibit mechanical properties that are different than the substrate. The interfacial strain that is present is sufficient to cause the layer to crack. It has been shown that glass that has undergone delamination has a surface chemistry that is depleted of Si, thereby supporting that delamination is an extension of weathering.[2]

EXPERIMENTAL PROCEDURE

A large, statistical design of experiments was created to test the impact of all of the possible factors that could compromise the chemical resistance of glass vials used to contain pharmaceutical products. These factors included:

- Depyrogenation -- or thermal exposure in a continuous oven-- to remove any carbonaceous debris and to sterilize the vials prior to filling.
- Exposure to terminal sterilization via autoclaving (123°C for 15 minutes – representative cycle for the pharmaceutical industry).
- The kind of glass vial used, including ammonium sulfate treatment that is commonly employed to reduce surface alkalinity (preventing alkaline elements from being leached into the drug product). Table I. contains the chemical compositions of the vials that were used in this study.
- The chemical nature of the liquid being contained.
- Exposure times and temperatures for filled vials.

Table I. Composition of glass vials used in this study.

	Vendor A	Vendor B	Vendor B
	(wt. %)	(wt. %)	(wt. %)
SiO_2	80.5	75.0	75.0
B_2O_3	12.6	10.5	10.5
Al_2O_3	2.2	5.0	5.0
Na_2O	4.2	7.0	7.0
K_2O	<0.1	–	–
CaO	<0.1	1.5	1.5
Ammonium sulfate treated	Yes	No	No
Vial interior coated with CVD CVD	No	No	Yes

Phase I of the study looked only at depyrogenation times and temperatures on the interior surface of the glass vial for Type I glass vials (three different types of vials were selected, see Table I for description). Type I glass is a borosilicate glass, but does not have the same composition as Pyrex™. Although the depyrogenation temperatures were well below the glass transition temperatures for the glass that was studied, it was included in the study to present a comprehensive analysis of all possible factors. After the completion of Phase I, it was concluded that the depyrogenation temperature (250 and 350 °C) and time (from 0-18 hours) had no impact on the interior surface chemistry as monitored by dynamic secondary ion mass spectroscopy (D-SIMS). It should be noted that the oven used for depyrogenation was continuous and the environment did not contain significant moisture. Had significant moisture been present, it is possible that the unit operation of depyrogenation could have had an impact on the interior vial surface chemistry. Table I shows the various factors that were investigated in Phase II of the investigation.

Table II. Experimental conditions used to test glass stability.

Factor	Specific Levels for Each Factor
Vial Types	Type I (no ammonium sulfate), Type I (ammonium sulfate), Type I (CVD coating of SiO_2)
Depyrogenation Temperature, °C	250, 350 (time fixed at 4 hours)
Terminal Sterilization Cycles	0, 2
Test Solutions	glutaric acid solution (equimolar with MTA formulation), hippuric acid (equimolar with MTA formulation), drug product liquid formulation

All test solutions produced similar results in that delamination was detected. Because of space limitations, only data from the glutaric acid-containing vials will be discussed.

EXPERIMENTAL CONDITIONS

The tests used for this investigation and the rationale behind their selection have been published elsewhere; thus, they will not be presented here.[2,3] Additionally, the rationale behind such testing was clearly recognized by Dimbleby, who published her work in the early part of the 20th century.[5-8]

EXPERIMENTAL RESULTS

Scanning Electron Microscopy

Analysis of the vial interior via scanning electron microscopy continues to be one of the defining tests to assess glass compatibility with pharmaceutical liquids. Figure 3 shows four different scanning electron micrographs of glass that has undergone severe delamination. Note the presence of the glass flakes on the interior surface and the pits that are present in the flakes.

pH Measurements

At elevated pH values, glass dissolution is the dominant mechanism, particularly dissolution of the silicate glass forming network. Under suitable conditions, the dissolved species can react to form silicic acid, which will lower the pH of the solution.[16]

Figure 4 shows a plot of pH vs. time for vials that received no terminal sterilization.

Concentration of Si as Measure by ICP-OES

Figure 5 is a plot of silicon concentration vs. time for vials containing glutaric acid stored at 60°C. Note that for Vendor A vials, the concentration of Si changes over time, Vendor B vials do not demonstrate this behavior. Also worth noting is that the dissolution of silica in S-g glass powder* stored at 50°C in water is less severe by an order of magnitude (as a comparator).

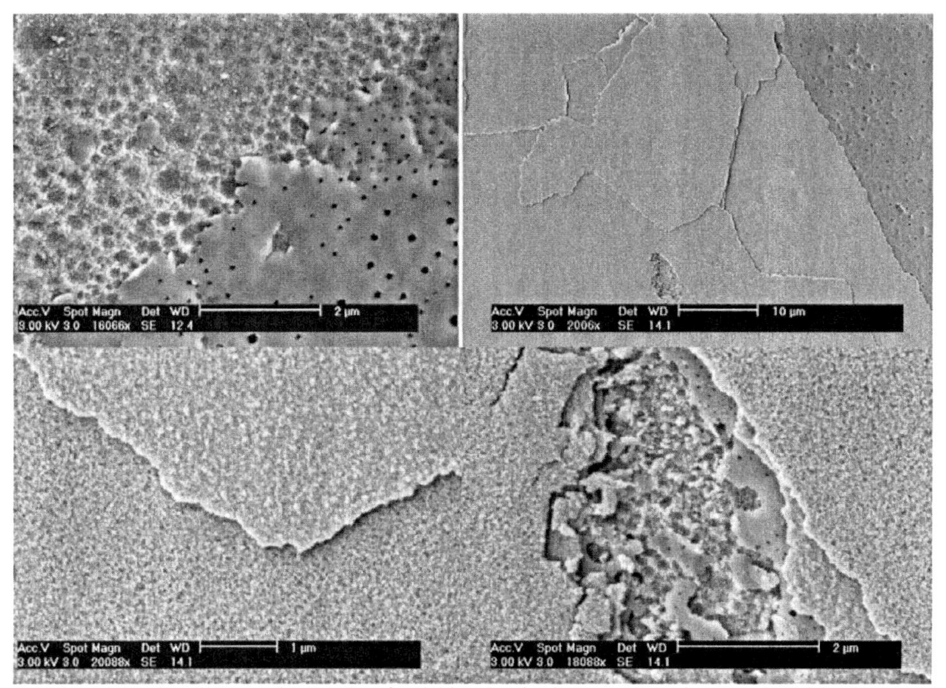

Figure 3. Scanning electron micrographs of a vial interior that has experienced severe glass delamination

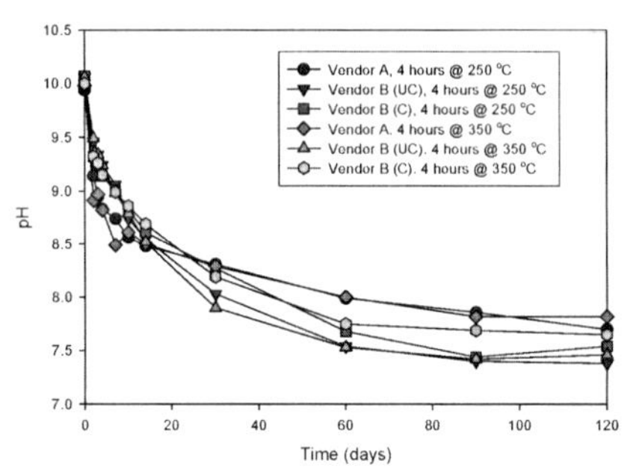

Figure 4. Plot of pH vs. time for three different glass types. These vials were not exposed to terminal sterilization via autoclaving. Vials were stored at 60 °C. (Vendor A = uncoated, ammonium sulfate treated Type I vial; Vendor B (UC) = uncoated, non-ammonium sulfate treated Type I vial; Vendor B (C) = Type I vial with interior coating of pure SiO_2).

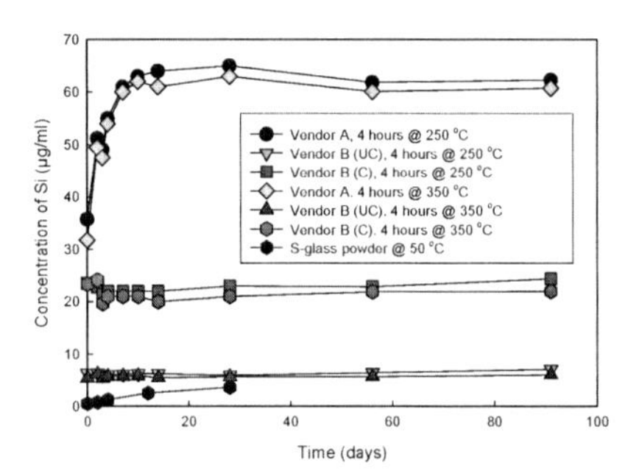

Figure 5. Plot of silicon concentration as a function of time for vials receiving no terminal sterilization via autoclaving, and stored at 60 °C. (Vendor A = uncoated, ammonium sulfate treated Type I vial; Vendor B (UC) = uncoated, non-ammonium sulfate treated Type I vial; Vendor B (C) = Type I vial with interior coating of pure SiO_2).

DISCUSSION

The data presented in this paper support the conclusion that glass delamination is an extension of glass weathering. Silica is dissolved from the vial interior. This is confirmed by dynamic SIMS data (not presented here) of the vial interior,[2] and enters into solution, as evidenced by the ICP data. The glass left behind, whether an actual glass or a gel layer, does not have the same mechanical properties as the Type I substrate. These flakes spall into the liquid, creating a visible particulate in the drug product liquid. It is hypothesized that the pits shown in Figure 3 play a key role in the delamination process, as these pits are the remnants of a Na-B rich phase that separated during the conversion of the glass tubing into a vial.

The observed differences between vendors can be traced back to the manufacturing processes used by each manufacturer. Vendor A used higher temperatures in the manufacture of the vial, which allowed for higher productivity. In doing so, the glass was phase separated, which increased the surface alkalinity. Ammonium sulfate treatment was required to qualify the glass for pharmaceutical use. The pits that resulted, along with discontinuities, made the glass more prone to delamination. Vendor B, on the other hand, uses lower processing temperatures in vial production, does not phase separate the glass, does not need to apply ammonium sulfate, and therefore does not produce a pitted surface on the vial interior.

CONCLUSIONS

Based upon information in the published literature, glass delamination can be described by the same mechanism as weathering. The affected surface is depleted of silicon, which creates a layer that has dissimilar mechanical properties from the bulk substrate, causing the outer surface to spall. Tests other than the visual observation of flakes can provide more timely indications of glass incompatibility with pharmaceutical liquids.

REFERENCES

[1]Ennis, R.D., et al., *Glass Vials for Small Volume Parenterals: Influence of Drug and Manufacturing Processes on Glass Delamination.* Pharmaceutical Development Technology, 2001. **6**(3): p. 393-405.
[2]Iacocca, R.G. and M.A. Allgeier, *Corrosive attack of glass by a pharmaceutical compound.* Journal of Materials Science, 2007. **42**(3): p. 801-811.
[3]Iacocca, R.G., et al., *Factors Affecting the Chemical Durability of Glass Used in the Pharmaceutical Industry.* AAPS PharmSciTech, 2010. **11**(3): p. 1340-1349.
[4]Dimbleby, V., *Glass for Pharmaceutical Purposes.* J. Pharm. Pharmacol, 1953. **5**: p. 969-989.
[5]Dimbleby, V., *Chemical durability of glass.* Science school review, 1937. **18**: p. 476-489.
[6]Dimbleby, V., H.S.Y. Gill, and W.E.S. Turner, *Some effects of storage on the chemical durability of glass containers.* Journal of the Society of Glass Technology, 1935. **19**: p. 231-43.
[7]Dimbleby, V., *Notes on some methods used in the analyses of glasses.* Journal of the Society of Glass Technology, 1927. **11**: p. 153-66.
[8]Blackmore, H.S., V. Dimbleby, and W.E.S. Turner, *A rapid method of testing the durability of glassware.* Journal of the Society of Glass Technology, 1923. **7**: p. 122-129.
[9]Lavoisier, in *Memoires de L'Academie des Sciences.* 1770. p. 73-90.
[10]Branda, F., et al., *Weathering of a Roman Glass: A new Hypothesis for Pit Formation on Glass Surfaces.* Glass Technology, 1999. **40**(3): p. 89-91.
[11]Gillies, K.J.S. and A. Cox, *Decay of Medieval Stained Glass at York, Canterbury, and Carlisle: Part 2. Relationship between the Composition of the Glass, its Durability and the Weathering Products.* Glastech. Ber., 1988. **61**(4): p. 101-107.
[12]Gillies, K.J.S. and A. Cox, *Decay of Medieval Stained Glass at York, Canterbury and Carlisle.* Glastechnische Berichte, 1988. **61**: p. 75-84.
[13]Rogers, P., D. McPhail, and J. Ryan, *A Quantitative Study of Decay Processes of Venetian Glass in a Museum Environment.* Glass Technology, 1993. **34**(2): p. 67-68.
[14]White, W.B., *Theory of Corrosion of Glass and Ceramics*, in *Corrosion of Glass, Ceramics and Ceramic Superconductors: Principles, Testing, Characterization, and Applications*, D.E. Clark and B.K. Zoitos, Editors. 1992, Noyes Publications: Park Ridge, NJ. p. 2-28.
[15]McIntyre, N.S., G.G. Strathdee, and B.F. Phillips, *Secondary Ion Mass Spectrometric Studies of the Aqueous Leaching of a Borosilicate Glass.* Surface Science, 1980. **100**: p. 71-84.
[16]Smets, B.M.J., *On the Mechanism of the Corrosion of Glass by Water.* Philips Tech. Rev., 1985. **42**(2): p. 59-64.

*Corresponding author: Iacocca_Ronald_G@Lilly.com

MICROSTRUCTURAL PHASE SEPARATION AND DELAMINATION IN GLASS FOR PHARMA APPLICATIONS

Patrick K. Kreski and Arun K. Varshneya
Saxon Glass Technologies, Inc.
Alfred NY 14802 USA

ABSTRACT

Flaking or peeling of inner glass container wall, termed "delamination", has been observed in Type I borosilicate glasses due to long term storage of some parenteral drugs and has been the reason for several drug recalls in recent years. While there may be different types of delamination phenomena, it is suggested that the dominant precursor to delamination is the well-known microstructural phase separation in borosilicate glasses and the subsequent preferential dissolution of the lesser chemically durable phase by corrosive medicines. Hence, processing of glass during container forming or minor composition changes that tend to keep glass out of the immiscibility dome should help reduce such issues. Introduction of a high surface compression in the glass container by chemical strengthening may also have beneficial effect by requiring a higher tensile stress for crack propagation during peeling.

INTRODUCTION

Over the past two years numerous parenteral drug recalls have been attributed to glass particulates or glass flakes.[1] Although no immediate adverse effects on any patients were noted, the recalls were issued voluntarily to reduce such possibility. These recalls spanned various parenteral formulations and various Type I borosilicate glass compositions and, as such, the nature of glass particulate generation and attributes of these particulates are expected to have some variation from case to case. In some instances, the alleged particulate matter might have come from a fractured glass chip, however, in others it clearly resulted from some reaction of the drug formulation with glass interior wall over time.

A recent study by Iacocca and Allgeier[2] showed the appearance of flakes after 30-day storage at 40°C in Type I borosilicate glass vials containing an initially unbuffered pH 8.2 formulation to simulate parenteral (injectable) solutions and suspensions. The flakes or lamellae were 10 to 15 micron in size and ~200 nm thick, and were determined to be high-silica glass phases using D-SIMS (Figure 1). Delamination was found to occur after 8 weeks storage at 30°C. In addition to the glass

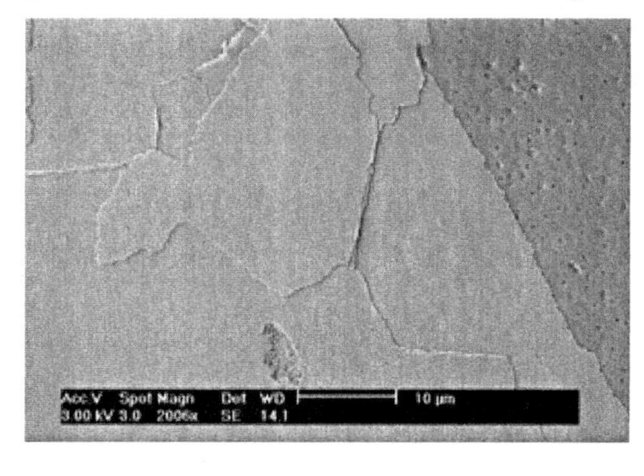

Figure 1. Scanning electron micrograph of the interior of a Type I borosilicate glass vial after 30-day storage at 40°C with sodium bicarbonate unbuffered formulation of pH 8.2 (After Iacocca & Allgeier[2] and R. G. Iacocca, presented at the Rx-360 Symposium on Glass Delamination, May 25, 2011, Arlington VA).

composition, treatment temperature, time, and pH of the solution that usually control the chemical corrosion, interior glass surface pretreatment by $(NH_4)_2SO_4$ solution and differences due to manufacturing process used for tube-forming (Danner versus Vello) were also examined. Iacocca and Allgeier suggested that aqueous solution diffuses into the glass surface defects where the ionic concentration builds up with time, creating a highly corrosive environment eventually leading to spalling and delamination (Figure 2). There have been other studies in the past specifically on the subject of delamination in Type I borosilicate glasses due to interaction with parenteral solutions. See for example, Ennis et. al.[3] for more references on the subject.

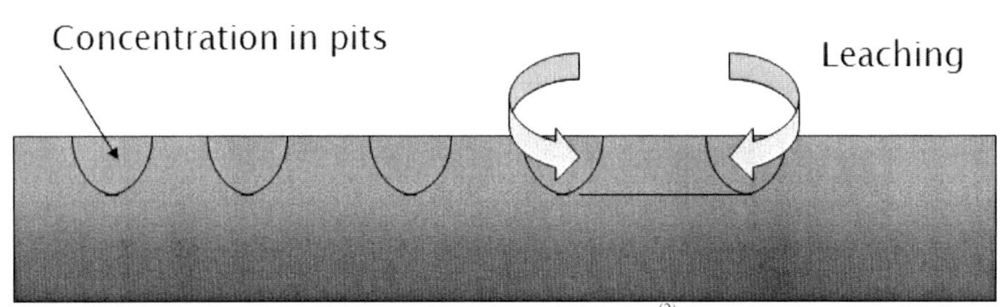

Figure 2. Delamination schematic proposed by Iacocca & Allgeier.[(2)] Corrosion process creates pits at pre-existing surface defect sites. The corrosive liquid eventually attacks to create subsurface cracks.

The science of metastable liquid-liquid immiscibility of glass leading to microphase separation has been established by Cahn and Charles.[4] The objective of this paper is to suggest that microstructural phase separation from liquid-liquid metastable immiscibility, commonly known to occur in sodium borosilicate glasses, is the root cause of the delamination process. Hence, suitable techniques to mitigate such phenomena would be (1) use of processing temperatures that avoid the immiscibility dome, (2) composition adjustments, and (3) the introduction, perhaps, of surface compression, for instance, by chemical strengthening that reduces the effective tensile stress causing flaking.

EXPERIMENTAL PROCEDURE

Specimens of Schott Fiolax® 50-mL vials[5] were obtained and immersed in a mild HF etchant at 60°C in order to accelerate corrosion of both the inside as well as the outside glass surface. Vials were washed, cooled, depyrogenated (three distilled water rinse and dry cycles, then exposure to 330°C for 13 minutes) and examined using optical microscopy and SEM.

RESULTS

The etched vials clearly showed regions of fogginess in a ring form (halo) at the upper and lower portion (Figure 3A). The location of these rings coincided with the "stress rings" otherwise known to occur in vials during the tube-conversion process (Figure 4). An optical micrograph of the lower ring is shown in Figure 3B indicating extensive phase separation. Higher magnification SEM images of the same region showed pitting indicative of droplet-type phase separation (Figure 5). Away from the foggy region, no phase separation was visible in the SEM (Figure 6).

Figure 3. (A) Optical micrograph displaying two foggy halos, one below the shoulder and one above the heel, of a 50 mL vial after mild etching. Vial height is approximately 7.5 cm. (B) Optical micrograph of a region displaying a foggy halo after mild etching.

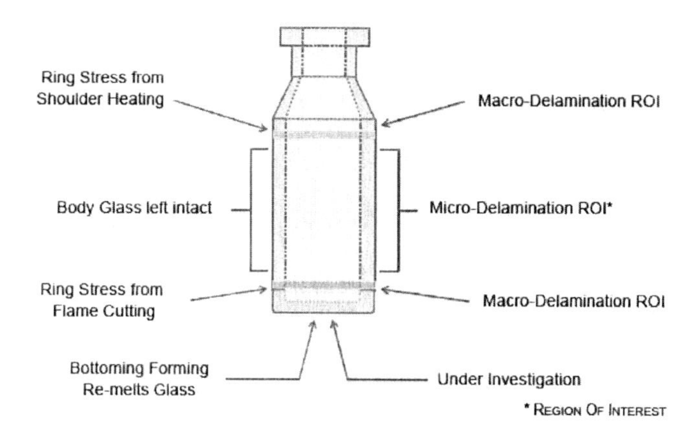

Figure 4. Sources of delamination in tube-converted Type I "medium expansion coefficient" borosilicate glass vials (After Don Kraus, Gerresheimer, Presented at the Rx-360 Symposium, May 25, 2011, Arlington VA).

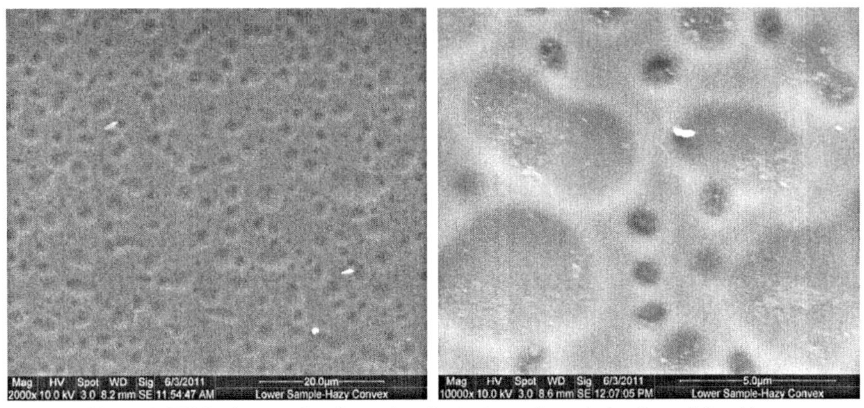

Figure 5. SEM micrographs of a region displaying a foggy halo after mild etching.

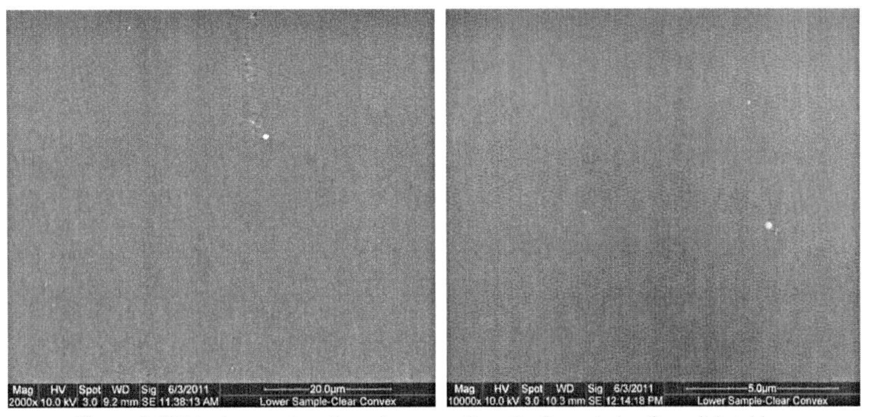

Figure 6. SEM micrographs of a region away from the foggy halo after mild etching.

DISCUSSION

The nominal chemical composition of a typical commercial Type I borosilicate such as Schott Fiolax® is:[5] 7.1 Na_2O · 1.7 CaO · 3.1 Al_2O_3 · 9.5 B_2O_3 · 78.6 SiO_2 (mol%). If we combine the alumina and boric oxide contents, and combine the alkali and alkaline earth contents, then Fiolax® can be seen to fall (Figure 7) near the 700-750°C immiscibility domes cut in a (pseudoternary) Na_2O-B_2O_3-SiO_2 liquid-liquid immiscibility phase diagram.[6] This implies that glass forming temperatures above 750°C yield a single-phase microstructure; however, below 750°C after some extended holding time may yield droplet phase separation. The present approximation makes prediction of the compositional nature of the droplet phase (alkali borate-rich or silica-rich) difficult, but SEM images suggest an alkali-rich phase was removed by etching (Figure 5). Note, phase separation in Pyrex®-type low-expansion coefficient Type I borosilicate glasses usually occurs by a similar separation of sodium borate-rich phase in a silica-rich matrix. Because the sodium-rich phase is the higher thermal

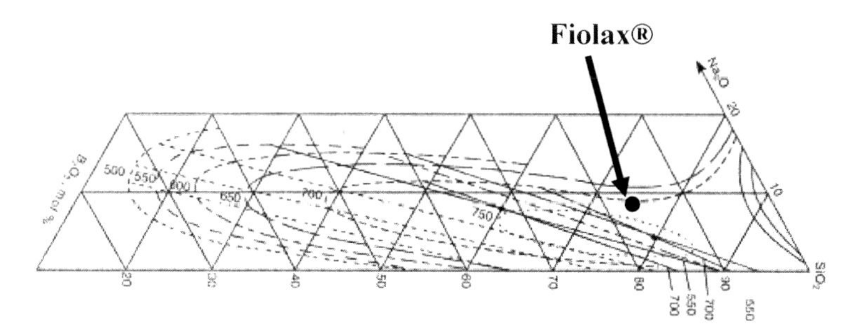

Figure 7. Type I borosilicate glass such as Schott Fiolax® is located near the 700-750°C immiscibility domes in a (pseudoternary) Na_2O-B_2O_3-SiO_2 diagram (modified from Mazurin[6]).

expansion phase, cooling of the glass would result in the generation of tension in the sodium borate-rich phase while compression develops in the silica-rich phase. It can be argued that the subsequent preferential dissolution of the sodium borate-rich phase will result in swelling of the silica-rich phase and eventual flaking of this phase due to tensile incompatibility with the underlying substrate. Swelling of glass during dissolution of the sodium borate-rich phase is a known issue in the Vycor® process and is alleviated by a judicious choice of the starting composition.[7]

Although experiments were not conducted with the intent of developing glass flakes, it is believed that development of fogginess from leaching of an alkali-rich phase is a precursor to delamination. Hence, several techniques could be applied to mitigate possibilities of delamination:

(1) Conversion of vials from glass tubing in a way that the temperature regime for liquid-liquid metastability is rapidly passed. This would imply that much of the internal surface of the glass vials is either processed at higher temperatures, or at significantly lower temperatures, than immiscibility to minimize growth opportunity of phase separation. Molded vials which have shorter dwell times should show reduced delamination.

(2) Composition adjustment. Increasing alumina content in glass may shrink the immiscibility dome.[8]

(3) Since delamination is the action of a tensile stress to separate two surfaces, addition of a large magnitude surface compression, such as inside glass wall chemical strengthening, has the potential to impede crack propagation.

(4) Another possibility is the deposition of an inert coating (such as amorphous silica) on the interior of the container. However, bonding of the coating to the substrate itself has to survive shelf-life.

SUMMARY

(1) Microstructural phase separation from liquid-liquid immiscibility known to exist in sodium borosilicate glasses is argued to be the primary precursor to glass delamination observed in Type I borosilicate glass containers for many parenteral drug formulations.

(2) Glass processing procedures that reduce dwell time in the immiscibility regime should reduce opportunities for delamination. Likewise, compositional modifications that shrink the immiscibility dome should also be favored.

(3) Introduction of surface compression by chemical strengthening should counteract tensile stresses that tend to peel flakes away from the substrate.

REFERENCES

[1]See, for example: "MedWatch The FDA Safety Information and Adverse Event Reporting Program," US Food and Drug Administration, Accessed July 2011.Available at: http://www.fda.gov/Safety/MedWatch/SafetyInformation/default.htm.
Recalls for glass particulates or glass flakes (Company, Approx. Date(s)): Baxter International Inc. 05/17/2010; Lundbeck Inc. 07/30/2010; Amgen 09/24/2010; Sandoz 10/29/2010; American Regent 12/24/2010; Cumberland Pharmaceuticals Inc. 12/30/2011; American Regent 02/03/2011, 02/04/2011, 03/15/2011, 03/16/2011, 04/26/2011, 05/05/2011, 06/06/2011, 06/15/2011.

[2]R.G. Iacocca and M. Allgeier, "Corrosive attack of glass by a pharmaceutical compound", *J. Mater. Sci.*, **42**, 801-811 (2007).

[3]R.D. Ennis, R. Pritchard, C. Nakamura, M. Coulon, T. Yang, G.C. Visor, and W.A. Lee, "Glass vials for small volume parenterals: influence of drug and manufacturing processes on glass delamination," *Pharma. Dev. & Tech.*, **6**[3] 393-405 (2001).

[4]J.W. Cahn and R. J. Charles, "The initial stages of phase separation in glasses", *Phys. Chem. Glasses*, **6**[5] 181-191 (1965).

[5]"Schott Fiolax®," (2010) Product Literature, Schott North America, Inc. Accessed June 2011. Available at <http://www.schott.com/tubing/english/download/schott-tubing_brochure_fiolax_english.pdf>.

[6]O.V. Mazurin in *Phase Separation in Glass*, Fig 59, p155. Elsevier Science Publishing Co. Amsterdam, Netherlands, 1984.

[7]H. P. Hood and M. E. Nordberg, "Borosilicate glass," U.S. Pat. 2,221,709, Nov. 1940.

[8]W.-F. Du, K. Kuraoka, T. Akai, and T. Yazawa, "Study of Al_2O_3 effect on structural changes and phase separation in Na_2O-B_2O_3-SiO_2 glass by NMR", *J. Mater. Sci.*, **35**,4865-4871 (2000).

*Corresponding author: kreski@saxonglass.com

SURFACE AND INTERFACE MODIFICATION OF SILICATE GLASS VIA SUPERCRITICAL WATER

Shingo Kanehira
Center for the Promotion of Interdisciplinary Education and Research, Kyoto University
Yoshida-Honmachi, Sakyo-ku, Kyoto 606-8501, Japan

Kazuyuki Hirao
Department of material chemistry, graduate school of engineering, Kyoto University
Katsura, Nishikyo-ku, Kyoto 615-8510, Japan

Takahiro Maruyama and Tsutomu Sawano
Advanced Materials Innovation Center, Shiohama, Yokkaichi, Mie 510-0851, Japan

ABSTRACT

The reaction between supercritical water and silicate glasses was analyzed using various surface analysis methods. The alkali silicate glasses with a diameter of 1 inch were treated over 300°C inside the chamber and cooled into room temperature after the treatment. The surface of the glasses was severely etched especially over 400°C, which is near the supercritical point of water; 374°C, 22.1MPa. Not only the formation of pores but also the precipitation of alkali in the glass were observed from the image of secondary electron microscopy (SEM) and EPMA analysis. An optical interferometry microscope revealed that the average roughness after the treatment dramatically increased depending on the increase of treatment temperature. The polishing test of the treated glass indicated that the mechanical properties at the surface of the treated glass dramatically decreased due to the reaction between supercritical water and glass.

INTRODUCTION

The analysis of reaction between water and glasses is very important to a wide variety of fields. Alkali silicate glasses have an amorphous structure composed of Si-O tetrahedron and some of them react with water, resulting in the trapping of H_2O molecular or OH^- ion in the structure called hydration. The first step in the reaction of water with alkali silicate glass is generally assumed to be the exchange of alkali ions in the glass with hydrogen-bearing ions from the water as follows:[1,2]

$$Na^+(glass) + 2H_2O = H_3O^+(glass) + Na^+ + OH^- \tag{1}$$

The alkali ions, in case of sodium in equation (1), diffuse in the glass structure. In addition, the surface of the glass also dissolves into the water, resulting in the formation of specific structure at the surface. A previous report showed that diffusion area of sodium in the soda-lime glass was about ~ 0.6μm after hydration under 90°C over 500h.[1] It is said that a new layer after the hydration reaction formed at the surface and it affects the durability of the glass. These effects were characterized using resonant nuclear reaction.[2]

Supercritical fluid has many interesting characteristics such as low dielectric constant, low kinematic viscosity, high reactivity, and high diffusivity into solids. The supercritical point in water exists in 374°C, 22.1MPa. There are a few reports about the reaction between supercritical water and glass fiber[3] or glasses.[4,5] These reports contain the analysis of etched area in the fiber structure and the formation of porous structure, however, the detailed explanation and additional information is needed. Our interest focused on the reaction between supercritical water and commercial silicate glasses. Here, we treated commercial silicate glasses under the supercritical water and the microstructure or mechanical properties near the surface was also examined using various methods in detail.

EXPERIMENTAL PROCEDURE

We used commercial silicate glasses composed of 71% SiO_2, 13% Na_2O, 9% CaO, 4% MgO, 2% Al_2O_3, 1% K_2O in our experiments. The surface of the sample glass was mirror finished, and its diameter and thickness were about 25mm (1inch) and 2mm, respectively. The main chamber of 300ml volume composed of Hastelloy alloy was surrounded with coil heater, and the temperature inside was controllable up to 450^0C. The temperature and pressure inside the chamber was confirmed using a K-type thermocouple and pressure gage. At first, distilled water was poured into the chamber and the glass sample was put under the water. The amount of the water was changed between 40 and 60ml to control the dependence of pressure and temperature inside the chamber. We increased the temperature to 300~450^0C at 10°C/min and maintained for 0 ~2h, resulting in the accomplishment of supercritical water treatment. After the treatment, the chamber was cooled using a fan and opened up after cooling. The treated sample was rinsed by distilled water and ethanol for 10 min. in a ultrasonic bath, and sufficiently dried in an oven for 24h.

The surface of the glass sample after the treatment was observed using a field-emission secondary electron microscope (FE-SEM, JSM6700F, JEOL, Japan). Distribution of alkali ions was analyzed using electron probe micro analysis (EPMA; JXA-8100, JEOL, Japan). An average surface roughness after the treatment was measured using white light interferometry (NewView 7300, Zygo Corporation, USA). The abrasive test of the treated sample was performed using lapping test apparatus (LGP-15S-I, Lapmaster SFT Corporation, Japan.) The slurry for the test was commercial CeO_2 for polishing (SHOROX-A20). Three samples were fixed at the attachment of the tester and stress of 9.8kPa was continuously applied during the test.

RESULTS AND DISCUSSIONS

Figure 1 indicates a plot of pressure as a function of temperature inside the chamber when the amount of initial water changed from 40ml to 60ml. The solid curve shows a theoretical calculation, which is called a vapor pressure curve. The pressure in the chamber increased along the vapor pressure curve under ~350^0C, however, it deviated from the calculation curve and increased linearly as the temperature increased over 350^0C. The result indicated that the water in liquid completely converted into another phase such as gas or supercritical phase. The amount of initial water affected the change of pressure inside. The 40ml volume of water needs over 450^0C to realize the supercritical phase, however, the 60ml volume of water can realize the phase at 400^0C. Therefore, we fixed the amount of water at 60 ml in the treatment experiment.

Figure 2 (a) shows photograph of the glass substrate before and after the supercritical water treatment at 4500C/30MPa. The color of the sample was apparently changed from transparent into opaque and white color after the supercritical water treatment. The opaque area corresponds to the corrosion area by the supercritical water and white area indicates a precipitation of alkali. FE-SEM micro-structure photo of the same glass substrate was also shown in Fig. 2(b). The initial smooth surface of the glass sample converted into the irregular one due to the reaction between the glass and the supercritical water. Not only the pore but also the precipitation of alkali were also observed at the surface of the glass. Generally, the alkali in glass material can diffuse freely by addition of the external field such as electrical field, thermal, and so on. A severe supercritical water atmosphere triggered the diffusion of alkali in the glass substrate.

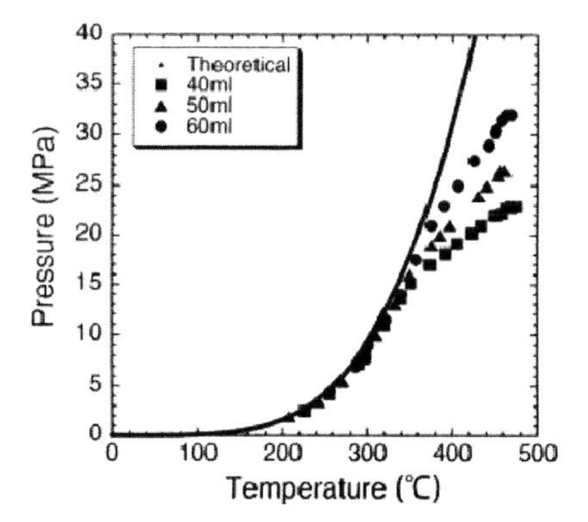

Figure 1. Pressure dependence inside chamber as a function of temperature.

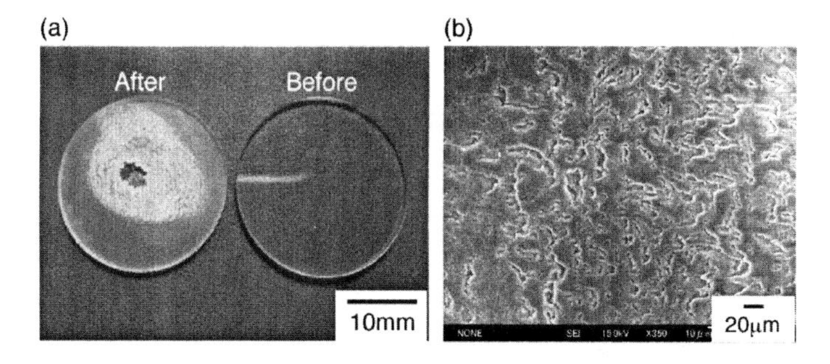

Figure 2. (a) Photograph of glass substrate after the supercritical water treatment. (b) Microstructure of the treated glass substrate at surface observed by FE-SEM.

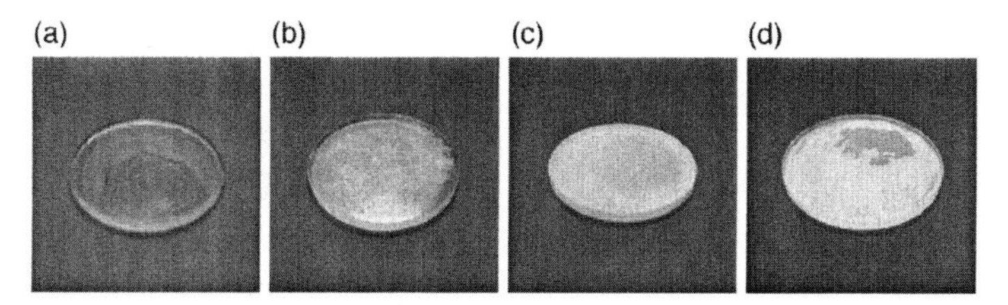

Figure 3. Photographs of glass substrates after the supercritical water treatment at (a)300°C, (b)350°C, (c)400°C, and (d)450°C, respectively.

Figure 3 shows the glass substrate after the supercritical water treatment at (a)300°C, (b)350°C, (c) 400°C, and (d)450°C, respectively. There was no holding time in the process of treatment inside the chamber. The treated glass at 300°C has a transparency; however, it gradually turned opaque over 350°C. The glass surface over 400°C was extensively etched by the supercritical water. The temperature of 300-350°C, and 400-450°C corresponded to gas and supercritical phase, respectively. Therefore, the supercritical water has a high reactivity towards the glass substrate rather than that of vaporized water.

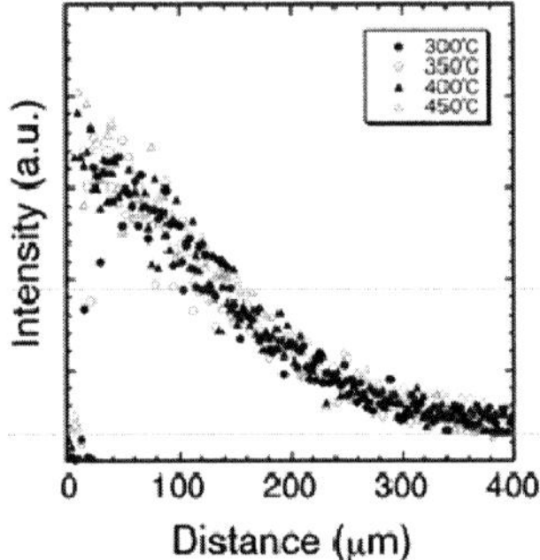

Figure 4. Distribution of sodium ions after the vapor or supercritical water treatment at (a)300°C, (b)350°C, (c)400°C, and (d)450°C, respectively. The measurement was performed near the surface of the sample glass; zero indicates the surface of glass.

Figure 4 shows distribution of sodium ions near the surface of the sample glass after the treatment at various temperatures. The sodium ion aggregated at the surface and its concentration gradually decreased as the distance from the surface increased. Therefore, the white area at the surface shown in Fig. 3 is due to the segregation of alkali in the glass. Many papers suppose theoretical models about the reaction between alkali glass and water. In case of our glass sample, multiple cation such as sodium, potassium, calcium, and magnesium may complicate the diffusion mechanism. The difference among previous reports is that the diffused distance (~ 300μm) is much longer than that of ones, probably due to an ability of strong diffusion of supercritical water. Now we are investigating the model of supercritical water attack against the glass.

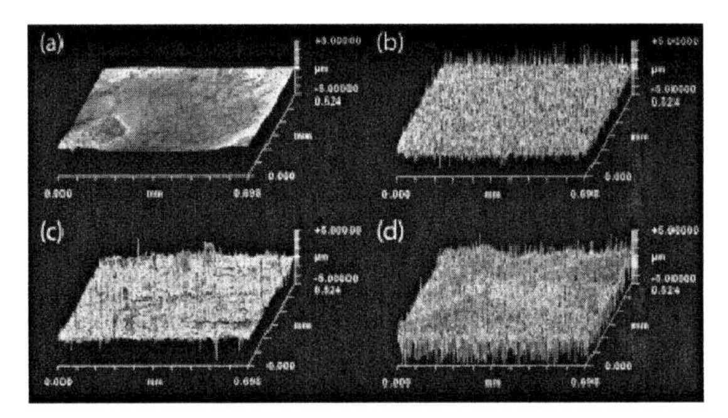

Figure 5. Profile of glass surface after the vaporized water or supercritical water treatment at (a) 300°C, (b) 350°C, (c) 400°C, and (d) 450°C, respectively.

Figure 5 indicates profiles of the glass surface after the vaporized water and supercritical water atmosphere. The glasses were treated at (a) 300°C, (b) 350°C, (c) 400°C, and (d) 450°C, respectively. The glasses were washed carefully by ultrasonic washer attached with a distilled water bath before the observation to ignore the effect of alkali precipitated at the surface. The average roughness of the glass was evaluated from the area of $525 \times 700 \mu m$.[2] In case of Fig. 5(a), the average roughness of the treated glass was almost the same as the virgin glass, however, it linearly increased when the treatment temperature increased over 350°C. A previous paper reported that the porous structure which had an interconnected framework was observed after the hydrothermal treatment at 450°C for 3h in Pyrex glass.[4] The surface morphology of our treated glass is different in case of the previous report under the same treatment conditions, therefore, the porous structure tends to strongly depend on the glass components.

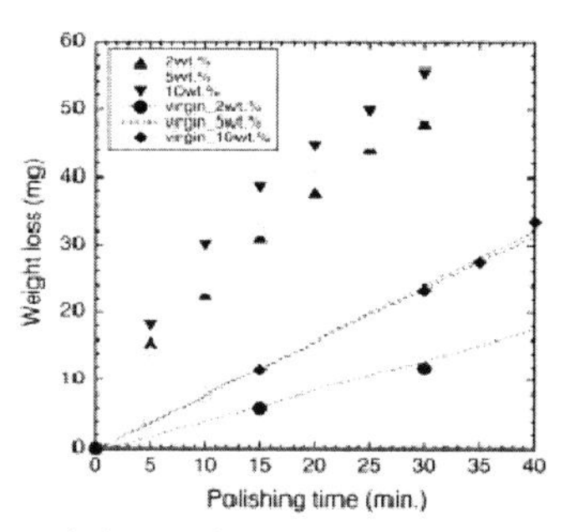

Figure 6. Weight loss of glasses after polishing treated by supercritical water at 400°C as a function of polishing time. The change of virgin glass is also plotted in the same figure. The amount of CeO_2 abrasive powder was changed from 2 to 10wt.%

Figure 6 indicates weight loss of glass sample after polishing as a function of holding time. The weight loss of the virgin glass substrate is also plotted in the same figure. In case of the virgin glass, the loss linearly increased when the holding time increased. In addition, the loss also increased when the amount of CeO_2 increased, however, it saturated over the 5wt.% CeO_2. These results indicated that the glass surface was filled with CeO_2 slurry at 5wt.%, and the additional slurry did not work for the abrasive of the glass substrate. On the contrary, the loss of the glass treated with supercritical water was three times higher than that of the virgin glass. In addition, the increase of CeO_2 slurry from 5 to 10wt.% resulted in the growth of the weight loss of treated glasses. The difference between virgin glass and treated one was probably due to the difference of the surface morphology. The slurry tended to penetrate the treated glass surface with a porous structure; therefore, the additional slurry was needed to act as an effective abrasive. The increase of weight loss corresponds to the decrease of mechanical properties of the glass substrate. That is, the mechanical properties of the glass substrate treated with supercritical water dramatically decreased due to the formation of porous and soft structure near the surface of the glass substrate. The existence of the soft layer near the glass surface is under investigation. The glass surface after the abrasive test showed a good flatness as same as the initial glass.

CONCLUSIONS

We have treated the commercial glass substrate under the vaporized or supercritical water inside the chamber. The average roughness of the glass substrate linearly increased due to the formation of porous structure via the reaction between glass and supercritical water. The sodium in the glass aggregated at the surface after the supercritical water treatment. The mechanical properties of the treated glass were dramatically decreased evaluating from CeO_2 abrasive test due to the structural change near its surface.

REFERENCES

[1]W.A. Lanford, K. Davis, P. Lamarche, T. Laursen, R. Groleau, and R.H. Doremus, Hydration of Soda-lime Glass, J. Non-Cryst. Solids, **33**, 249-66 (1979).

[2]R.H. Doremus, Y. Mehrotra, W. A. Lanford, and C. Burman, Reaction of Water with Glass ; Influence of a Transformed Surface Layer, J. Mat. Sci., **18**, 612-22 (1983).

[3]V.N. Bagratashvili, A.N. Konovalov, A.A. Novitskiy, M. Poliakoff, and S.I. Tsypina, Reflectometric Studies of the Etching of a Silica Fiber with a Germanium Silicate Core in Sub- and Supercritical Water, Russian J. Phys. Chem. B, **3**, 1154-64 (2009).

[4]F.A. Sigoli, S. Feliciano, M.V. Giotto, M.R. Davolos, and M. Jafelicci Jr., Porous Silica Matrix Obtained from Pyrex Glass by Hydrothermal Treatment: Characterization and Nature of the Porosity, J. Am. Ceram. Soc., **86**, 1196-201 (2003).

[5]F.A. Sigoli, Y. Kawano, M.R. Davolos, and M. Jafelicci Jr., Phase Separation in Pyrex Glass by Hydrothermal Treatment: Evidence from Micro-Raman Spectroscopy, J. Non-Cryst. Solids, **284**, 49-54 (2001).

*Corresponding author: kane_w01@hotmail.co.jp

DEMANDS AND ACHIEVEMENTS IN CURRENT GLASS CONTAINER STRENGTHENING

C. Roos[1] and G. Lubitz[2]

[1]International Partners in Glass Research, Bülach, Switzerland
[2]Vetropack Holding, Bülach, Switzerland

ABSTRACT
In the sector of glass-container production, innovation is heavily dominated by cost-issues and return on investment. This leads often to more evolutional and incremental changes in manufacturing than to revolutionary ones. In this paper the interaction of demands especially focused on glass strengthening and hence glass-container light-weighting are discussed. It is explained how these interactions should be overcome and where considerable achievements have been made possible and still can be made. Emphasis is put on three ways of increasing the useable glass strength, namely preserving the initial strength, restoring the initial strength and "adding" an additional strength to the glass-container.

INTRODUCTION

The glass container manufacturing process comprises of the typical glass manufacturing procedures used at most industrial glass productions: a) mixing of raw materials and cullet, b) melting and conditioning of this batch mixture, c) forming of the glass, d) annealing, and e) inspection. Some special characteristics of the glass-container manufacturing are the high level of external cullet that is being used and the multitude of inspection techniques. The former is important both for recycling and energy saving reasons. The latter is especially important as glass-containers come in direct contact with food and beverages. Hence demands to ensure highest quality is put on glass-containers, both from the direct customer (filler) and consumer of the final product. Also, the glass-container manufacturing has to compete with much simpler and less elaborate container production techniques, such as PET-bottle forming and can- or cardboard box manufacturing. Combined with the aspects that the glass industry in general is quite cost intensive and involves long payback-times, a certain conservative attitude applies to most glass manufacturers and therefore also to glass-container manufacturers.

A special aspect in glass-container manufacturing is the forming process which is here only briefly described. Detailed descriptions are available elsewhere.[1] The forming process comprises of cutting a glass gob and forming a parison out of this gob, either by blowing or by pressing. In a subsequent step this parison is blown into the shape of the final bottle. Because of these rather complex processes and due to the very competitive market situation there are continuous demands for improvements. The interaction of these demands can be complex as shown in figure 1.

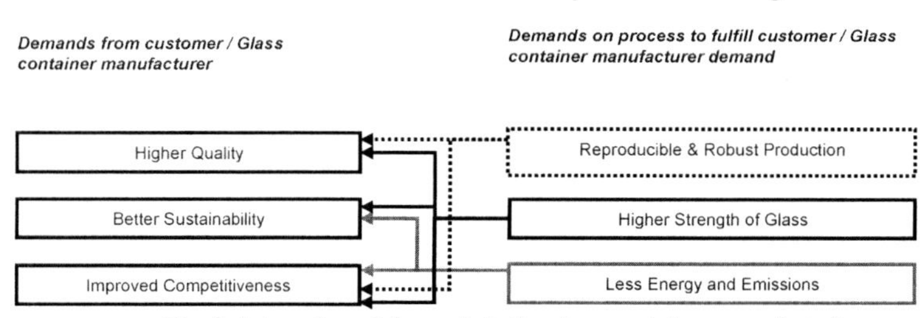

Fig. 1: Interaction of demands in the glass-container manufacturing

From customer and manufacturer point of view, a yet higher quality of the glass, and even although glass-containers are in for a really 100% cradle to cradle recycling, better sustainability are demanded. Additionally, improved competitiveness is important. These demands again trigger certain demands on the process-side: More reproducible and robust production, higher strength of the glass articles and lesser energy consumption and emissions during production.

STRENGTHENING AND LIGHT-WEIGHTING, HOW?

Focusing on the demand for higher strength of glass-containers, there are basically three ways to increase the strength of glass:

a) Preserving the initial strength

Preserving the initial glass strength can be achieved by protecting the glass (e.g. with a coating) or by optimizing the process to avoid damages. As strength of glass is a question of surface damage, all of this certainly has to be done before the glass has been damaged.

b) Restoring the initial strength

The initial glass-strength can be restored by e.g. defect healing through tempering near Tg or by coatings. Logically, this approach has to be applied after the glass has been damaged.

c) Applying an additional strength

In this case the glass strength is increased by "giving" additional strength to the article (e.g. by ion exchange, coatings, thermal toughening, etc). Applying an additional strength can be done before or after the glass has been damaged.

Improvements to strengthening and light-weighting of glass-containers in the past have mainly been achieved through more robust and reproducible processes. These improvements aimed largely on preserving the initial strength. No "active" strengthening methods have been applied permanently in real manufacturing. Various techniques are applied in production and help, as mentioned, to preserve the initial glass strength and enable a significant light-weighting of containers. There are such revolutionary developments as the independent section machine (IS machine) in 1924 and the narrow-neck press-blow process in 1976. Other more evolutionary improvements also helped to increase the glass-container strength such as parallel mould opening (1977), servo controlled IS machines (2000), vacuum application to reduce the settle wave, gob measurement and gob control systems, direct lehr loading, and innovative handling materials. But what stands behind this is the all-encompassing questions "how to achieve a considerable strength increase in glass-containers?"

Basically when considering the breakage distribution of glass a Weibull statistic is applied. In contrast to metals which follow a Gaussian breakage curve the Weibull distribution is asymmetric. Different ways of assessing the strength distribution of glass have been discussed.[2] Transferring the Weibull breakage distribution to glass-containers, three regions of such a distribution can be identified (Fig. 2). A so called "low-strength tail" (I) is evident, which comprises breakage at a low level because of e.g. unstable processes, improvable cleanliness, uncontrolled or not-reproducible parameters or improper alignment. This region is subject to a good-manufacturing-practice which has to be applied and controlled by the individual glass-container manufacturer.

There is a second region (II) in which breakage is caused by individual and process-inherent manufacturing technique. In every glass-manufacturing there is a process where the glass eventually will come in contact with another material. Because of these contacts, primarily at the surface, unavoidable damages are induced in the glass. This region is considered as most important for applied research, preferably in coalition between research parties, glass-container manufacturers and possible suppliers.

Thirdly, there is a region (III) where the glass strength is already considerably high, but in most cases still far away from the theoretical value of glass strength of more than 30 GPa.[3] Here breakage is caused by so called "fundamental factors", such as glass chemistry, network modifier interaction, crack initiation and growth, Griffith flaws and so on.[4] This region is a matter of basic research on the above mentioned topics and many researchers are currently aiming at improving this fundamental glass-strength.[5]

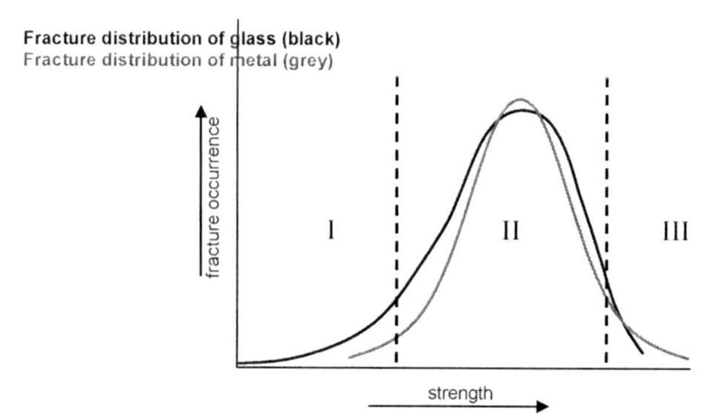

Fig. 2: Fracture distribution of glass compared to metal

If two breakage curves are considered, which have the same average value (Fig. 3) for each curve the relevant glass-container is designed with respect to the lowest strength values. Designing a glass-container with respect to its demanded strength means general glass thickness, or more correctly, glass thickness distribution and hence weight designing. As mentioned before, the quality demands on glass-containers are very high, each manufacturer designs its glass thickness distribution for the containers with respect to the lowest breakage values that occur. This also means that given the curves in figure 3, not first and foremost the, average glass-strength is limiting the glass-thickness, but the scattering of breakage values. A more narrow breakage distribution (series B in fig. 3) would tremendously help to decrease the glass weight, by simply allowing a production process closer to the average strength value, without the need to "overdesign" glass-container thickness.

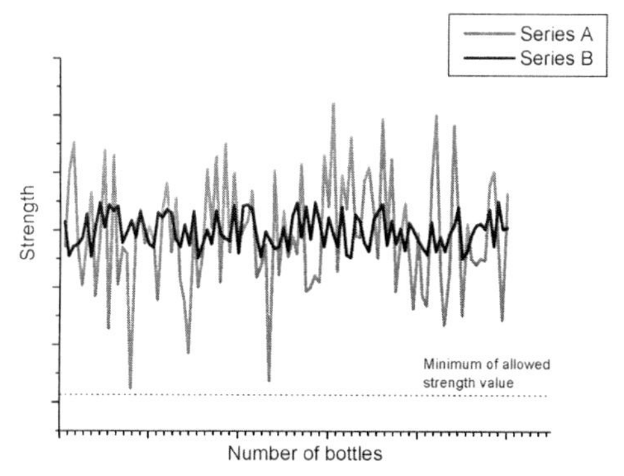

Fig. 3: Comparison of high and low scattering breakage distributions (both curves have the same mean value)

This leads to the point that basically two approaches are necessary for a considerable step in light-weighting a glass-container: Increasing the mean strength, and also of equal importance, minimize scattering of the strength values. Due to competitive reasons for alternative packaging materials, it is needless to say that if bottles still break after a strengthening, it is (nearly) not possible to pass the costs for the strength increase to the customer. Therefore strengthening should lead to light-weighting, costs savings and to decreasing emissions. These benefits have to weigh up the costs of whatever treatment or technique is applied to increase the strength of the glass-container. An exception might be valid for returnable glass-containers where more return-trips might justify a higher price.

EXAMPLES FOR APPROACHES

In the following section examples are given where strength increasing for glass-containers can be achieved. Referring to the previous three ways of glass strengthening, these examples are divided into preserving or restoring the initial strength or adding strength to the glass.

Benchmark for Strength Losses in Production

A recent benchmark in IPGR has been conducted to evaluate the potential of strength decrease in the handling process of the container. For this, containers were carefully sampled from different locations in the process and manually without any further contact or damage sources transported to the lehr, annealed and burst pressure strength tested. For this benchmark a special guideline with sampling, handling and material instructions has been developed and applied.

The earliest possibility to sample a final-shape container is after the take-out has placed the container over the dead-plate but before the container has been released onto the dead-plate or has been contacted by the pusher. This sampling location is referred to as "Handling 1" in the subsequent discussion. The next sampling location, "Handling 2", was just after the container has been wiped onto the dead-plate by the pusher, but before the container has travelled a significant distance. For the third sampling, stated as "Reference", the containers have been transported to the lehr in the standard way. All containers have been recollected just after the annealing lehr, before the cold-end coating was applied. Additionally, containers were collected and inspected after the cold-end coating and inspection, referred to as "Production mean sect." For each location a total of 160 containers were sampled and tested.

The focus of that benchmark was put on outside handling damages. Therefore burst-pressure strength was tested. For the evaluation, only containers with a light-weight index of < 1 were considered as suitable. The light-weight index is defined as

$$L = 0.44 \times \text{Bottle weight (g)} / \text{Overflow Capacity (ml)}^{0.77} \quad (1)$$

All the above mentioned containers were sampled from one section, alternating cavities, in a time of less than 20 minutes and with a 5 minute safety period away from a swabbing cycle. All containers were taped for crack origin analysis before burst pressure testing.

As an example, the evaluation on a one-way, green glass bottle with 345ml, 175g and light-weight index L = 0,86 is shown in figure 4. It is obvious that from "Handling 1" to "Production mean, Sect. 10" approximately 8 bar of strength is lost. Considering a total average strength of the container of 40bar, this equals a strength loss due to handling of 20%. In other words, 20% of strength in this case can be gained by applying good manufacturing practices such as e.g. proper alignment, handling, cleanliness and proper contact-material selection.

In the presented case it is interesting to see, that the strength did not decrease due to packing of the containers (compare "production mean" and "packing mean"). This was analysed by randomly testing 80 containers of the complete production, while another 80 containers were packed and unpacked again after 24 hours and tested. Critical crack-growth can certainly not be determined by this approach which focused only on handling.

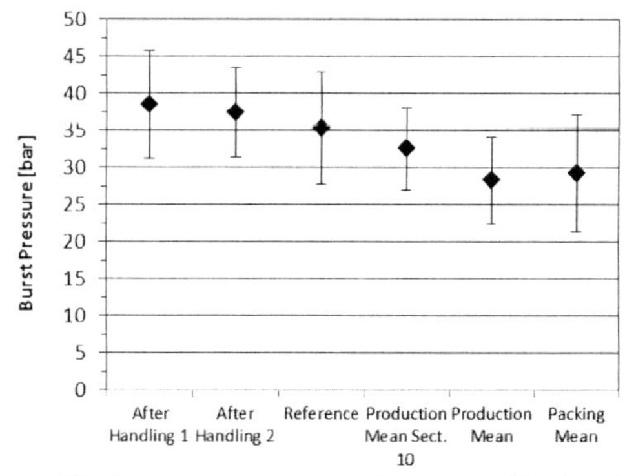

Fig. 4: Burst pressure strength over sampling locations

Investigating these breakages more closely the following has been found: With respect to the breakage origin, the most dominant differences between the sampling locations are found at body and knurling breakages. As can be seen from figure 5 the level of knurling breakages increases with each sampling location more downstream. This gives direct optimization potential for the bottom contact materials and cleanliness at bottom contacts, as there are dead-plate, conveyor-belt and secondary conveyor belt. Additionally this points out that differences in relative speed such as a) when the container is wiped onto the conveyor belt, b) the conveyor belt is changing into the secondary conveyor and c) the container is pushed into the lehr by the lehr loader can be considered as crucial. Additionally particles from rust or glass breakages can strongly affect the strength when brought onto the container, especially between the bottom knurling as tensile stresses during loading are present between the stippling.

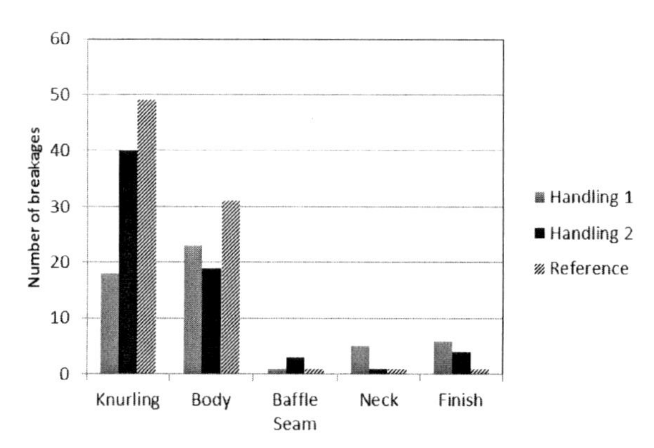

Fig. 5: Distribution of breakage locations

Secondly, as can be seen from figure 5, the level of body breakages is significantly higher for reference containers. Looking at the body breakage more closely and analyzing these breakages by height at the container, it can be seen that a significant number of breakages for the reference containers occurs at the 100 – 120mm height (fig. 6). This subsequently led to the source of the breakage, namely the contact position of the upper edge of the secondary conveyor transfer which induced defects to the container body.

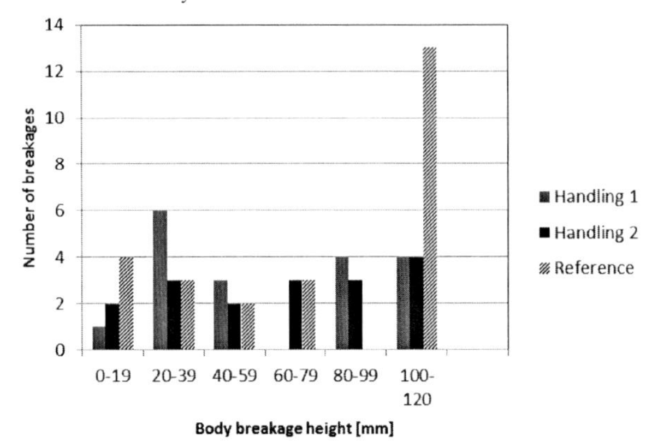

Figure 6: Height distribution of body breakages

Glass Contact Interactions

Investigations are also started to better understand the interaction of glass at temperatures above or near Tg with contact materials. In figure 7 the diffusion of elements between a container glass and cast iron is shown, measured by wavelength dispersive X-ray spectroscopy (WDX). For that, diffusion couples have been prepared and heat treated at 800°C for 72h. It can be seen that Sodium (Na) is depleting from the glass and is diffusing into the metal. Also Iron (Fe) is diffusing into the glass and is enriched at the very near surface layer. These diffusions can have a multitude of effects on the glass at that stage of the process. Leaching of sodium from the glass to the mould increases the viscosity of the surface-near glass and hence leads to stabilizing the parison. The depletion of sodium also makes the glass more prone to contact damage as it becomes more rigid and "less healing". The iron enrichment at the surface, on the other hand, can have different effects. It influences the redox-conditions, sticking, viscosity, and surface conditioning. It might also induce surface crystallization under certain circumstances. Yet, this is still subject to current investigations. But what is emphasized here is that interactions between glass and contact material are crucial, unavoidable and yet not fully understood.

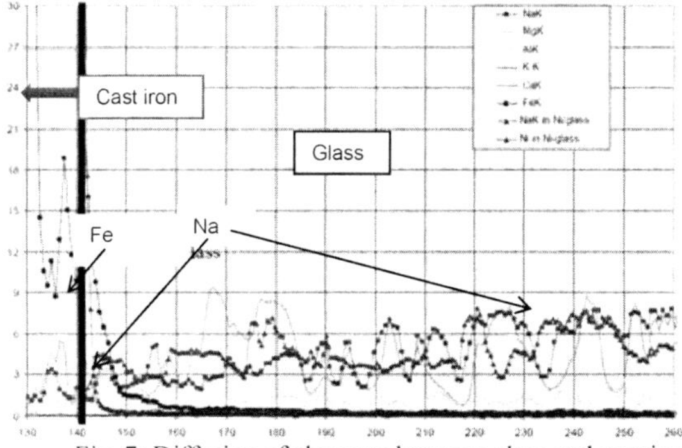

Fig. 7: Diffusion of elements between glass and cast iron

The former comment is even more valid, when referring to figure 8, where locations of surface defects and locations with no defects have been investigated more closely by X-ray photoelectron spectroscopy (XPS). At places of defects, more iron diffused into the glass than on reference positions with no defects. This is also the subject of current research and shows how complex these surface interactions can be and how much still needs to be investigated.

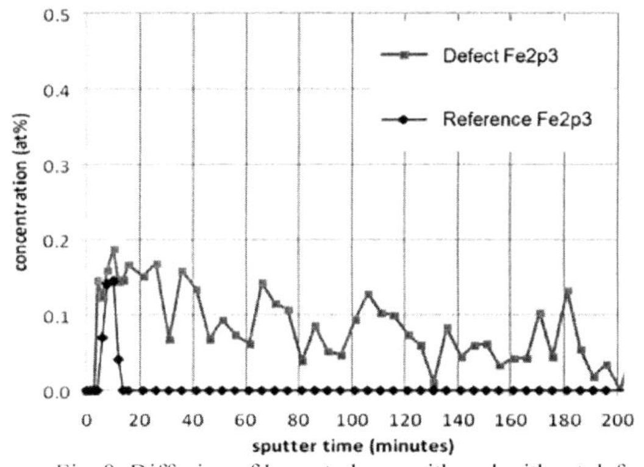

Fig. 8: Diffusion of Iron at places with and without defects

HF Treatment on Glasses

Another aspect of strength increasing, as mentioned before, is restoring of the initial strength. In figure 9 the effect of a special hydrofluoric acid (HF) etching on the burst pressure strength is shown. Both, a green glass and a flint glass of the same hydrolytic class have been subjected to an HF etching, ensuring no changes in optical appearance. Whereas for the green glass no significant differences could be observed, the flint glass experiences an increase in burst pressure strength and more important a significant decrease in standard deviation of the breakage values. Obviously in the flint glass a defect blunting has taken place. It is remarkable that this treatment is dependent on the glass colour and probably on redox conditions in the glass.

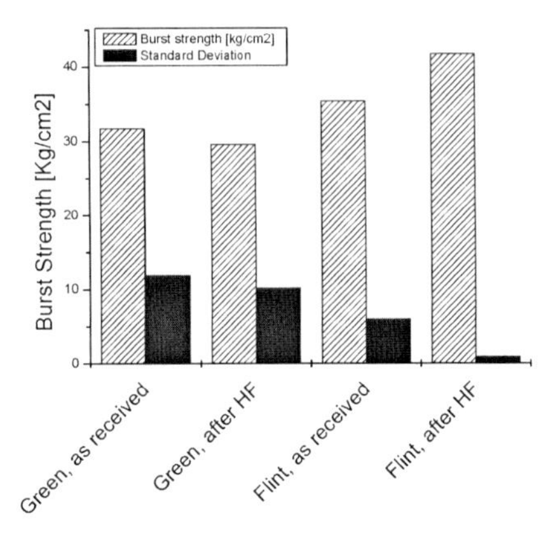

Fig. 9: Influence of etching on the burst pressure strength of different glass-types.

Certainly it is quite rarely appropriate for environmental and health reasons to apply a HF treatment. But then it is to conclude that even more research should be focused on alternative solutions that have the same effect and are less harmful. All the above refers to what has been described in the previous section and reveals the tremendous potential that is available for increasing the useable glass-strength by decreasing the scattering of fracture values.

Polymer Coating

The third possibility of increasing the glass-strength, as by adding an "additional" strength to the container, was the subject of numerous research throughout the past decades. For instance, in 1994 IPGR developed a polymer coating for containers that enabled a strength increase of about 25% in bursting strength as well as colouring options. In figure 10 it is demonstrated that the weight of a 330ml bottle could be decreased from 155g to 120g while keeping the same burst pressure strength. This technique has been implemented into production by Wiegand-Glas, Germany and Nihon Yamamura Glass, Japan. The process is still in use by Nihon Yamamura Glass as a colouring technique.

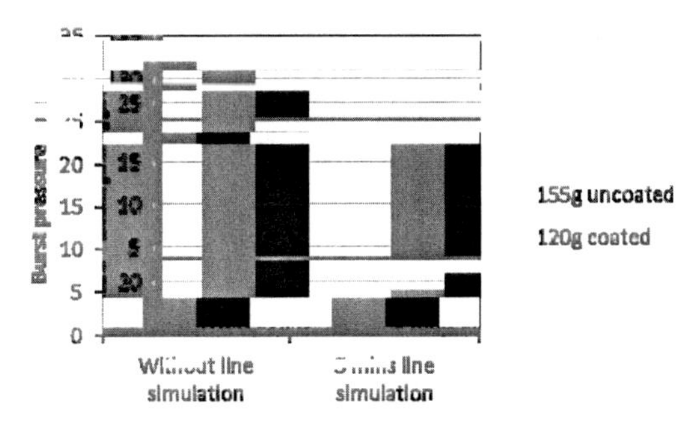

Fig. 10: Effect of polymer coating on burst pressure strength of a 330ml container

Techniques such as ion-strengthening, heat treatment, surface crystallization and others have been investigated to increase the strength of glass-containers. Unfortunately none of these techniques have really been implemented into the production process permanently. In the authors opinion there are two reasons for that. Firstly, in many cases researchers did not really understand the needs of the industry. This led to the point that at a certain stage of the development project a so-called "killer"-criteria came up which had not been considered earlier and could not be overcome. Often these "killer"-criteria involved cost, speed or recycling issues. As a second reason, in many cases the industry was not willing to take the risk to step into a new technology. This is because, as explained earlier, most companies are using a conservative approach to innovation which seems from their point of view more appropriate to deal with the high investment costs and the long payback times involved in the manufacturing of glass-containers.

CONCLUSION

In this paper insight into the needs and demands of the glass-container industry has been given, with focus on strengthening of glass that can enable light-weighting of glass-containers. An introduction to possible glass strengthening approaches has been presented and conclusions are drawn to enable the next big leap in strengthening of glass-containers. It has been shown that strength increase, and hence light weighting, in the past was solely achieved by so-called "passive" strengthening methods that aimed to restoring and preserving the initial glass strength. It also has been shown that, although considerable improvements have taken place, there is still room for improvement by good-manufacturing-practice (GMP) and by more robust and reproducible processes. Here it has to be emphasized that there is still a lack in glass-container manufacturing for thoughtful and sensible loops to enable a proper "measure, control and feedback" of parameters. It has been explained that interactions in the forming process between glass and contact materials are not yet fully understood. Here understanding and hence seizing this potential yields great room for improvement as well.

Because of the competitive nature of glass-container manufacturing, active strengthening methods (and hence light weighting) can only be motivated by internal benefits, until the bottle is virtually "unbreakable". It has been explained why, although a lot of developments and investigations have been conducted in the past decades, no permanent implementation of "active" strengthening methods into the production process was enabled. To enable this in the future, it is vital to make researchers understand the real needs and thinking of glass-container manufacturers. Additionally it is important for the glass-container industry to be open to innovative developments which would incorporate a certain amount of courage to take the risk and step into a new technology.

Finally and probably most important it is stated that from the authors point of view an all-encompassing approach is needed for a considerable step in glass-container strengthening and light-weighting. Increasing the mean strength and minimizing scattering of the strength values by keeping in mind the leverage of preserving and restoring the initial strength and adding additional strength to the glass. This also leads to the opinion that not solely one coating, one treatment step, or one new handling material can enable the next "big leap".

ACKNOWLEDGMENTS

The authors would like to thank Dr. Frits Gubbels, TNO, Netherlands and Dr. M. Kilo, Fraunhofer ISC, Germany for providing the XPS and WDS measurements and the HF etchings, respectively. Also, the support of all IPGR member companies in carrying out the strength benchmark is highly appreciated.

REFERENCES

[1]A. Schaeffer, M. Benz-Zauner, eds., Glass Hollowware, Deutsches Museum (2010)
[2]W.C. LaCourse; J. Can. Ceram. Soc. **52** (1983) 49-53
[3]A.K. Varshneya; Internat. J. Appl. Glass Science **1** (2010) 1-12
[4]S. Karlsson, B. Jonson, C. Stalhandske; Glass Technol. – Europ. J. Glass Sci. Techn. **A51** (2010) 41-54
[5]C.R. Kurkjian, P.K. Gupta, R.K. Brow; Internat. J. Appl. Glass Science **1** (2010) 27–37

*Corresponding author: christian.roos@ipgr.ch

THE CHEMISTRY OF CHEMICAL STRENGTHENING OF GLASS

Arun K. Varshneya and Patrick K. Kreski
Saxon Glass Technologies, Inc.
Alfred NY 14802 USA

ABSTRACT

Alkali-containing inorganic glass products can be readily strengthened by immersion in a bath of molten alkali salts at temperatures lower than the strain point of glass. The larger bath alkali ions gradually exchange with the smaller host alkali ions of the glass resulting in a stuffed surface which leads to practical strengthening. Chemical strengthening technology of glass is greatly affected by glass composition chemistry such as soda lime silicate, borosilicate and aluminosilicate. Likewise, the chemistry differences of the "tin" and the "air" surfaces influence the strengthening behavior of float glass. With progressive exchange, the bath chemistry alters due to increasing impurities and needs to be brought under control for a successful product technology. The physics of surface compression development has been described in detail earlier. In this review, we discuss the effects of variations in the chemistry of glass and the immersion bath.

INTRODUCTION

Chemical strengthening was first invented in 1962 by Kistler[1] and Acloque *et. a.l.*[2] independently. An alkali-containing glass is immersed in an electrically heated bath (tank) containing molten alkali nitrate salt (usually) at temperatures well below the glass annealing point. An exchange between the host alkali ions of glass and the larger invading ions from the salt occurs (see Figure 1). The resultant stuffing in a near-rigid atomic glass network leads to the development of high surface compression and some balancing interior tension depending upon the diffusion depth and product wall thickness. It is the high surface compression which causes glass product strengthening since flaws usually occur on the outside surface due to handling. Thus soda lime silicate glasses (SLS) can be readily strengthened by immersion in KNO_3 bath at temperatures ~450°C; and lithium aluminosilicate glasses can be strengthened by immersion in $NaNO_3$ salt bath at temperatures ~400°C.

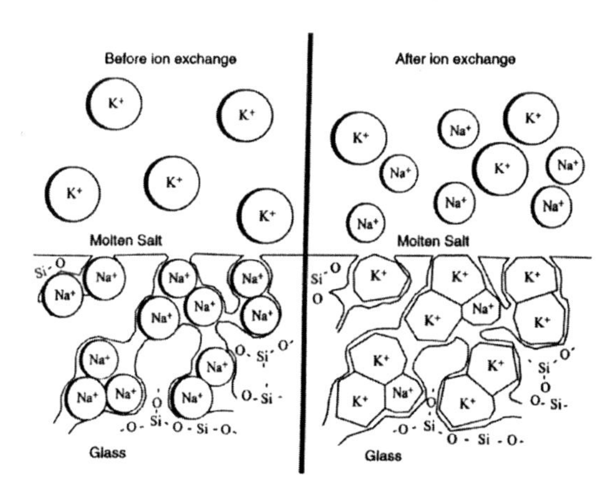

Figure 1. Alkali ions 'stuffing' in glass by ion exchange (redrawn from Nordberg, et. al.[5]).

The advantages of this process in comparison to other glass strengthening techniques (namely, thermal tempering) are: (i) introduction of relatively high surface compression, (ii) no measurable optical distortion, (iii) thin plates, even 100 μm thin, can be strengthened, and (iv) products having irregular geometry can be strengthened so long as the surface can be contacted by the molten salt. Disadvantages of chemical strengthening are (a) specifically limited to alkali-containing glass, (b) low case depth for SLS glass which makes the glass susceptible to weakness from handling flaws in severe situations, and (c) high cost due to extended bath immersion time.

Potential strength increase by chemical strengthening is dependent upon the initial flaw condition of the glass and the compression magnitude and case depth generated by the ion exchange. Compression magnitude and case depth are controlled by the glass chemistry and the ion-exchange medium chemistry. Oxide glass products in the consumer market belong to several major glass families, notably, high silicas, soda lime silicates, sodium borosilicates, alkali aluminosilicates, and lead-alkali silicates. Because the type and the concentration of the alkali ion vary from glass to glass, it is clear that bath chemistry requires careful selection for an optimized strengthening process. Likewise, most glass surfaces differ from the bulk composition depending upon the glass forming technique utilized.

Over the five decades since the discovery, we have learned much about the process of glass chemical strengthening. There are still several lessons yet to be learned. These have been discussed by the author recently.[3] Additionally, the physics of ion-exchange strengthening in glass has also been reviewed recently by the author.[4] The objective of this paper is to review the chemistry of the glass chemical strengthening process. Attention is paid specifically to the effects of (1) the chemistry of glass composition, (2) chemistry non-uniformity due to glass processing and (3) the exchange bath chemistry.

GLASS COMPOSITION CHEMISTRY

Pure silica or high silica glasses do not strengthen simply because there is insufficient concentration of alkali ions present to develop significant surface compression. By the same token, low expansion borosilicates (of the Pyrex®-type) and lead-alkali silicates do not strengthen well.[5] Our unpublished data suggests that Pyrex®-type borosilicates require as long as 3 days or so immersion in KNO_3 bath at ~450°C to develop some beneficial effects; their exchangeable alkali ions are presumably (nearly) inaccessible due to liquid-liquid phase separation characteristics. In lead-alkali silicates, the large lead ions act to block interdiffusion kinetics.

The medium expansion Type I borosilicate glasses, such as Kimble N51A or Schott Fiolax®, which are used as packaging for parenteral drugs in pharmaceutical applications and for cosmetics, can be strengthened reasonably well. Surface compressions of around 300 MPa and a case depth of 20 – 35 microns can be obtained by immersion in KNO_3. Increased sensitivity to glass breakage during transportation and during customer use has resulted in successful introduction of chemically strengthened autoinjectors.[6]

Glasses that strengthen well are the soda lime silicates and the alkali aluminosilicates. In common SLS glasses, the Na_2O content is generally 13 to 15% and the phase microstructure of glass allows exchange with invading K^+ ions from a KNO_3 bath. Soda lime silicates can be readily exchanged to develop roughly 350 MPa surface compression and as much as 700 MPa in some specially designed compositions with added alumina. High magnitude surface compression is particularly useful for impact resistance of thin display windows in portable electronic devices such as mobile phones. One limitation in the soda lime silicates is the rather short case depth, generally no more than 50 μm. Attempts to obtain deeper penetration of K^+ typically results in the relaxation of the beneficial surface compression. For thin display windows, this limitation is actually beneficial in that deeper penetration can result in the development of a large internal tension which, in turn, would cause undesirable frangibility upon failure. The work of Sinton et. al.[7] showed that, while increasing the total

MgO+CaO content reduces case depth, the replacement of CaO by MgO in SLS glasses increases case depth. Up to 90 μm case depths in ordinary SLS glasses were demonstrated without significant degradation of strength (Figure 2A). Molecular dynamics simulations of Pedone, et. al.[8] have suggested Mg^{2+} is incorporated into the silica network in four-fold coordination, rather than a modifying role like Ca^{2+}. MgO substitution for CaO decreases network packing density and opens alkali diffusion pathways. Sinton et. al.[7] also showed that case depth could be increased by replacing some Na_2O by K_2O (Figure 2B). This is consistent with higher interdiffusion coefficients observed in mixed alkali compositions,[9] but this substitution is at the expense of some amount of surface compression.

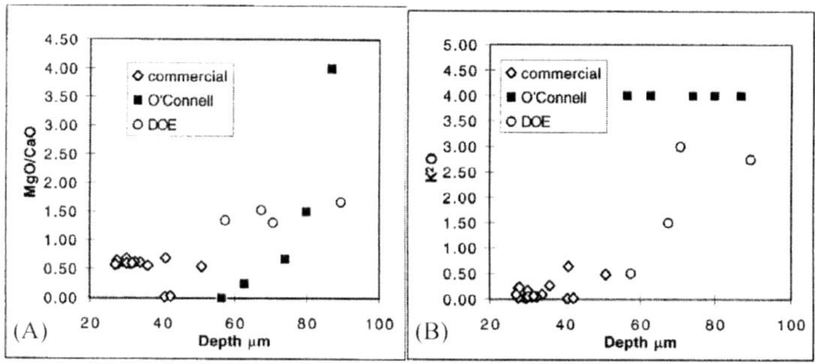

Figure 2. Variations in case depth with (A) MgO/CaO ratio and (B) K_2O content in the host glass (from Sinton, et. al.[7]).

Larger case depths and greater compression magnitudes are obtainable in alkali aluminosilicates. Specially designed formulations of alkali aluminosilicates can be ion exchanged for surface compression of 300 MPa – 1 GPa and case depth of ~50 μm to 1 mm. Nordberg et. al.[5] have shown that modulus of rupture (MOR) of K^+-exchanged Na_2O-Al_2O_3-SiO_2 glasses, where Na_2O was 10 or 20%, increased steadily with Al_2O_3 additions from 5 to 35 wt% when tested after 150-grit abrasion (Figure 3). This increase of abraded MOR resulted from improved compression magnitude and slower rate of compression relaxation for equivalent ion-exchange processing conditions. Alumina enters the Na_2O-SiO_2 network as $[AlO_4]^-$ occupying an alkali ion for charge balance, therefore each $[AlO_4]^-$ removes one non-bridging oxygen (NBO) from the silicate network.[6] Upon accommodating a stuffing alkali ion, the alkali site associated with a $[AlO_4]^-$ group presumably exhibits less plasticity[4] than a NBO site leading to larger observed compression magnitudes. This is evident in stress profile measurements by Kim[10] of K^+-exchanged $25Na_2O\cdot xAl_2O_3\cdot(75-x)SiO_2$ (mol%) glass for alumina/alkali ratios of zero to 0.6 where surface compression magnitude increased with increasing alumina content. In this same work, case depth was observed to decrease with alumina addition. Although Na^+ mobility has been shown to increase with alumina/alkali ratio of zero to one,[11] alkali interdiffusion is controlled by the slower diffusing species[6] therefore the shorter case depth is due to greater activation energy for K^+ diffusion. This is presumably caused by less plastic accommodation of the K^+ ion by the $[AlO_4]^-$ alkali site (i.e. greater elastic deformation of the site). Therefore, a trade-off is present between compression magnitude and case depth when working with alkali aluminosilicate glasses for chemical strengthening. Nordberg et. al.[5] have also shown that 0 to 20 wt% addition of ZrO_2 improves 150-grit abraded MOR of Na_2O-ZrO_2-SiO_2 glasses. Like Al_2O_3, additions of ZrO_2 presumably cause reduced network plasticity.

Thin yet very strong glass windshields for use in automobiles were marketed in the late 1960s by General Motors. These utilized Corning code 0317 "Chemcor" glass which is a predecessor to

Corning's current Gorilla® glass. Saint Gobain's "SOLIDION®" glass is a cross between SLS and aluminosilicate which, after the Na^+ (glass) ↔ K^+ (bath) exchange, develops ~480 MPa of surface

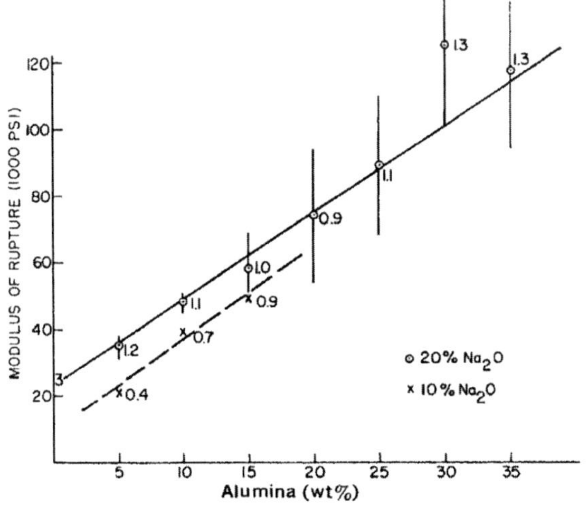

Figure 3. Abraded (150-grit SiC) modulus of rupture versus alumina content in the host glass (from Nordberg, *et. al.*[5]).

compression with a case depth of ~200 μm. The glass in laminated form is used for aircraft cockpit windshields. On the other hand, PPG's "Herculite®" glass for aircraft cockpit windshields is a lithium aluminosilicate glass laminate where the Li^+ (glass) ↔ Na^+ (bath) exchange is carried out at relatively low temperatures (~375°C) for about 24h to yield a surface compression of ~375 MPa and a case depth of ~330 μm. The deeper case depth allows superior protection from abrasion due to dust storms, hail stones, and impact by flying birds. Perhaps the most remarkable surface compression magnitude with high case depths are obtained for an alkali aluminosilicate glass strengthened by immersion in a mixed alkali salt bath. This glass, marketed as Ion-Armor™ by Saxon Glass Technologies, Inc., has a MOR of nearly 1 GPa resulting from surface compression of ~1 GPa with case depth of ~1 mm.[12-13]

It is usual to observe a subsurface compression maximum in alkali silicate and soda lime silicate glasses[14] (Figure 4). Stress relaxation on the surface is also observed in Na_2O-Al_2O_3-SiO_2 glasses,[15] though normally only after extended treatment times.

CHEMISTRY DIFFERENCE FROM GLASS PROCESSING

The most important forming-related chemistry is that of float glasses where the tin diffuses into the glass surface in contact with the tin bath. This chemistry difference has been well-documented[16] and can lead to varying properties between the two sides. Shown in Table I are our recent results of Na^+ (glass) ↔ K^+ (bath) exchange at three different temperatures for 1.1 mm window glasses (intended for cell phone application). It is clear that the tin side has lower surface compression and lower case depth. Imbalance of surface compression and case depth between the tin and air sides causes geometric warp that is insignificant for 1.1 mm thickness, but is notable for 0.4 mm thickness. Techniques must be developed to mitigate warp of chemically strengthened float glasses for their widespread adoption as protective windows for ever increasing display panel and television sizes.

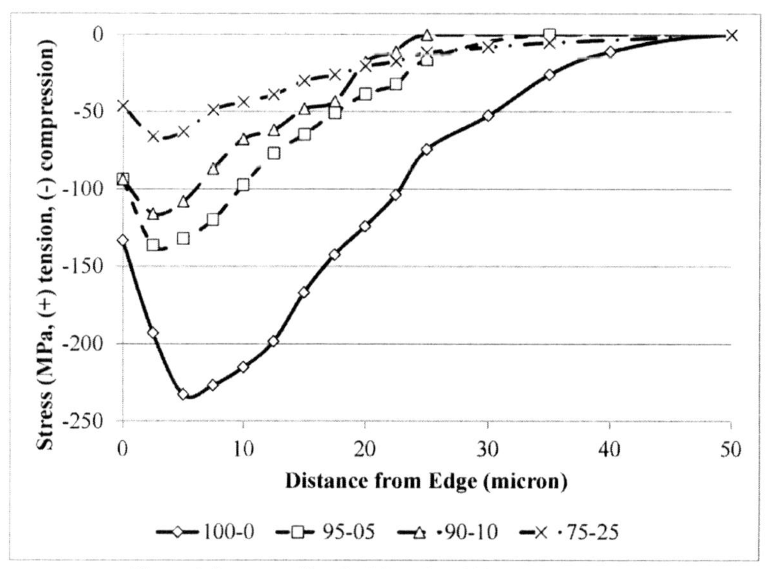

Figure 4. Stress profiles for $23Na_2O \cdot 77SiO_2$ glass immersed in various KNO_3-$NaNO_3$ salt bath compositions (mol%).

Table I. Ion-exchange strengthened SLS air versus tin surface compression and case depth measured via ellipsometry.

	1.1 mm Substrate			
Exchange	Surface Compression (MPa)		Case Depth (micron)	
Temperature	Tin	Air	Tin	Air
A	488	501	19.8	23.4
B	453	466	23.6	27.3
C	382	394	28.2	31.8

SALT BATH CHEMISTRY

Salt bath chemistry is perhaps the most important obstacle in a "stable" manufacturing technology for commercial products. For instance, during the production of chemically strengthened SLS glass windows by immersion in a KNO_3 salt bath, it should be realized that the Na^+ ions exiting from glass gradually weaken the potency of the salt bath. The thermodynamic activity of the K^+ ions on the glass surface is given by the product of chemical concentration and activity coefficient, where activity coefficient is often less than one. As Na^+ impurity level increases, the K^+ activity coefficient may be further reduced to tenths or hundredths. This is demonstrated in the work of Wakabayashi[17] by the equilibrium mole fractions of K^+ exchanged into two sodium silicate glasses (Figure 5), where N_K represents the quantity of K_2O in the glass or molten nitrate. As N_K in the molten nitrate decreases, i.e. as Na_2O increases, the equilibrium quantity of K_2O in the glass drastically decreases. Further examples of such experiments are shown in Figure 4 where a $23Na_2O \cdot 77SiO_2$ glass was ion exchanged in mixed alkali nitrate baths. Addition of 5 mol% $NaNO_3$ as a "contaminant" to the bath nearly halves the magnitude of the subsurface compression maximum.

Soda lime silicate glasses are particularly sensitive to alkaline earth impurities in the bath. The work of Zhang et. al.[18] shows that as little as 0.005% Ca^{2+} impurity can reduce the bend strength by a factor of nearly three, while alkaline earth impurities of larger ionic radius have similar effects at higher concentrations (Figure 6). These impurities are presumed to adsorb to the glass surface blocking access to alkali diffusion pathways.[19]

Recently published work of Xu et. al.[20] suggests that minor additions of Al_2O_3, KOH, K_2SiO_3, K_3PO_4 or K_2CO_3 could improve bath quality (Table II and Figure 7). They found a combination of 0.5 wt% Al_2O_3 and 0.5 wt% KOH apparently produced favorable results.

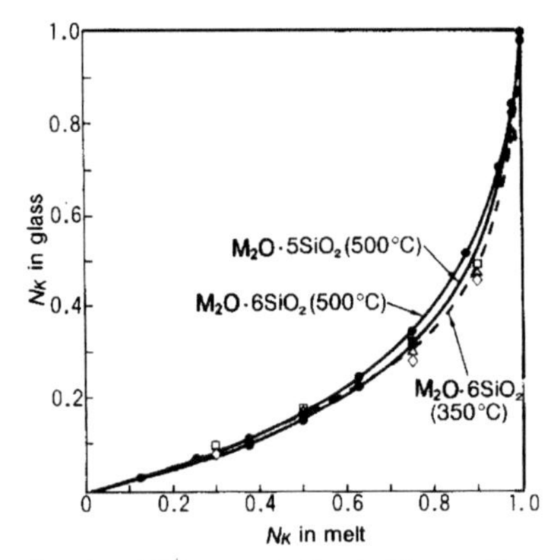

Figure 5. Equilibrium fractional K^+ concentration in binary silicate glasses as a function of fractional K^+ concentration in the melt (salt bath). (From Wakabayashi[17]).

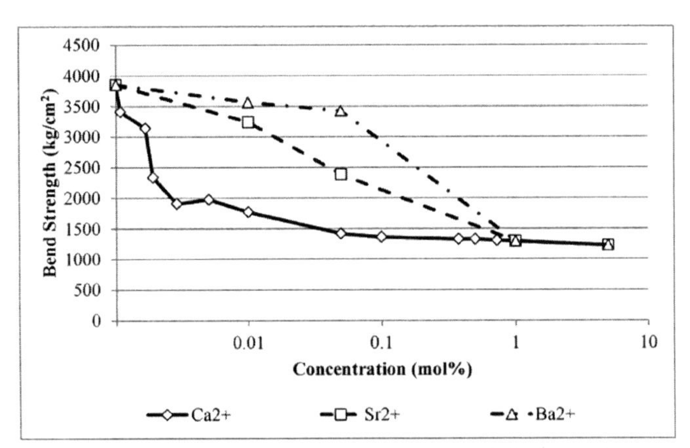

Figure 6. Bend strength versus concentration of alkaline earth impurities in the salt bath (reproduced from Zhang, *et. al.*[20]).

Table II. Salt bath compositions used by Xu, et. al.[20]

	Composition of molten salt (wt%)					
Sample No.	KNO_3	Al_2O_3	KOH	K_2SiO_3	K_5PO_4	K_2CO_3
1	100					
2	99	0.5	0.5			
3	97	0.5	0.5	2		
4	97	0.5	0.5		2	
5	97	0.5	0.5			2
6	100*					

*Analytical purity KNO_3 without additive.
All others used commercial grade KNO_3.

SUMMARY

(1) Of the available commercial glasses, soda lime silicate and alkali aluminosilicate glasses strengthen the best. Surface compression can range from 300 MPa to almost 1 GPa. However, the case depths vary widely. For most SLS glasses, one should only expect ~35 micron case depth. However, case depths as much as 1 mm can be obtained in alkali aluminosilicate glasses.

(2) Air and tin surfaces of float glass can display different interdiffusion and compression development characteristics during chemical strengthening. In thin glasses used for personal mobile communication devices, such specimens could become warped after ion exchange. Warp reduction techniques need to be developed.

(3) Parts per million levels of alkaline earth impurities in KNO_3 baths can have a strong poisoning effect on the strengthening of soda lime silicates. Small additions of Al_2O_3 and KOH to the bath could be beneficial.

REFERENCES

[1]S.S. Kistler, "Stresses in glass produced by nonuniform exchange of monovalent ions", *J. Am. Ceram. Soc.*, **45**[2] 59-68 (1962).

[2]P. Acloque and J. Tochon, "Measurement of mechanical resistance of glass after reinforcement", pp 687-704 in *Colloquium on Mechanical Strength of Glass and Ways of Improving It*, Sept 25 -29, 1961, Florence Italy. Published by Union Scientifique Continentale du Verre, Charlroi, Belgium 1962 pp 1044, 1962.

[3]A.K. Varshneya, "Chemical strengthening of glass: Lessons learned and yet to be learned", *Int. J. Appl. Glass Sci.*, **1**[2] 131-142 (2010).

[4]A.K. Varshneya, "The physics of chemical strengthening of glass: Room for a new view", *J. Non-Cryst. Solids*, **356**, 2289-2294 (2010).

[5]M.E. Nordberg, E.L. Mochel, H.M. Garfinkel, and J.S. Olcott, "Strengthening by ion exchange", *J. Am. Ceram. Soc.*, **47**[5] 215-219 (1964).

[6]A.K. Varshneya, in *Fundamentals of Inorganic Glasses*, 2nd ed. p129 and p398 *ff.* Soc. Glass Tech. Sheffield, UK 2006.

[7]C.W. Sinton, W.C. LaCourse, M.J. O'Connell, "Variations in K^+-Na^+ ion exchange depth in commercial and experimental float glass compositions," *Mater. Res. Bull.*, **34**[14-15] 2351-2359 (1999).

[8]A. Pedone, G. Malavasi, M.C. Menziani, U. Segre, and A.N. Cormack, "Role of magnesium in soda-lime glasses: Insight into structural, transport, and mechanical properties through computer simulations," *J. Chem. Phys. C*, **112**[29] 11034-11041 (2008).

[9]A.K. Varshneya and M.E. Milberg, "Ion exchange in sodium borosilicate glasses," *J. Am Ceram. Soc.*, **57**[4] 165-169 (1974)
[10]S. Kim, "Mechanism of alumina effect on ion-exchange strengthening of glass," M.S. Thesis, Alfred University (1984).
[11]J.O. Isard, "Electrical conduction in the aluminosilicate glasses," *J. Soc. Glass Technol.*, **43**, 113T-123T (1959).
[12]Saxon Glass Technologies, Inc. <http://www.saxonglass.com>
[13]A.K. Varshneya and Ian M. Spinelli, "High-strength, large-case-depth chemically strengthened lithium aluminosilicate glass", *Am. Ceram. Soc. Bull.*, **88**[5] 27-33 (2009).
[14]A.Y. Sane and A.R. Cooper, "Stress buildup and relaxation during ion exchange strengthening of glass", *J. Am. Ceram. Soc.*, **70**[2] 86-89 (1987).
[15]W.J. Spoor and A.J. Burggraaf, "The strengthening of glass by ion exchange. Part 3. Mathematical description of the stress relaxation after ion exchange in alkali aluminosilicate glasses", *Phys. Chem. Glasses.*, 7[5] 173-177 (1966).
[16]See, for example, L. Colombin, A. Jelli, J. Riga, J.J. Pireaux, and J. Verbist, "Penetration depth of tin in float glass," *J. Non-Crystal. Solids*, **24**[2] 253-258 (1977).
[17]H. Wakabayashi, "The relationship between kinetic and thermodynamic properties in mixed alkali glass," *J. Non-Crystal. Solids*, **203**, 274-279 (1996).
[18]X. Zhang, O. He, C. Xu, Y. Zhang, "The effect of impurity ions in molten salt KNO_3 on ion-exchange and strengthening of glass", *J. Non-Cryst. Solids.*, **80**[1-3] 313-318 (1986).
[19]A.M. Kleshchevnikov, A.P. Milovanov, V.V. Moiseev, V.I. Nefedov, V.I. Portnyagin, and G.A. Shashkina, "An x-ray photoelectron spectroscopic study of the surface of sodium silicate glass after ion-exchange treatment," *Fizika Khimiya Stekla*, **9**[5] 622-8 (1983).
[20]J. Xu, B. Zhang, Y. Huo, and J. Ma, "Effect of additives in molten salt on ion exchange and strengthening of glass", *J. Chinese Ceram. Soc.*, **37**[5] 851-854 (2009).

*Corresponding author: Varshneya@alfred.edu

A STUDY OF SILICA GLASS FIBER STRUCTURE AND ELASTIC PROPERTIES, USING MOLECULAR DYNAMICS SIMULATIONS

Laura Adkins and Alastair Cormack
Alfred University
Alfred, NY, USA

ABSTRACT
Silica whiskers and fibers, with diameters ranging from a few Å to 20nm have been modeled using classical molecular dynamics simulations. The surface structure of square cross-sectioned fibers has been characterized in terms of structural defect species and their population as a function of depth from the surface. The behavior of a fiber under tensile strain is described, and the atomic scale structural origin of fracture is discussed.

INTRODUCTION

It is well-established that computational modeling is an excellent technique for probing the structure and properties of silica glasses. A number of atomistic simulations have also been conducted to examine the failure of silica glass under stress,[1-14] but the majority of these studies have used bulk glass, using periodic boundary conditions to avoid surface effects.[1,2,7-10,12,14] In these studies, failure was found to occur through the coalescence of nano-scale voids inherent in the structure of the glass.[1,7,14]

Although a significant body of work exists examining the structure of both bulk silica glass,[6,9,12,15-43] and silica surfaces,[44-61] using atomistic computational techniques, little work has been done to examine the structure of silica fibers. In this work, silica fibers were created, and their structure was examined. Through the use of a BlueGene/L supercomputer, we have been able to simulate fibers up to 20nm in cross-sectional diameter. After creating these fibers, they were strained under tension, to failure, and the fracture process was observed.

FIBER FORMATION

Creation

A standard computational process was used with the molecular dynamics package DLPOLY to form bulk glass samples of varying sizes containing 12,000, 19,500, and 599,040 atoms. In order to create a fiber from the simulated bulk structure, the periodic boundary conditions were manipulated. At least 20Å of extra space was added on either side of the box, in the x and y directions, leaving the z direction as continuously periodic. This produced individual fibers that were spaced far enough apart in the x and y directions to avoid any interactions between images, but still continuous in the z direction.

Annealing

After the fibers were created, the surface of the fibers was found to be very highly disordered. This disorder is characterized by the presence of unusual structural defect species in the surface of the sample, that are not found in the bulk structure, such as non-bridging oxygens (NBO) and concomitant Q_2 & Q_3 units, and under-coordinated silicon atoms, Si.[3] In order to allow some of these defect species to re-coordinate back into the surface of the fiber, the sample was heated to 2,500K before it was cut, and then after the fiber was cut, it was held at that temperature for 250,000 timesteps to model an annealing process. The concentration of defect species in the sample as it was cut and annealed is shown in Table I.

From Table I., it is clear that a great deal of disorder is added to the system, specifically when the fiber is formed from the bulk. Although most of the defect species are recombined back into the structure during the annealing process, a small portion of the disorder remains in the final sample. It was found primarily in the outer region of the fiber, defining a surface layer, as described in the next section.

Table I. Defect species in the square fiber at various stages of the forming process

Forming Step	NBO	Q_3	Q_2	$Si^{(3)}$
Bulk Glass Sample	0.3 %	0.8 %	0.1 %	0.4 %
Hot Bulk Sample	0.6 %	1.9 %	0.3 %	1.2 %
Hot Cut Sample	4.0 %	10.6 %	2.3 %	7.3 %
Hot Annealed Sample	0.6 %	2.1 %	0.1 %	1.1 %
Cooled Final Fiber	0.7 %	2.2 %	0.2 %	1.3 %

DEFINING SURFACE THICKNESS

In the simulations of silica surfaces, the surface of a sample has typically been defined by the presence of unusually coordinated species.[62] The depth of a surface structure is determined by the variation in population of these species from the bulk structure. One surface feature that has been noted in simulated surfaces is a bond-angle of 120°. An SiO_4 tetrahedron has bond-angles of 109° between each set of O-Si-O atoms. When an O-Si-O bond-angle of 120° is found in silica glass, it is usually associated with three-coordinated silicon atoms. Trioni et al.[59] observed such a peak in the bond-angle distribution within 5Å of the surface. Similarly, Wilson and Walsh[48] found that the peak at 120° was not noticeable beyond 5.29Å from the surface. Mischler, Kob, and Binder,[45] in an ab-initio MD simulation, noted that a shift back to a bond-angle around 109° from more strained values was only apparent when they examined a sample which had a surface depth greater than 5Å. Based on three-coordinated silicon atoms, two-membered silicon rings, and non-bridging oxygen in the structure, among other unusual species, Du and Cormack[62] determined that the surface layer extended 5-10Å into the sample.

The concentration of OH that would be formed from the hydroxylation of surface defect species was found to be 2.6 sites/nm^2 using temperature programmed static SIMS.[63] Simulations by Pantano and colleagues[46] suggested concentrations of 3.9 OH/nm^2 while Du and Cormack[62] suggested that a range of OH concentrations from 4.5 to 2.5 sites/nm^2 were possible, depending on the species that are assumed to react to form Si-OH.

Material surfaces can also be defined by the variation in density from the bulk value.[50] Using a depth profiling technique, Yuan and Cormack found that silica glass had a surface depth of approximately 10-15Å. The surfaces in their study were fracture surfaces, produced from the tensile failure of a bulk glass, and were expected to be rougher than the clean, cut, outer surface of the fibers studied in this work.

To define the surface depth in the fibers simulated in this work, 0.5Å thick radial slices were taken from the center outward, mimicking the shape of the cross section, and the number of defect species found in each slice was noted. For the purposes of defining a surface layer, the defect species included non-bridging oxygen, triply-bridged oxygen, any silicon atom not connected to four oxygens, and any Q species other than Q_4.

Figure 1 shows the variation in the number of defect species for three fibers, from which it can be deduced that the approximate depths of the surface layers are 9.3±0.2, 9.5±0.6, and 8.4±0.7Å, for the 12,000-, 19,500-, and 599,040-atom fibers, respectively. The variation in surface depth of these samples is small, and largely within a single standard deviation. This does not suggest a trend based on sample size. Since the fibers were free to move in the sample box during the annealing, both the

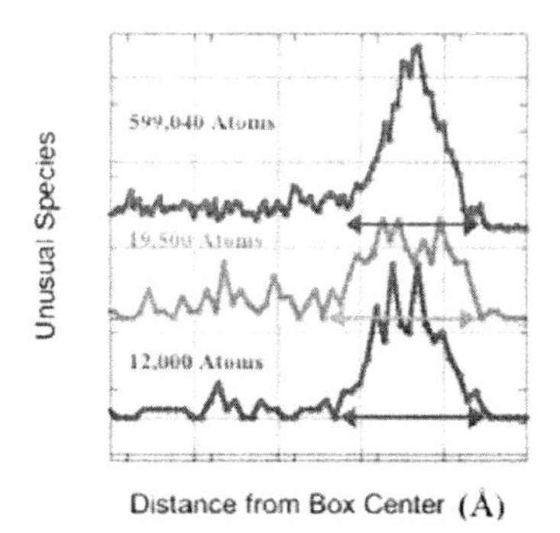

Figure 1. The number of unusual species forming the surface of the three differently-sized fibers with square-cross-sections is shown here. Surfaces were found to be 8.4±0.7, 9.5±0.6, and 9.3±0.2Å thick, respectively, for fibers of increasing size.

12,000- and 19,500-atom fibers shifted slightly. The resulting fibers still retained their box cross-sectional shape but were angled away from the initial box formed parallel to the x and y axes. Thus the radial slices of the boxes, which are taken parallel to the axes, provided a slightly wider surface layer than would have otherwise been found. Since the 599,040-atom fiber did not shift away from the original parallel position to the axes, the surface structure is slightly thinner.

The concentration of these defect species in the glass, including small-membered rings, if fully converted to OH in a wet atmosphere, would lead to a surface with 2.7 OH/nm^2 for the 19,500-atom square fiber. This agrees well with the results of the experimental work referenced earlier that suggested an OH concentration of 2.6/nm^2.

In addition to a spike in defect species population, low-density areas may also be used to define the surfaces. The density of the 19,500-atom sample, given as function of the depth into the box, is shown in Figure 2. The surface thicknesses obtained this way seen for the samples were found to be 5.7, 8.7, and 11Å, for the 12,000-, 19,500-, and 599,040-atom samples, respectively. Although there is a trend with sample size, it quickly reaches an asymptote and so the surface region is not expected to increase much further with increasing fiber size beyond 20nm.

STRAINING THE SILICA FIBER

After the silica fiber was formed, it was strained under tension until failure, to examine the fracture process as it occurred. Rather than apply stress and measure the resultant strain, as is commonly done experimentally,[64-67] computationally, the simulation box and atoms were strained, and then the stress of the system was calculated from the new atomic positions, forces and velocities. Strain in a single dimension is calculated as the change in length of the sample in that direction divided by the original length, and is reported as a percent change. In the straining process, the simulation box is expanded along the z-direction, followed by a scaling of the z coordinate of each atom. Interatomic forces are calculated for these new positions and the atoms are allowed to relax (in all three directions). Since the simulation box contains empty space in the x and y directions, the fiber can contract naturally in the unstrained directions, as a response to the strain applied in z.

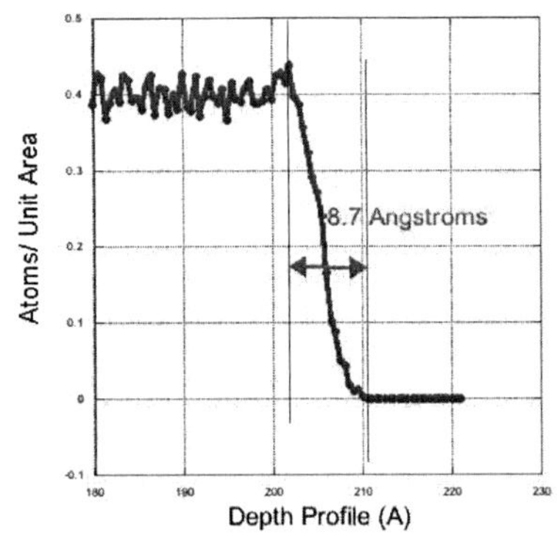

Figure 2. The density of the sample, as a function of depth profile, for the 19,500-atom fiber.

The samples were strained at a rate of 0.001/picosecond. Although this strain rate is fast relative to experimental strain rates, it is one of the slowest that has been used in computational simulation of fracture of silicate glasses.[1] The use of a fast strain rate is dictated by the computational resource constraints. Fracture of the fiber typically occurred within 300,000 timesteps, and the runs were allowed to continue for a total of 600,000 timesteps, the equivalent of 1.2 nanoseconds in real time. With a timestep of 2 femtoseconds, to run these calculations for a single second of real time would require 5×10^{14} iterations of the calculations, and is too computationally expensive to be feasible using current technologies.

As the systems were strained, they deformed elastically before failing in a brittle manner. A graph of the stress-strain curve for the 19,500-atom fiber is shown in Figure 3 and the values of failure stress and failure strain found in this work are compared with the results of simulations of bulk glass and experimental study in Table II.

The stress at failure agrees reasonable well with the results of experimental study shown here, and is closer to the experimental values than the result of a similar study using bulk silica glass. Failure strain was lower than any values found either by the experimental study or the MD simulation of bulk silica, leading to a value of Young's modulus that was higher than expected. Although the stress-strain curve shows apparently linear behavior as the fiber is initially deformed, non-linear behavior is clearly noted as the strain of the system is increased and the fiber approaches failure. In concordance with this behavior, also noted by Gupta and Kurkjian,[68] Young's modulus was calculated by fitting a third-order polynomial to the curve, up to a strain of 13%.

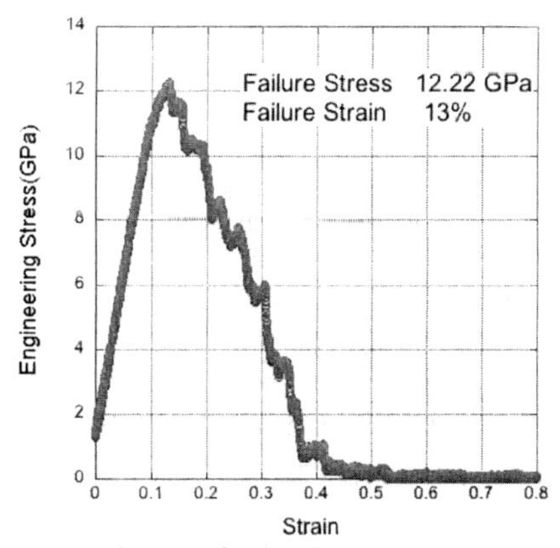

Figure 3. A stress-strain curve for the 19,500-atom square-cross-section fiber.

Table II. Failure stress, failure strain, and the elastic modulus for the fibers, previous bulk silica work, and experimental study

Property	MD Bulk Silica[1]	Experimental Fiber[65]	MD Fiber 19,500-atom Square
Failure Stress (GPa)	10.8	12.77	12.21 ± 0.04
Failure Strain (%)	15.3	17.73 ± 0.16	13.04
Young's Modulus (GPa)	69.9	71.99 ± 0.65	90

DISTORTION OF THE STRUCTURE UNDER STRAIN

Under strain, it is expected that the structure of the glass will undergo some distortion. As the simulation box is elongated (i.e. strain is increased), the Si-O bond-lengths increase from an average of 1.618Å to 1.636Å at the failure strain, and then return to the initial value, as the fibers are broken and the two parts separate completely. Failure of the fiber is initiated at approximately 12-14% strain, and so the bond-lengths begin to decrease, indicating elastic recovery, after this point.

The changes that occur to the intertetrahedral bond-angle are somewhat different. At zero strain, the intertetrahedral bond-angle of 150° matches the peak position seen by Du and Cormack in their surface structures of silica glass.[62] As the strain increases, the average Si-O-Si angle increases only slightly to 151.6°; however, subsequent relaxation results in a substantial lowering of the final Si-O-Si bond-angle to below its initial value. This is probably due to the formation of small rings on the freshly formed fracture surface.

With failure of the 19,500-atom square fiber, the number of two- and three-membered rings increases from approximately 0.8% to 1.5%, partially accounting for the bond-angle shift, because these rings have much tighter bond-angles than those found in five- and six-membered rings, typically around 95-100° for two-membered rings, and between 120-140° for three-membered rings. The rest of the shift in Si-O-Si bond-angle is accounted for by the formation of triply-bonded oxygen on the new fracture surface. The concentration of triply-bonded oxygen found in the initial fiber was 0.19%, but after failure the concentration increased to 0.49%. The Si-O-Si bond-angles associated with such an oxygen species are between 110-125°.

Although the area of the fracture surface is small when compared with the external surface of the fiber, it contains many more small-membered rings and TBO than the fiber surface because the fracture occurred at room temperature, and the species on the fracture surface have less energy with which to reconnect back into the structure in favorable ways. These small, strained bond-angles in the two- and three-membered rings shift the average bond-angle of the system to lower values. As the fracture process proceeds, the population of these rings and TBO increases and the average bond-angle decreases, remaining lower than in the unstrained fiber.

THE ORIGIN OF FRACTURE

To examine origins of fracture, the 19,500-atom structure was separated into 10Å thick sections of the left, front, right and back surfaces, the center core, and the full fiber, and then snapshots of the fracture process were taken for visual examination, as shown in Figure 4.

From regions A and B, it is clear that the fracture process began at the surface of the fiber. As the strain increased, the crack expanded from the surface at A (on the right panel), through the bulk at C, to region D, on the opposite surface of the fiber, to region A. Further analysis is underway to define the exact origin of failure and the role of the structure of the glass in this process. Preliminary results indicate that failure is initiated from surface structural defects such as a three-coordinated silicon or triply bridged oxygen in the structure.

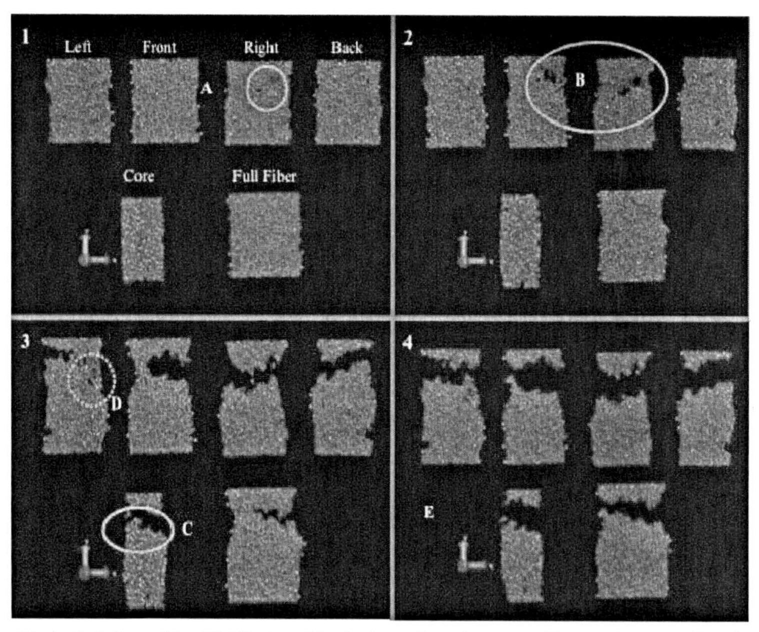

Figure 4. Failure of the 19,500-atom fiber is shown here, in stages. The panels are numbered in sequential order and show; A) Initial void formation, B) Cracks propagating in the surface of the fiber, C) Failure in the core of the fiber is completed, D) Part of the surface of the fiber is still intact, at the same time C is observed, E) Failure of the fiber is complete.

CONCLUSIONS

Silica glass fibers of different sizes have been created and characterized using atomistic computer simulations. A distinct surface layer has been formed in each sample studied, and can be identified by the presence of structural defect species and variations in sample density. For characteristic samples, containing 19,500 and 599,040 atoms, the surface layer was approximately 9-10 Å thick. Under uniaxial tension, the 19,500 atom fiber broke at values of stress and strain comparable to experimental data, and the fracture process was noted to begin at the surface of the fiber, move through the bulk, and complete on the other side. Initiation of failure appears to be linked with high strain sites on the surface associated with surface structural defects.

REFERENCES

[1]Pedone, A.; Malavasi, G.; Cristina Menziani, M.; Segre, U.; Cormack, A. N. *Chem. Mater.* **2008**, *20*, 4356.

[2]Nomura, K.-I.; Chen, Y.-C.; Weiqiang, W.; Kalia, R.K.; Nakano, A.; Vashishta, P.; Yang, L.H. *J. Phys. D: Appl. Phys.* **2009**, *42*.

[3]Rountree, C. L.; Bonamy, D.; Dalmas, D.; Prades, S.; Kalia, R.K.; Guillot, C.; Bouchaud, E. *Physics and Chemistry of Glasses: European Journal of Glass Science and Technology Part B* **2010**, *51*, 127.

[4]Rountree, C.L.; Prades, S.; Bonamy, D.; Bouchaud, E.; Kalia, R.; Guillot, C.*J. Alloys Compd.* **2007**, *434-435*, 60.

[5]Simmons, J.H. *J. Non-Cryst. Solids* **1998**, *239*, 1.

[6]Wang, J.; Omeltchenko, A.; Kalia, R.K.; Vashishta, P. *Proceedings of the 1996 MRS Fall Meeting, December 2, 1996 - December 6, 1996* **1997**, *455*, 267.

[7]Muralidharan, K.; Simmons, J.H.; Deymier, P.A.; Runge, K. *Papers from the Michael Weinberg Symposium* **2005**, *351*, 1532.

[8]Ochoa, R.; Swiler, T. P.; Simmons, J.H. *J. Non-Cryst. Solids* **1991**, *128*, 57.

[9]Simmons, J.H.; Swiler, T. P.; Ochoa, R. *J. Non-Cryst. Solids* **1991**, *134*, 179.

[10]Swiler, T.P. *Glass Res.* **2002**, *11*, 10.

[11]Swiler, T.P.; Simmons, J. H.; Wright, A.C. *J. Non-Cryst. Solids* **1995**, *182*, 68.

[12]Chen, Y.-C.; Lu, Z.; Nomura, K.-I.; Wang, W.; Kalia, R.K.; Nakano, A.; Vashishta, P. *Phys. Rev. Lett.* **2007**, *99*.

[13]Van Brutzel, L.; Rountree, C.L.; Kalia, R.K.; Nakano, A.; Vashishta, P. *Nanophase and Nanocomposite Materials IV, November 26, 2001 - November 29, 2001* **2002**, *703*, 117.

[14]Swiler, T.P.; Simmons, J.H.; Wright, A.C. *J. Non-Cryst. Solids* **1995**, *182*, 68.

[15]Kerrache, A.; Teboul, V.; Guichaoua, D.; Monteil, A. *Proceedings of the 4th Franco-Italian Symposium on SiO2 and Advanced Dielectrics, September 16, 2002-September 18, 2002* **2003**, *322*, 41.

[16]Stebbins, J.F.; McMillan, P. *J. Non-Cryst. Solids* **1993**, *160*, 116.

[17]Vessal, B.; Amini, M.; Catlow, C.R.A. *J. Non-Cryst. Solids* **1993**, *159*, 184.

[18]Zheng, L.; An, Q.; Fu, R.; Ni, S.; Luo, S.-N. *J. Chem. Phys.* **2006**, *125*.

[19]Rino, J.P.; Gutierrez, G.; Ebbsjo, I.; Kalia, R.K.; Vashishta, P. *Proceedings of the 1996 MRS Fall Symposium, November 27, 1995 - December 1, 1995* **1996**, *408*, 333.

[20]Huff, N.T.; Demiralp, E.; Cagin, T.; Goddard III, W. A. *J. Non-Cryst. Solids* **1999**, *253*, 133.

[21]Della, V.R.G.; Venuti, E. *Phys. Rev. B: Condens. Matter* **1996**, *54*, 3809.

[22]Leonforte, F.; Tanguy, A.; Wittmer, J.P.; Barrat, J.L. *Phys. Rev. Lett.* **2006**, *97*.

[23]Vinh, L.T.; Hung, P.K.; Hong, N.V.; Tu, T.T. *J. Non-Cryst. Solids* **2009**, *355*, 1215.

[24]Yuan, X.; Cormack, A.N. *J. Non-Cryst. Solids* **2001**, *283*, 69.

[25]Benoit, M.; Ispas, S.; Jund, P.; Jullien, R. *European Physical Journal B* **2000**, *13*, 631.

[26]Soules, T.F. *J. Non-Cryst. Solids* **1984**, *73*, 315.

[27]Soules, T.F. *J. Non-Cryst. Solids* **1982**, *49*, 29.

[28]Takada, A. *6th Pacific Rim Conference on Ceramic and Glass Technology, September 11, 2005 - September 16, 2005* **2006**, *197*, 179.
[29]Takada, A. *Nippon Seramikkusu Kyokai Gakujutsu Ronbunshi Journal of the Ceramic Society of Japan* **2008**, *116*, 880.
[30]Mitra, S.K. *Philosophical Magazine B: Physics of Condensed Matter; Electronic, Optical and Magnetic Properties* **1982**, *45*, 529.
[31]Takada, A.; Richet, P.; Catlow, C.R.A.; Price, G. D. *J. Non-Cryst. Solids* **2008**, *354*, 181.
[32]Takada, A.; Richet, P.; Catlow, C.R.A.; Price, G. D. *Physics of Non-Crystalline Solids 10, October 15, 2004 - October 15, 2004* **2004**, *345-346*, 224.
[33]Pedone, A.; Gianluca, M.; Menziani, C.M.; Serge, U.; Cormack, A.N. *Chem. Mater.* **2008**, *20*, 4356.
[34]Kuzuu, N.; Nagai, K.; Tanaka, M.; Tamai, Y. *Japanese Journal of Applied Physics, Part 1: Regular Papers and Short Notes and Review Papers* **2005**, *44*, 8086.
[35]Kuzuu, N.; Yoshie, H.; Tamai, Y.; Wang, C. *Glass Science for High Technology. 16th University Conference, August 13, 2004 - August 15, 2004* **2004**, *349*, 319.
[36]Yamahara, K.; Okazaki, K.; Kawamura, K. *J. Non-Cryst. Solids* **2001**, *291*, 32.
[37]Lee, B.M.; Munetoh, S.; Motooka, T.; Yun, Y.-W.; Lee, K.-M. *Theory, Modeling and Numerical Simulation of Multi-Physics Materials Behavior, November 26, 2007 - November 30, 2007* **2008**, *139*, 101.
[38]Takada, A.; Richet, P.; Atake, T. *J. Non-Cryst. Solids* **2009**, *355*, 694.
[39]Berthier, L. *Physical Review E - Statistical, Nonlinear, and Soft Matter Physics* **2007**, *76*.
[40]Liang, Y.; Miranda, C.R.; Scandolo, S. *High Pressure Research* **2008**, *28*, 35.
[41]Rountree, C.L.; Prades, S.; Bonamy, D.; Bouchaud, E.; Kalia, R.; Guillot, C. *J. Alloys Compd.* **2007**, *437-435*, 60.
[42]Chen, Y.-C.; Nomura, K.-I.; Kalia, R.K.; Nakano, A.; Vashishta, P. *Phys. Rev. Lett.* **2009**, *103*, 155506.
[43]Malavasi, G.; Menziani, M. C.; Pedone, A.; Segre, U. *J. Non-Cryst. Solids* **2006**, *352*, 285.
[44]Bakaev, V.A.; Steele, W.A.; Bakaeva, T.I.; Pantano, C.G. *J. Chem. Phys.* **1999**, *111*, 9813.
[45]Mischler, C.; Kob, W.; Binder, K. *CCP 2001, September 5, 2001 - September 8, 2001* **2002**, *147*, 222.
[46]Leed, E. A.; Pantano, C.G. *J. Non-Cryst. Solids* **2003**, *324*, 48.
[47]Branda, M.M.; Montani, R. A.; Catellani, N. J. *Surf. Sci.* **2000**, *446*, L89.
[48]Wilson, M.; Walsh, T.R. *J. Chem. Phys.* **2000**, *113*, 9180.
[49]Garofalini, S.H. *J. Non-Cryst. Solids* **1990**, *120*, 1.
[50]Cormack, A.N.; Yuan, X. *The Glass Researcher* **1999**, *9*, 1.
[51]Abbas, A.; Delaye, J.-M.; Ghaleb, D.; Calas, G. *J. Non-Cryst. Solids* **2003**, *315*, 187.
[52]Wang, C.; Kuzuu, N.; Tamai, Y. *J. Non-Cryst. Solids* **2003**, *318*, 131.
[53]West, J.K.; Latour Jr, R.; Hench, L.L. *J. Biomed. Mater. Res.* **1997**, *37*, 585.
[54]Branda, M.M.; Montani, R.A.; Castellani, N. J. *Surf. Sci.* **1995**, *341*, 295.
[55]Bakaev, V.A.; Steele, W.A. *J. Chem. Phys.* **1999**, *111*, 9803.
[56]Bakaev, V.A.; Steele, W.A.; Pantano, C.G. *J. Chem. Phys.* **2000**, *114*, 9599.
[57]Ma, Y.; Foster, A.S.; Nieminen, R.M. *J. Chem. Phys.* **2005**, *122*, 1.
[58]Webb, E.B.; Garofalini, S.H. *J. Non-Cryst. Solids* **1998**, *226*, 47.
[59]Trioni, M.I.; Bongiorno, A.; Colombo, L. *J. Non-Cryst. Solids* **1997**, *220*, 164.
[60]Levine, S.M.; Garofalini, S. H. *Defects in Glasses.* **1986**, *61*, 29.
[61]Pantano, C.G. *61st Conference on Glass Problems, October 17, 2000 - October 18, 2000* **2001**, *22*, 137.
[62]Du, J.; Cormack, A.N. *J. Am. Ceram. Soc.* **2005**, *88*, 2532.
[63]D'Souza, A.S.; Pantano, C.G. *J. Am. Ceram. Soc* **2002**, *85*, 1499.
[64]Lower, N.P.; Brow, R.K.; Kurkjian, C.R. *J. Non-Cryst. Solids* **2004**, *344*, 17.

[65]Lower, N.P.; Brow, R.K.; Kurkjian, C.R. *J. Non-Cryst. Solids* **2004**, *349*, 168.
[66]Andersons, J.; Joffe, R.; Hojo, M.; Ochiai, S. *Compos. Sci. Technol.* **2002**, *62*, 131.
[67]Varshneya, A.K. *Fundamentals of Inorganic Glasses*; Academic Press: San Diego, 1994.
[68]Gupta, P.K.; Kurkjian, C.R. *J. Non-Cryst. Solids* **2005**, *351*, 2324.

*Corresponding author: Cormack@alfred.edu

EFFECT OF GLASS COMPOSITION ON SILANOL CONTENT: A STUDY OF GREEN VERSUS SOLAR GLASS

Sefina Ali and Dan Bennett
AGC R&D

ABSTRACT

As an initiative to reduce vehicle cabin temperature; the composition of automotive glazing is changing from standard green to solar glass. Solar glass has greater infrared radiation absorbance characteristics while maintaining a high visible light transmittance. Transmission values are obtained using UV-VIS-NIR spectroscopy. The metal oxides responsible for infrared reflectance will be determined using SEM-EDS/WDS. It is well understood that non-bridging cations alter the structure of silicate glass. FTIR spectroscopy will be employed to determine the effect of metal oxides on silanol concentration. The concentration of the silanol functional group is critical for adequate silane coupling in post processing. Differences of radiation transmittance, elemental composition, and silanol content of solar glass versus standard green glass shall be discussed.

INTRODUCTION

As automakers strive to produce more energy efficient vehicles, one area of focus is the vehicle cabin temperature (VCT). Reduction of VCT reduces the demand for the air conditioning system and thereby increases the efficiency of the vehicle. It is known that the transmittance of infrared radiation into the vehicle cabin will increase the internal temperature.[1-2] Reduction of the infrared radiation transmittance is therefore critical. The area of the vehicle most responsible for infrared transmittance is the window assemblies; windshield, side windows and back window.

Various methods by which infrared transmittance can be reduced through window glazings have been discussed. The uses of electrochromic films or coatings have been shown to reduce infrared transmittance by reflectance.[3-6] Another method involves altering the glass composition to include heavy metal modifiers which reduce infrared transmittance through absorbance.[7-10] The use of iron oxide as the infrared absorbing medium is of particular interest.

Analysis of iron oxide modified glasses is extensive; however the primary focus is on infrared absorbance and bulk property analysis. Sakaguchi describes the variability of Fe^{2+}infrared absorption peak position for varying compositions of alkali alkaline-earth silicates.[9] It has been determined that glasses doped with iron results in surfaces highly enriched with Fe^{2+} ion.[11] The surface composition was determined with x-ray diffraction, Auger electron spectroscopy and energy dispersive x-ray spectroscopy techniques.

Composition analysis of silicate based glasses has been investigated previously; primarily focusing on the nature of the silicon-oxygen bond.[11-15] Devine and Pisciella have studied the effect of temperature and composition on the shifting of the Si-O-Si peak with FTIR analysis.[13-15] A correlation between the electronegativity of the modifying cations and the shift of the Si-O-Si peak was elucidated.

Commonly the study of glass surface chemistries utilize FTIR and thermogravimetric analysis methods.[16-18] FTIR analysis of the Si-OH bond represented by a broad peak at 3700-3400 cm^{-1} or thermal dehydroxylation monitored by weight loss is used to determine silanol concentration.

Silane coupling agents of the general formula R_1-Si-$(OR_2)_3$ are utilized in the automotive industry to adhere the glass substrate to an organic adhesive. The mechanism of silane coupling to a glass surface is well understood and is summarized by Plueddemann.[19] The effectiveness of a silane coupling agent is dependent upon the concentration of the silanol functional group at the coupling interface. A low concentration of silanol on the surface will result in decreased coupling with the organic molecule. The structure, reaction rates, and hydrolysis of silane coupling agents bonded to silica surfaces have been determined with FTIR and Raman spectroscopy.[20-21]

In general, correlations between surface science and bulk composition regarding silicate based glasses are limited. The effect of glass modifying heavy metal oxides on surface silanol concentrations for soda-lime silicate glass is desired and shall be discussed herein.

EXPERIMENTAL PROCEDURE

UV-VIS-NIR Transmittance

Samples of monolithic green glass (sample ID 14401) and solar glass (sample ID 21005) were obtained from Asahi Glass Company. The samples were cut to 100x100 mm sections and washed with distilled water and then air dried. UV-VIS-NIR transmittance measurements were conducted with a Varian Cary Win 5000 spectrophotometer. Samples were analyzed in the spectral range of 2500 - 250 nm with a grating change over at 800 nm. A spectral band width of 5.00 nm was utilized and the samples were scanned at a rate of 600 nm/min. A NIST spectral reference standard G-8661 was used as an external standard. The UV-VIS-NIR sources are CIE Illuminant A and Illuminant C. Illuminant A simulates incandescent lamp light and Illuminant C simulates average daylight.

The transmittance of ultraviolet (T_{uv}) and visible (T_{vis}) radiation is reported as a normalized Riemann sum of the spectra ranges of 300-400 and 380-760 nm respectively. The normalization factor was taken from ASTM E308-90 specification.[22]

SEM-EDS/WDS Elemental Analysis

Small sections (2.0 x 0.5 mm) were cut from monolithic 100x100 plaque samples. The sections were rinsed clean with methanol and a conductive carbon coating of 35 nm was applied using a Cressington 108A Carbon Coater. The thickness of the carbon coating was determined by the green-blue color on a polished brass slug adjacent to the glass section in the coating chamber.[23]

Elemental analysis was conducted with an Oxford Inca 250+ energy dispersive detector and an Oxford Inca Wave 500 wavelength dispersive detector coupled to a JEOL JSM7001F field emission scanning electron microscope. All samples were analyzed at 5000 times magnification with a detector working distance of 10 mm. The electron accelerating voltage was set at 15KeV. All x-ray fluorescence scans were measured for 60 seconds at 12KCPS in a 9.6E-5Pa vacuum atmosphere.

Semi-quantitative elemental analysis was completed by energy dispersive spectroscopy utilizing copper as an external reference. Weight percents of each element in the spectrum were calculated by the Inca software using the Cliff-Lorimer constants.[24]

Comparative quantitative elemental analysis was conducted with wavelength dispersive spectroscopy. A lithium fluoride (220) Bragg crystal was used to select the wavelength for the analysis of iron. The Bragg angle of the detector (2θ) was optimized with iron standards. A five point, detector response calibration curve utilizing: blank background, Cr Diopside, Biotite, Kaersitite, and pure iron wire was made. The counts per second (CPS) of x-ray's fluoresced were recorded for each sample analyzed.

FTIR Spectroscopy

The 0.5mm x 2.0mm glass samples were crushed to dust particles using hydraulic press in a steel cup. Weight percentages of ~ 1, 5 and 10% of each glass sample were weighed analytically and mixed homogenously with potassium bromide in a mortar and pestle. Transmittance windows were achieved by pressing the mixture with 8 tons of pressure.

The mid-infrared spectrum was collected from 4000-400 cm^{-1}. The spectral data was obtained using the Thermo Scientific FTIR, Nicolet Avatar 360. The interval spacing of the scan was 1.928 cm^{-}

[1] with the gain set to 2 and the resolution set to 4.0. A blank KBr window was used as a background for all scans and automatic atmospheric correction was used. The sample chamber was purged with nitrogen. No software manipulations were used including baseline correction or spectral smoothing. Each spectrum is an average of 64 scans.

RESULTS AND DISCUSSION

Glass samples of identical dimensions were analyzed. Visually the samples are very similar. The only noticeable difference is a color; solar glass has a darker green hue (Figure 1).

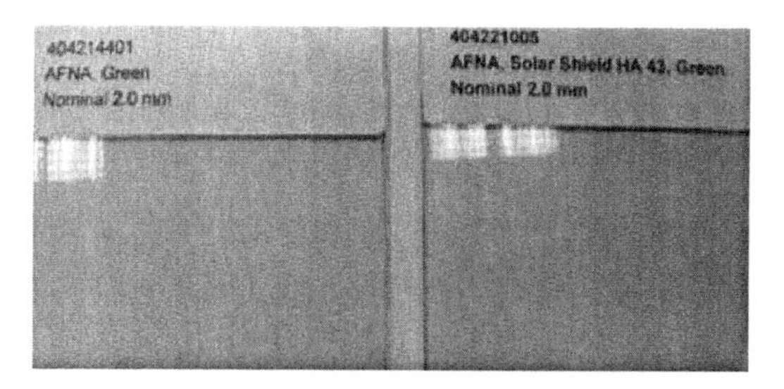

Figure 1. Samples as received

UV-VIS/NIR transmittance of each sample indicates that the green glass sample has higher transmittance of radiation as expected (Figure 2). The spectra from the entire region 2500-250nm, (Figure 2), shows the largest difference in transmittance lies in the near infrared range (750-2250 nm) while the transmittance in the visible region remains similar. This is expected as the objective of the solar glass is to reduce an automobile's interior temperature by reducing infrared radiation transmittance.

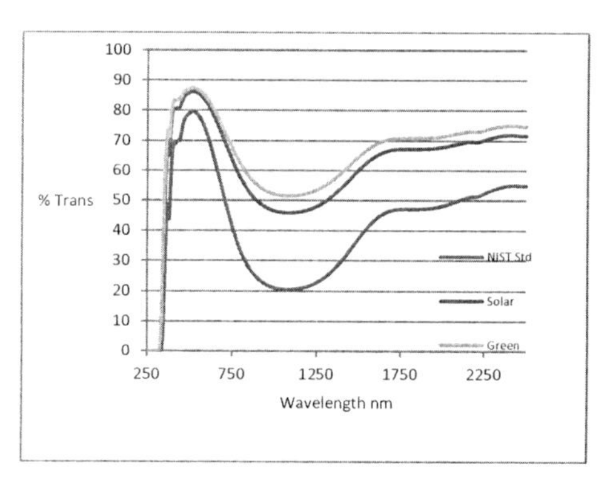

Figure 2. UV-VIS-NIR Spectra

The transmittance values for both Illuminate C and Illuminate A as well as the normalized percentage transmittance values of each radiation are listed in (Table 1). The transmittance of ultraviolet radiation region decreased from 62.2% to 56.3% and the visible light transmittance is decreased modestly from 85.3% to 83.6% for the solar glass sample compared to the green sample.

Table I. Varian Cary Win 5000 UV-VIS/NIR spectrophotometer data

Samples	(ILL. C)	(ILL. A)	% T_{sol}	% T_{uv}	% T_{Ir}
	%T	%T	300-2000nm	300-400nm	800-2100nm
2.0mm Solar	84.29	83.60	66.86	56.34	51.97
2.0mm Green	85.83	85.20	70.34	62.19	56.88
NIST Standard	76.04	74.74	49.44	39.41	27.49

Elemental analysis was used to determine the metal oxide responsible for the decrease in transmittance in the near-infrared region. Semi-quantitative analysis was achieved with energy dispersive spectroscopy (EDS). The x-ray fluorescence spectrum is shown (Figure 3).

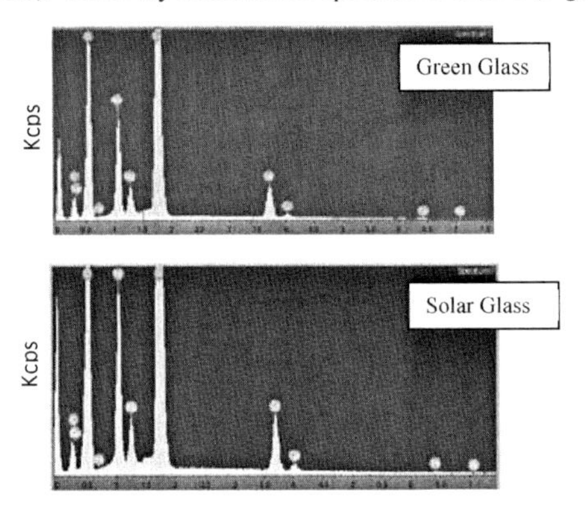

Figure 3. X-ray Fluorescence spectra (EDS)

The two samples vary in composition; however the iron content of the solar glass is three times that of the green glass composition (Table II). This indicates that the increased iron oxide content results in decreased near-infrared transmittance. Silicon, sodium, magnesium and calcium are all glass forming oxides whose weight percentages are consistent with soda lime silica glass (Table II). Iron oxide is a glass network modifying oxide.

Table II. Glass composition; Weight percentage (Energy Dispersive Spectroscopy)

	O	Na	Mg	Si	Ca	Fe
Green	48.04	9.06	2.21	34.76	5.63	0.24
Solar	46.29	8.84	1.81	35.45	6.94	0.77

Wavelength dispersive x-ray spectroscopy (WDS) was then used to determine the relative differences of iron oxide in each sample. The x-ray spectrum of iron is given (Figure 4); left: green glass and right: solar glass. The analysis of iron was conducted at a wavelength of 1.940 nm.

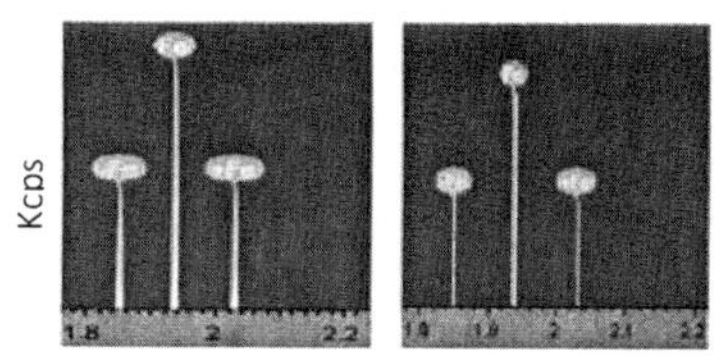

Figure 4. X-ray fluorescence spectra (WDS)

The detector response to the x-rays fluoresced was determined with a five point calibration curve using a blank background, Chromium Diopside, Biotite, Kaersitite and pure iron wire. The calibration points are an average of five spectrums with different sampling locations (Figure 5).

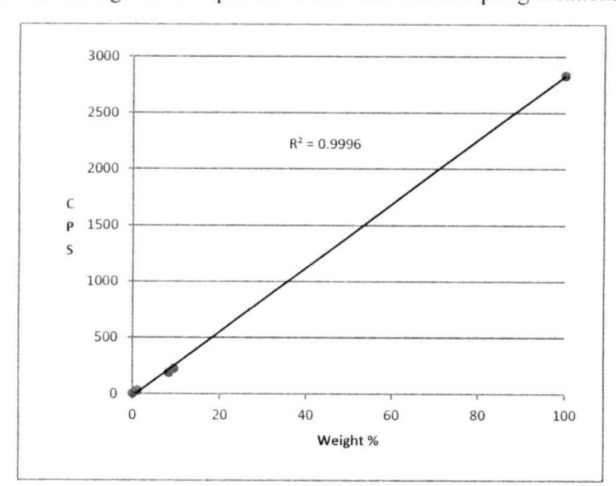

Figure 5. Detector response calibration plot

The intensity of the x-ray fluorescence is given in counts per second (CPS) instead of CPS/nA. The beam current was stable during the analysis with a value of 4.2±0.08 nA and was calculated into the intensity value. The detector response to the x-rays fluoresced by iron was found to be linear in nature. The matrix effect of the calibration samples did not affect the detector response. The calibration data is given in Table III.

Table III. Detector Response; Calibration Data

	Sample	Weight %	X-Ray CPS
1	Bkd Blank	0	4.2
2	Cr Diopside	1.08	30.4736
3	Biotite	8.33	189.837
4	Kaersutite	9.51	226.938
5	Iron Wire	99.995	2834.16

The glass samples were analyzed utilizing the same experimental parameters; the following x-ray fluorescence data was obtained (Table IV).

Table IV. Iron x-ray fluorescence of glass samples

	Green Glass	Solar Glass
Spectra	CPS	CPS
1	8.40	24.90
2	9.24	21.17
3	9.66	24.49
4	8.40	19.51
5	9.66	24.49
Average	9.07	22.91
Std Dev.	0.64	2.43

The intensity of the iron x-rays in the solar glass sample is 2.5x greater than the intensity of the green glass sample. Intensity in CPS of x-rays fluoresced is related to the weight percentage of the responsible element. Using the linear relationship of the detector response calibration curve a relative weight percent of iron can be determined. It was determined that the composition of the solar glass is $0.920 \pm 0.08\%$ iron and green glass is $0.299 \pm 0.02\%$ iron.

The effect of the iron on the surface functional groups was determined with FTIR spectroscopy. The silanol functional group is clearly resolved at 3400 cm^{-1}. Transmittance analysis of the samples as received did not offer favorable results. The samples at 2.0 mm in thickness did not resolve any useful peaks (Figure 6).

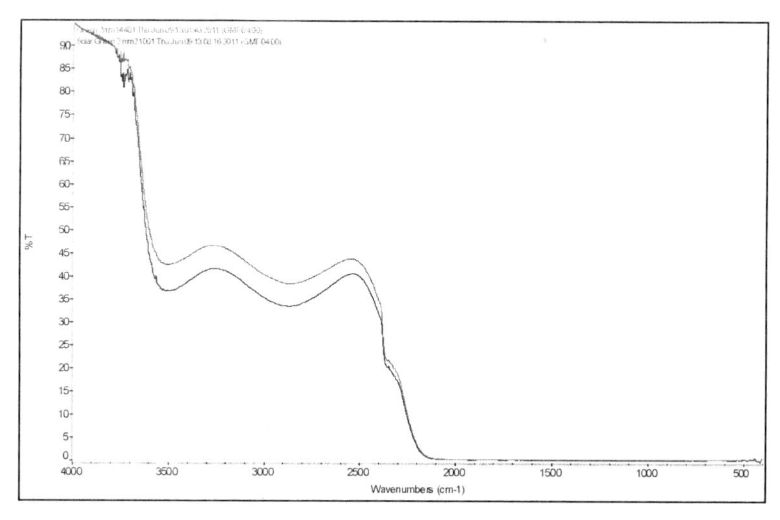

Figure 6: Transmittance of 2.0 mm samples

The 2.0 mm thickness of the samples resulted in absorption of the infrared radiation in the region of 400-2100 cm^{-1}. The glass samples were crushed to dust and mixed with KBr at 5% by weight. The transmittance of the samples was increased due to the lower concentration, smaller path length (Figure 7).

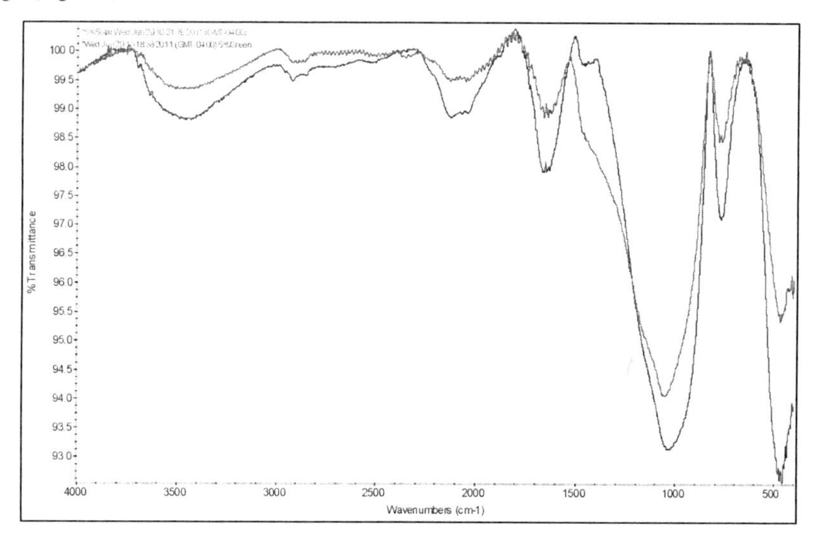

Figure 7: Transmittance of 5% by weight in KBr

The peak at 3450 cm^{-1} correlates to the bending of the silanol group (Si-OH). The large peak at 1050 cm^{-1} correlates to the stretching of the Si-O-Si functional group. The intensity of the Si-O peak is much larger than the silanol peak. The bulk of the glass matrix is made of a network of SiOx bonds. The silanol groups will only be on the surface of the glass samples. FTIR analysis indicates that the intensity of the silanol peaks for the green and the solar glass samples are not the same. However, the reference peak (Si-O at 1050 cm^{-1}) has the same intensity. The reference siloxane reference peak serves two purposes. The ratio of the intensity of the silanol peak to the siloxane peak will be used to determine percent change in silanol concentration. Also, it is expected that the intensity of the siloxane peak will not change with a small increase in iron oxide concentration. Therefore, the siloxane peak serves as an internal reference for overall sample concentration. Similar siloxane intensities indicate precise sampling of the glass specimens.

There is a noticeable peak shift of the Si-O peak for each sample. The green glass sample has a peak maximum at 1050 cm^{-1} while the green sample has a peak maximum at 1030 cm^{-1}. This is in agreement with previous studies which determined the peak shift effect of heavy metal cations on the Si-O-Si bond.[13] The increase in iron concentration shifts the peak to lower wave numbers by strengthening the Si-O-Si bond.

It was postulated that an increase in iron oxide may decrease the concentration of silanol functional groups on the surface due to the nature of iron oxide migrating to the surface of the glass. This was confirmed by the FTIR analysis, in which a 3 fold increase in iron oxide concentration resulted in a 50% decrease in the intensity of the silanol peak. Therefore the manufacturing process may be affected by the composition change. Glass adhesion performance utilizing silane coupling agents may be affected by changing from green to solar glass formulations and will require further adhesion evaluation testing. A summary of the properties comparing the two glass samples (Table VI).

Table VI. Differences by comparison summary.

Analysis Results	Green Glass	Solar Glass
Ultra violet radiation	62.1%	56.3%
Light Transmittance (T_{vis})	85.3%	83.6%
Weight Percent of Iron	0.299% Fe	0.920% Fe
Silanol Peak Area (FTIR 3750-2980 cm-1)	486.51	289.85
Siloxane Peak Area (FTIR 1400-800 cm-1	2298.81	2309.35
Siloxane Peak (FTIR)	1030 cm^{-1}	1050 cm^{-1}
Peak Area Ratio Silanol/Siloxane (FTIR)	0.211	0.126
% Change in Silanol Concentration	-	-50.45%

CONCLUSION

Comparison of radiation transmittance, elemental composition, and silanol content of solar glass versus standard green glass evaluated. The solar glass sample has a lower ultraviolet and near-infrared radiation transmittance due to the increased iron oxide concentration than the green glass sample. The effect of this increase in non-network modifying cation on the glass surface functional groups was found to be significant. The intensity of the silanol peak (Si-OH) absorption in FTIR decreased by more than 50% for the solar glass sample. Therefore, the performance of glass adhesion with silane coupling agents would be expected to decrease for the solar glass composition. Further studies are needed to correlate the effect of heavy metal cation concentration on silanol concentration. Specifically, glass samples with higher concentrations of cation modifiers as well as other cations (chromium, manganese, copper, cobalt) will be studied.

ACKNOWLEDGMENTS

The authors would like to thank and acknowledge the following advisors and supporters for their considerable input and resources on this project. Thank you to the Washtenaw Community College advisor, Prof. Tracy Schwab for her valued support and input. Special thanks to AGC R&D Inc. for providing essential funding, facilities, instrumentation, and commitment to this endeavor. Many thanks to the AFPG-9 Committee, including Dr. A. Varshneya for their meaningful student award scholarship. To IMI-NFG, thank you for the generous travel support scholarship–listed as NSF Grant No. DMR-0844014.

REFERENCES

[1]R.J. Goldstein, Heat transfer – a review of 1999 literature. J. of Heat and Mass Transfer 44 (2001).
[2]J. Yellott, Calculation of solar heat gain through single glass. Solar Energy 7 (1963).
[3]N. Jaksic, A feasibility study of electrochromic windows in vehicles. Solar Energy Materials 79 (2003).
[4]A. Taleb, Design of windows to reduce solar radiation transmittance. Solar and Wind Technology, 5 (1988).
[5]B. Jelle, Performance of an electrochromic window based on polyaniline, Prussian blue and tungsten oxide. Solar Energy Materials and Solar Cells, 58 (1999).
[6]S. Alamri, The temperature behavior of smart windows under direct solar radiation. Solar Energy Materials and Solar Cells. 93 (2009).
[7]Y. Ozeki, Effects of spectral properties of glass on the thermal comfort of car occupants. Elsevier Ergonomics Book series 3 (2005).
[8]R. Brow, A survey of energy and environmental applications of glass. J. of European Ceramic Society 29 (2009).
[9]K. Sakaguchi, Compositional dependence of infrared absorption of iron-doped silicate glasses.
[10]J. Combes, Proceedings in the 17th International Congress Glass. 1995.
[11]P. Patsalas, Surface properties and activity of Ge-Ni-B ternary glasses. J. of Alloys and Compunds (2007).
[12]L.T. Zhuravlev, The surface chemistry of amorphous silica. Colloids and Surfaces 173 (2000).
[13]R. Strens, IR study of cation ordering and clustering in some (Fe, Mg) amphibole solid solutions. Chemistry Communications 15 (1966).
[14]P. Pisciella, FTIR spectroscopy investigation of the crystallization process in an iron rich glass. J. European Ceramic Society 25 (2005).
[15]R. Devine, Ion implantation and radiation induced structural modifications in amorphous SiO2. J. Non-Crystalline Solids, 152 (1993).
[16]R. Iler, The chemistry of Silica: Solubility, Polymerization, Colloid and Surface Properties, and Biochemistry, Wiley-Interscience, New York, 1979.
[17]N. Yaroslavsky, Zh. Fiz. Khim. 24 (1950).
[18]T. Yoshida, FTIR study on the effect of OH content on the damage process in silica glasses irridated by hydrogen. Nuclear Instruments and Methods in Physics Research 218 (2004)
[19]E. Plueddemann, Adhesion through silane coupling agents. J. of Adhesion 2 (1970).

[20]P. Shih, Raman studies of the hydrolysis of silane coupling agents, Materials Science and Engineering 20 (1975).
[21]H. Ishida, Fourier transform infrared spectroscopic study of the silane coupling agent porous silica interface. J. Colloid and Interface Science 64 (1978).
[22]ASTM Standard E308-90 Computing the Colors of Objects by using the CIE system 1990.
[23]S.B.J. Reed Electron Microprobe Analysis 1st Edition (1975).
[24]F. Rousessac Chemical Analysis – Modern Instrumentation Methods and Techniques. Wiley Interscience (2007).

Corresponding author: Sefina.Ali@us.agc.com

COATING METHODS FOR THE STRENGTH INCREASE OF CONTAINERS: NEW RESULTS ON NANO PARTICLE ALUMINA COATINGS

K.Czarnacki,[1] J.Wasylak,[2]

[1]Pol-Am-Pack S.A. Oddział Huta Szkła "ORZESZE" w Orzeszu, ul. Gliwicka 59, 43-180 Orzesze, Poland.
[2]AGH - University of Science and Technology in Cracow, Faculty of Materials Science and Ceramics, al. Mickiewicza 30, 30-059 Cracow, Poland

ABSTRACT

For many years there has been scientific and industrial research into how to increase the mechanical resistance of glass packaging. One of the very important factors, which influences the increase of the mechanical resistance of glass packaging, was the development of the formation method in which we may include methods such as Blow & Blow, Press & Blow and Narrow Neck Press & Blow. Further factors influencing the correction of resistance parameters of packaging were the application of heat treatment, glass digestion, internal gasification and also different hot-end and cold-end coating. I will attempt various ways of the internal and external coating of the bottle and also the application of hardened polymer UV for cold end coating. The aim of this paper is to show the research results for the mechanical resistance of glass containers using a new method of hot-end coating by nano particles of alumina compounds. We will show the progress in the research results of internal pressure, vertical load resistance and impact resistance of glass packaging. These results may act as guidelines for future container weight and the decrease of wall thickness.

INTRODUCTION

Throughout the past 100 years we may observe progress in the production of lighter glass containers. One of the biggest milestones in this progress was the introduction of the automatic glass container machine by Michael Owens at the beginning of the twentieth century in 1905.[1] Even though the first bottles were heavier than containers produced by glass hand blowers, this patent started a revolution in the weight of containers causing a decrease in wall thickness.

A significant step in glass container production was the introduction of the IS machine in 1926[2] and thus a transition away from the use of carousel machines. Since the 1960's a compilation of a new forming method like Narrow Neck Press and Blow[3] and double coating has given the next significant step in the decrease in weight and wall thickness. Furthermore the transition of double gob to triple gob in the middle 1970's was the next step in this weight decreasing process. Moreover, changes and the development of the forming method, introducing hot end and cold end coating, gave increasing glass container mechanical resistance. In spite of the outflow time and continued development of the forming methods, research is still being carried out by many public and private institutions whose main goal is the weight reduction of glass containers. A further research example in this area could be the work of I.P.G.R. Group (International Partners in Glass Research) and also producers of glass and glass machines.[4] This Group presented to the market epoxy-coating protecting the external surface of glass containers causing greater mechanical resistance. For many years Wiegand Glass in Germany has used this technology which is called PCT (Polymer Coating Technology). A Further example introduced by IPGR is the use of the automatic swabbing of blank moulds in the AIS machine which produces a higher stability of the container forming process and also produces a greater stability in wall thickness. All packaging groups and glass-works are constantly working on the development of the weight decrease in glass containers in the various ranges. In 2007 WRAP (Waste & Resources Action Programme)[5] presented to the market a successful project of weight decrease. The development of mechanization and the development in the coatings application of the hot end and cold end gave a meaningful weight reduction in glass containers causing them to become more competitive on the market against plastic or aluminium containers.

COATING METHODS AND PROCEDURES

The ideal coating for a glass container should be defined in three areas:[6]

1. Coating material.
2. Deposition process.
3. Performance characteristics.

Currently used methods increasing the mechanical strength of glass containers:

1. Physical (hot end and cold end coating).
2. Chemical (phosphate, nitrate).
3. Chemical - physical (sulphate).
4. External (sleeve labels or pre-labels).

PHYSICAL TREATMENT

Hot End Coating (HEC)

The physical methods described below are the main aspects used in the glass container industry. Usually it is a combination of hot end and cold end coating. In order to complete the protection of glass containers both hot end coating and cold end coating should be used. The material which is used as a HEC should not only be theoretically operational but should also be financially viable. Tin compounds are mostly used as hot end coatings. The TC 100 coating system was invented by M&T Chemical Inc. Tin oxide as a hot end coating increased the mechanical resistance from 25 to 35 percent in contrast to tin chloride which was used up until that time. A great advantage of TC 100 is less cap corrosion and a decrease in maintenance costs.

According to Smay,[4] metal oxide used as a hot end coating should:

- create strong bonding between the thin layer of the glass and the organic layer,
- reduce sodium migration during the interaction between the organic layer and the glass surface,
- reduce the poisoning effect of alkali ions under the influence of weak Bronsted acid present on the glass surface.

Commercially we can get the HEC from a few coating producers. Arkema is one of the main suppliers of Cernicoat TC 100 (mono butyl tin trichloride - MBTC). A further company is Bohemi Chemicals which supplies Starin S (MBTC). Imacoat International also offers a hot end coating as Imacoat HE-100 (MBTC). Glasschem offers a product called Glasschem HOT 99 (MBTC).

Table I. Hot end coating - MBTC data

Product	Appearance	Density (g/ml, 25 ^{0}C)	Molecular wt. (g/mol)	Freezing point (^{0}C)	Boiling point 760 mm Hg (^{0}C)
TC 100	straw/brown colored liquid	1.72	282,20	- 63	190-200
STARIN S	clear up to slight yellow	1,70-1,73	282,17	-	102 (16 hPa)
IMACOAT HE-100	light yellow	1.70	-	- 63	193
GLASSCHEM HOT 99	-	-	-	-	-
GENERAL	brown colored liquid	1,72	282,30	- 63	190-200

Cold End Coating (CEC)

Polyethylene waxes are mostly used commercially in the glass container industry e.g. as cold end coatings. The preferred surface temperature of the glass articles on which the coatings are applied should be in between 80°C and 140°C. The ratio required to obtain the optimum performance of cold end coating is 1:100. Under specific conditions the ratio used may increase to 1:175. This can be determined after initial trials. The use of de-mineralised water is also advised for cold end coatings. Spray nozzles should have an adjustable flat spray angle and a diameter of 0.4-0.8 mm. The initial pressure of the solution should be approximately 1 bar and the pressure of the solution in the nozzle should be 4-6 bar. Too high a concentration or distribution of the solution may lead to visibility or drops. Companies deliver both hot end coatings and cold end coatings. Arkema is a supplier of Tegoglas products for cold end coatings and produces different types of Tegoglas for different purposes. The Tegoglas products include: Tegoglas RP 40LT, Tegoglas RP 40, Tegoglas RP 40C, Tegoglas 702, Tegoglas T5, Tegoglas 702, Tegoglas 80 OL, and Tegoglas 3000+. Cold end coating improves lubricity and scratch resistance, which improves durability during handling and transportation. Generally these coatings are particularly suitable for the following types of containers: thin-walled, lightweight glass containers (non-returnable containers), disposable, beer and other beverage bottles, returnable glass containers, food jars for hot-filling/re-cooling, flagons, pharmaceutical glass, liquor bottles, perfume bottles, wine, oil, food glass containers, unstressed glassware and glass tableware.

Table II. CEC Tegoglas family properties[7]

Application Area	TEGOGLAS					
	RP 40C	RP40	RP40LT	702	T5	OL80
High requirement of wet scratch resistance	+++	+++	+++	+	-	-
High requirement of dry scratch resistance	+++	+++	+++	+++	++	++
High resistance to washing and sterilization	+++	+++	+++	-	-	-
High smoothness	+++	+++	+++	+++	+++	+
Pharmaceutical	+(++)	+(++)	+(++)	+++	+++	+++
Surface temperature						
< 80 ^{0}C	-	+	++	++	++	-
80-130 ^{0}C	+++	+++	+++	+++	+++	+++
>130 ^{0}C	+++	++	+	-	-	+++
Glues						
Synthetic	+++	+++	+++	+++	+++	+++
Casein	+++	++	+			
Starch	++	++	+++	+++	+++	++
Dextrine	-	-	-	+/-	+/-	+/-
Self-adhesive labels	+++	+++	++	-	-	++
After confirmation of pharmaceutical companies, (++) depending on filling product +++ highly suitable, ++ suitable, - not suitable						

Some of these coatings were tested in PAPC and the results will be published in the next part of the publication. Bohemi-chemicals deliver cold end coatings. They produce: Polyglas D4218M, Polyglas D4662, and Polyglas EN 21. Suddeutche Emulsions - Chemie produce two types of Sudranol135, Sudranol VP 410. These coatings also protect containers during transportation and conveyance and create appropriate resistance to micro-cracking. These coatings could be used for the wine and spirit market, small glass containers and beer bottles.

Gas Surface Treatment

Glass treating by gas is the next coating method, which improves the physical-chemical properties of glass containers. This method is used in Glassworks "ORZESZE." Generally this method is used for pure-alcohol bottles exported on the Eastern markets. This method improves the hydrolytic class of containers due to the control combustion mixture which contains air and freon compounds. The temperature of the process should be higher than 540°C. That is why the coating unit should be as close as possible to the AIS machine. During the internal coating the freon coating reacts with the hot surface of the glass generating a completely new coating with a higher chemical resistance. Depending on the hydrolytic class of the glass containers, this gas treatment eliminates both glass obscuring and internal shivering.

Chemical and Chemical - Physical Strengthening Treatment

In the 1970's Yamamura Glass Co. Ltd.[8] started producing chemically strengthened glass containers. At first, potassium salts were tried but then Yamamura used a mixture of potassium chloride and potassium nitrate, both highly soluble in water.

The mixture allowed the possibility of achieving a coating of 15 μm giving a compressive strength of 100 Pa. Over the entire range of bottle weights, the chemical treatments achieved a 53% increase in bursting pressure strength. A further example[9] of the chemical strengthening of glass is applying potassium ions to the surface of the glass article with a temperature not lower than annealing point. This process should last 5 min. at a temperature around 150^0C. In one embodiment, the glass articles may by dipped in a salt bath to apply the ions. Alternatively the glass could be sprayed by molten potassium salt to apply the ions. Articles treated in this way will be strengthened by having increased surface stress or may contain less glass but have the same strength as articles not treated by this method.

Another coating method[10] involves simultaneous alkali removal and stress layer formation. The dealkalization of the inner surface of the glass container is achieved by using an alkali removing agent (SO_2,$(NH_4)_2SO_4$). The process lasts from 10 to 40 min. at a temperature of approximately 580-650^0C. In the next stage containers are cooled externally and internally by blowing cool air on them to remove alkali products in order to develop a compressive stress layer. To improve lubricity and anti-wear properties bottles are coated by a synthetic resin emulsion or surfactant (paraffin and fatty acid). Duraglass Research and Development Corporation[10] developed another process of strengthening glass bottles. Container moulds are sprayed with fatty acid like behemic, stearic, glutamic and combinations of stearic and bohemic acids. Bottle drop heights increased from 60 cm (control) up to 450 cm (stearic/bohemic applied at 594 ^{0}C. The temperature of the process is 480^0C-700^0C). The theory is that the fatty acids chemically react with the atoms on the glass surface, strengthening the tips of the micro-cracks and preventing further crack propagation when the surface is under stress.

External Treatment

The shrink sleeve application is a sample of external treatment. Different types of PET (polyethylene terephthalate) films are used as sleeve labels. This process consists of three steps: 1-product handling, 2-sleeve application, 3- shrinkage.

Glass bottles arrive at the machine infeed where they are separated by an infeed screw- or separator-system. From this system bottles are transported to the applicator head at a pre-determined pitch. Film reels are unwound creating a film web, which is transported to the cutting unit system in the applicator frame, where the film web is cut into sleeves. When the bottles arrive at the applicator head, they are inspected and the sleeves are applied to the products. When an inspection unit is installed, sleeve presence, sleeve position (sleeve height) and splices are automatically detected. When a faulty bottle is detected, it is rejected from the system before entering the shrink tunnel. Rejected products are collected. After the bottles are inspected they run through the shrink tunnel (hot air or steam) where the sleeves are shrunk. Finally they are transferred onto the downstream conveyor system of the glassworks. During the research test bottles obtained from Fuji Seal and labeled by the sleeves system were checked.

RESULTS

The Base Conditions of Hot End Treatment Including Nano Partical Aluminium Compounds.

The beer glass bottles 330 ml were the object of Pol-Am-Pack S.A. and AGH research. The bottles were produced in ten-section triple gob machines (Emhart AIS 10 section TG machines, AIS - automatic independent section machine, TG - triple gob). Our coating which includes alumina compounds nano particles was spread on the external container surface in prototype devices – hot end coating hoods. The temperature of coated glass bottles were near the softening point - more than 600^0C. The prototype devices were positioned next to the AIS forming machines. An alumina hydroxide $Al(OH)_3$ was used as one of the base coating substances. This powder, together with a conformal substance, was dry-grinding in the post and vibrating mill. The grain size of the produced nano powder was determined by the SEM method and by a Nanosizer – ZS, respectively. The size of the nano particles were approximately 50 - 300 nm. The new coating was spread on the container's external surface in prototype devices. After hot end coating, the bottles were then annealed in annealing lehrs and cold end coated. The thickness of the hot end coating (10 - 90 nm) was determined by the ellipsometry method. The measurements were made by a Syntech 400 ellipsometer, $\lambda = 400$ nm.

During the research, the possibility of using a few types of CEC was investigated. During the first step the key point was to choose coating which is accessible on the market. For this purpose some coatings from Tegoglas family were applied as RP 40 LT, RP 40 C, RP 40, 702, 3000+. Additionally CEC CC25 and Sudranol 135 were checked in the PAPC laboratory. Most of the cold end coatings applied in the lab except Tegoglas 3000+ contained polyethylene waxes as a base. The Tegoglas chemical composition allows the use of CEC for production tests. Further cold end coatings were UV-polyesters. These three resins are used in the Can Pack Group. The following samples were made: uncoated bottles - comparison sample/base sample, hot-end coating samples, cold-end coatings samples, hot end and cold end coating samples, in order to make a comparison analysis between SnO_2 and Al_2O_3.

The prepared set of samples from each section of the AIS machine were taken (24h production tests; average values).

The following physical parameters were determined:

- Micro-hardness: 2011 - 2'st prototype HEC device
- Impact resistance: 2010 - 1'st prototype HEC device, 2011 - 2'nd prototype HEC device
- Vertical load: 2010 - 1'st prototype HEC device, 2011 - 2'nd prototype HEC device
- Internal pressure: 2010 - 1'st prototype HEC device, 2011 - 2'nd prototype HEC device
- Slip angle

Chemical parameters:

- Hydrolytic class.

Micro-hardness was determined according to the Polish Standard: PN-70/B-13150 (1970-03-27) Glass; Methods of Tests; Determination of Micro Hardness of Glassware.[11] The Vickers method was used PMT-3. The average values of 24 h production tests are shown in figure 1.

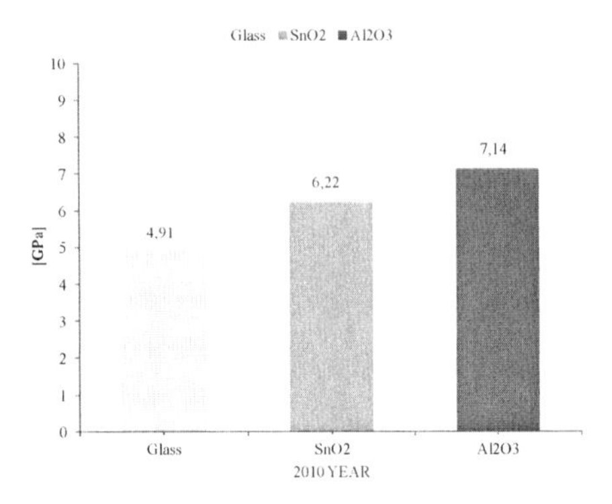

Figure 1. Micro-hardness, Vickers method, PMT-3 device, glass bottle 330 ml.

Impact resistance was determined according to the German Standard: DIN 52295 (1993-04) Testing of Glass; Pendulum Impact Test of Containers, Inspection by Attributes and by Variables.[12] The measurements were made on a AGR Impact Tester device (IT). The average values of a 24h production test are shown in figure 2.

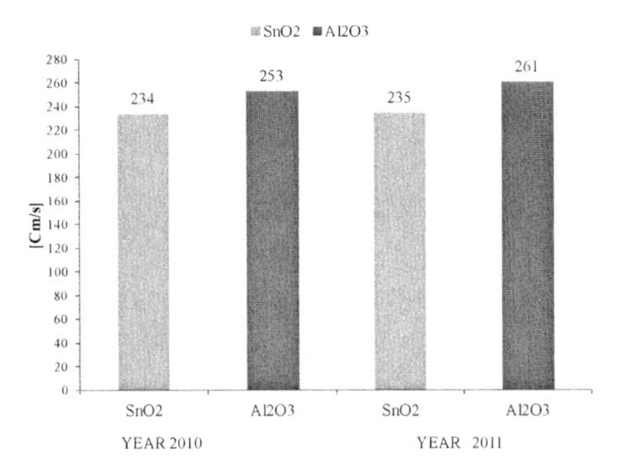

Figure 2. Impact resistance, glass bottle 330 ml.

Vertical load was verified according to the German Standard: DIN EN ISO 8113 Glass containers - Resistance to vertical load - Test method (ISO 8113:2004).[13] The measurements were made on an AGR Vertical Load Tester device (VLT). The average values of a 24h production test are shown in figure 3.

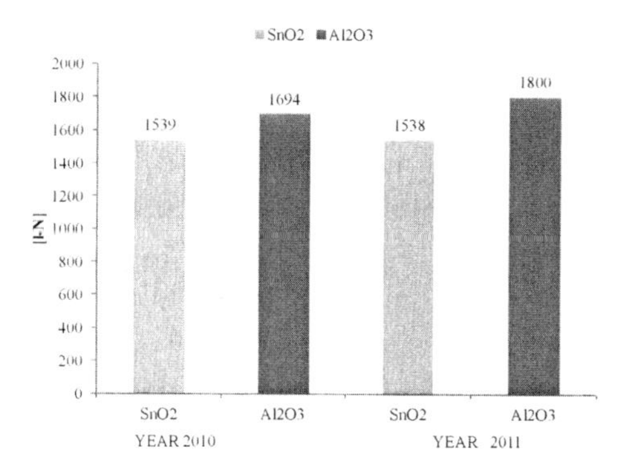

Figure 3. Vertical Load, glass bottle 330 ml.

Internal pressure was determined according to the German Standard: DIN EN ISO 7458 Glass containers - Internal pressure resistance - Test methods (ISO 7458:2004).[14] The measurements were made on an AGR Ramp Pressure Tester 2 device (RPT2). The average values of a 24h production test are shown in figure 4.

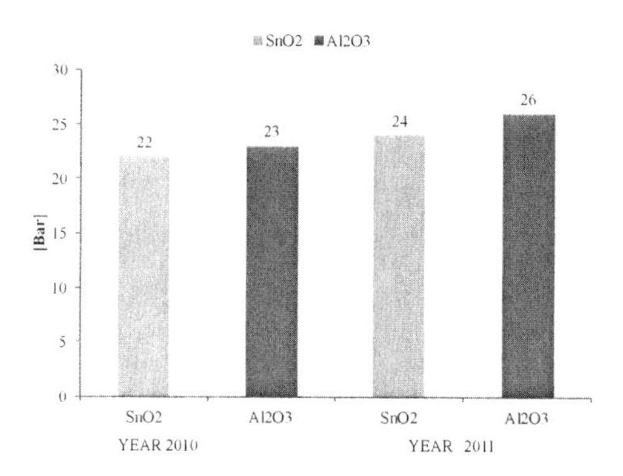

Figure 4. Internal pressure, glass bottle 330 ml

The measurements of the slip angle were made on an AGR Tilt Table device. The comparison between glass coated by TC-100 (SnO_2) versus new coating (Al_2O_3) and a few commercially available cold end coatings is shown in table 3. The average values of a 24h production test.

Table III. Slip angle, 24h production test, AGR tilt table, dry test, glass bottle 330 ml

Coatings Hot End / Cold End	Laboratory test	Production test
	Slip angle (^{0})/dry	Slip angle (^{0})/dry
TC100/ TEGOGLAS RP 40 LT	14	14
Al_2O_3/TEGOGLAS RP 40 LT	17	15
Al_2O_3/TEGOGLAS RP 40	18	16
Al_2O_3/TEGOGLAS RP 40 C	25	22
Al_2O_3/TEGOGLAS 702	24	20
Al_2O_3/GLASCHEM CC 25	22	20
Al_2O_3/SUDRANOL 135	14	15
UV RESIN 1	24	-
UV RESIN 2	19	-
UV RESIN 3	20	-

Hydrolytic class was according to the Polish Standard PN ISO 4802-1(2010) Glassware; Hydrolytic resistance of the interior surfaces of glass containers; Part 1: Determination by the titration method and classification.[15] (Table IV.)

Table IV. Hydrolytic class,

Sample	Base glass	Al_2O_3	Base glass	Al_2O_3	Gas coating	Al_2O_3	Gas coating	Al_2O_3
	V_{HCl} / 100 ml solution [ml/100 ml]	V_{HCl} / 100 ml solution [ml/100ml]	1 ml solution [c(HCl)=0,0 1mol/l] =310 µg Na_2O	1 ml solution [c(HCl)=0,0 1mol/l]=310 µg Na_2O	V_{HCl} / 100 ml solution [ml/100ml]	V_{HCl} / 100 ml solution [ml/100ml]	1 ml solution [c(HCl)=0,0 1mol/l] =310 µg Na_2O	1 ml solution [c(HCl)=0,0 1mol/l] =310 µg Na_2O
1	3,80	0,5	1085,40	135,20	0,74	0,43	232,50	145,70
2	3,40	0,5			0,71	0,46		
3	3,00	0,2			0,71	0,50		
4	3,70	0,3			0,70	0,42		
5	3,80	0,5			0,69	0,41		
6	3,20	0,3			0,72	0,40		
7	3,40	0,2						
8	3,60	0,4						
9	3,70	0,4						
10	3,70	0,5						
Avg.	3,53	0,38	1085,40	135,20	0,71	0,44	232,50	145,70

The Al_2O_3 coating prevents the loss of alkalines from the surface of the glass bottles, which causes a decrease of the hydrolytic class from H3 to H2.

CONCLUSION

Hot End Coating

These various methods of forming and coating were the next milestones in glass weight and wall thickness reduction. The implementation of a new hot end coating system in our glass plant gave better results of container mechanical resistance. The benefits of this new method of application were to give a better complete layer over the whole glass container surface. This new method of application allows the possibility of controlling, in a more effective way, the coating thickness. This completely new method of application will be developed in Glassworks "ORZESZE."

Cold End Coating

The cold end coating production test didn't give a 100 percent satisfactory result but helped us understand the key. The application of the Tegoglas family, CC 25 and Sudranol 135 increased the slip angle of produced bottles significantly. These tests help us to choose CEC coatings for disposable bottles. UV-polymers gave surprising results and this part of the research will also be developed in spite of problems with the application process.

ACKNOWLEDGMENTS

Special thanks to Pol-Am-Pack S.A. for the opportunity for research work in their glassworks "ORZESZE" and to the University of Science and Technology in Cracow, Faculty of Materials Science and Ceramics.

REFERENCES

[1]R. Chinella, Owens Illinois, "Lightweighting of Glass Containers", Ceram. Eng. Sci. Proc., (1989), 173-181.

[2]G.L. Miller and C. Sullivan, Machine-Made Glass Containers and the End of Production for Mouth-Blown Bottles", Historical Archaeology, Vol. [18], 2, (1984).

[3]H. Griffel, Heye-Glas "Lightweighting in the Glass Container Industry", Ceram. Eng. Sci. Proc., 8 [3-4], 156-170, (1987).

[4]P.J. Doyle, The British Glass Industry Research Association, "Recent Development in the Production of Stronger Glass Containers", Packaging Technology and Science (1988), 47-53.

[5]Demonstrating Glass Lightweighting , "Container Lite Final Report" WRAP. Retailer Innovation, (2007).

[6]F.V. Tooley, "The Handbook of Glass Manufacturing 3rd edition," Books for the Glass Industry Division Ashlee Publishing Co., Inc., New York, NY 10017, (1984).

[7]Atofina Deutschland GmbH, "Product Overview."

[8]C.G. Pantano, "The Role of Coating and Other Surface Treatments in the Strength of Glass," Department of Materials Science and Engineering Material Institute, The Pennsylvania State University.

[9]Nippon Taisan Bin K, Patent JP 58064248, "Surface Treatment of Glass Bottle to Improve Strength-Involves Simultaneous Alkali Removal and Compressive Stress Layer Formation", Derwent (1981).

[10]Sipe N.E., Noble J., Duraglass R&D Corporation, Patent US4039310, "Process of Strengthening Glass Bottles," Boulder (1977).

[11]Polish Standard: PN-70/B-13150; Glass; Methods Of Tests; Determination of Micro-Hardness of Glassware, (1970-03-27).

[12]German Standard: DIN EN ISO 8113; Glass containers - Resistance to vertical load - Test method (ISO 8113:2004).

[13]German Standard: DIN EN ISO 8113; Glass containers - Resistance to vertical load - Test method (ISO 8113:2004).

[14]German Standard: DIN EN ISO 7458; Glass containers - Internal pressure resistance - Test methods (ISO 7458:2004).
[15]Polish Standard PN ISO 4802-1; Glassware; Hydrolytic resistance of the interior surfaces of glass containers; Part 1: Determination by titration method and classification; (1994-12-30).

*Corresponding author: KCzarnacki@canpack.eu

EFFECT OF TiO_2 ADDITION ON THE DISTRIBUTION OF PHOSPHORUS ASSOCIATED WITH PHASE SEPARATION OF BOROSILICATE GLASSES

Y. Ohtsuki,[1] S. Sakida,[2] Y. Benino[1] and T. Nanba[1]
[1]Graduate School of Environmental Science Okayama University, Japan
[2]Environmental Management Center Okayama University, Japan

ABSTRACT

Behavior of phosphorus associated with phase separation and the effect of TiO_2 addition on the phosphorus distribution were investigated in Na_2O-B_2O_3-SiO_2 ternary glass system. After heat treatment, phase separation by spinodal decomposition was commonly observed. In the phase separation of TiO_2-free glasses, phosphorus was preferentially distributed into B_2O_3-rich phase, and after TiO_2 addition, phosphorus distribution into SiO_2-rich phase was also observed. With increasing TiO_2 content, the amount of phosphorus distributed into SiO_2-rich phases increased up to 90%. ^{31}P MAS NMR indicated that phosphorus in the glasses was mainly present in negatively-charged Q^2 units with terminal oxygen atoms, and in the TiO_2-containing glasses, Q^1 or Q^0 units containing P-O-Ti bond were confirmed. Despite the formation of negatively-charged Q^1 or Q^0 units, phosphorus was distributed into SiO_2-rich phase, which was due to the distribution of Ti^{4+} ions with higher coordination numbers acting as network modifiers.

INTRODUCTION

Phosphorus has been widely used in various applications, such as fertilizer, agrichemicals, pesticides, cement, toothpaste, food additives, and so on. However, Japan has long ceased the production of phosphorus natural resources and is almost totally dependent on imports of phosphate ores as phosphorus resources. Recently, phosphorus producing countries such as China and the USA limited the exports of phosphorus resources, and hence a stable supply of phosphorus resources becomes a critical issue in Japan.

The authors research group has developed a chemical recycling process of inorganic wastes by using phase separation of glass.[1,2] Inorganic wastes such as granulated blast furnace slags and high-temperature molten slags of municipal wastes, SiO_2, CaO and Al_2O_3 are present as major constituents. When B_2O_3 is added, the wastes are easily vitrified, and phase separation is induced by heat-treatments. In the phase separation, coloring species such as Fe ions as well as alkali and alkaline earth ions such as Na and Ca ions are preferentially introduced to B_2O_3-rich phase, and hence SiO_2-rich colorless and transparent glass is recovered by immersing the phase-separated slag glass into acidic solutions. In the municipal waste slags,[2] it was found that phosphorus was distributed to SiO_2-phase at a high rate of 60 - 70%, suggesting that phase separation of wastes was also available to the recovery of phosphorus, and phosphorus recovered from wastes was used as a phosphorus resource.

The authors examined the distribution of phosphorus due to phase separation. In the previous study,[3] a simple Na_2O-B_2O_3-SiO_2 system was chosen to reveal fundamental mechanisms of phosphorus distribution, in which it was found that phosphorus was completely distributed to B_2O_3-rich phase without additives other than P_2O_5, and the addition of Al_2O_3 initiated the phosphorus distribution to SiO_2-rich phase. In the Na_2O-B_2O_3-SiO_2 glass, aluminum was selectively distributed to SiO_2-rich phase, and in the municipal waste slag glass, however, aluminum was oppositely distributed to B_2O_3-rich phase. In the municipal waste slag glass, simultaneous distribution of titanium to SiO_2-rich phase was observed[2] implying that titanium also had a promoting effect of phosphorus distribution to the SiO_2-rich phase. In this study the effect of TiO_2 addition on the phosphorus distribution has been investigated in the phase separation of ternary Na_2O-B_2O_3-SiO_2 systems, and the mechanism of phosphorus distribution was also discussed based on the structural analyses.

METHODS AND PROCEDURES

Table I shows the glass compositions prepared in the present study, and the glasses were prepared by a conventional melt-quenching method. The raw materials of reagent grade Na_2CO_3, B_2O_3, SiO_2, Na_3PO_4 and TiO_2 were mixed thoroughly, and melted in a Pt crucible at 1400°C for 30 min. By pressing the melts with iron plates, glass plates were obtained. Differential thermal analyses (DTA) were carried out to determine the glass transition and crystallization temperatures, T_g and T_x.

Table I. Nominal compositions (mol%), T_g and T_x (°C) of the samples.

No.	Na_2O	B_2O_3	SiO_2	P_2O_5	TiO_2	T_g	T_x
1	10	40	50	-	-	469	-
2	10	40	47	3	-	450	-
3	10	40	46	3	1	426	-
4	10	40	47	3	3	420	-
5	10	40	44	3	5	412	663
6	10	40	37	3	10	412	549

To induce phase separation, the glasses were heat-treated at 500 - 660°C for 16h. After the heat treatment, the glasses were immersed in nitric acid of 1.0 N for 24h at 90°C to dissolve B_2O_3-rich phase. The insoluble residues were collected with vacuum filtration, and after being rinsed in water and dried in an oven, silica-rich glasses were recovered.

X-ray diffraction (XRD) measurement with a Cu-Kα radiation was carried out to confirm the crystallization. Chemical compositions of the specimens, such as the glasses before phase separation, the solutions and insoluble residues after immersion in nitric acid were determined by inductively-coupled plasma (ICP) emission spectroscopy and x-ray fluorescence (XRF) spectrometry. Microscopic texture of the glasses after phase separation was observed by scanning electron microscope (SEM). ^{29}Si, ^{11}B, and ^{31}P MAS-NMR measurements were performed with Varian Unity Inova 300 spectrometer to investigate the local structures around Si, B, and P, respectively. For the ^{29}Si measurements, 0.2 mol% of Fe_2O_3 was added to the glass batches in order to shorten the relaxation time. The respective NMR measurement conditions of ^{29}Si, ^{11}B and ^{31}P measurements were as follows: magnetic field = 7.05 T, frequency = 59.6 MHz, 96.2 MHz and 121.4 MHz, sample spinning speed = 5.0 kHz (all), pulse duration = 2.5 μs, 0.6 μs and 3.3 μs, repetition time = 1.0 s, 1.0 s, 5.0 s, chemical shift second standard = poly dimethyl siloxane (PDMS), BPO_4 and $NH_4H_2PO_4$. Infrared absorption spectra were recorded on an FT-IR spectrometer in the 4000 - 400 cm^{-1} range using KBr pellets containing powdered glasses.

RESULTS AND DISCUSSION

SEM photographs of the heat-treated glasses are shown in Fig. 1. Typical textures associated with spinodal decomposition are commonly observed in the presence or absence of P_2O_5 or TiO_2. In the DTA curve of the glasses No. 5 and No. 6, an exothermic peak due to crystallization was observed (See T_x in Table I.). As shown in Fig. 2, anatase-type TiO_2 crystal was precipitated in the heat-treated glasses with TiO_2 content ≥ 5 mol%.

Figure 1. SEM photos of glass surfaces heated at 600°C for 16h.

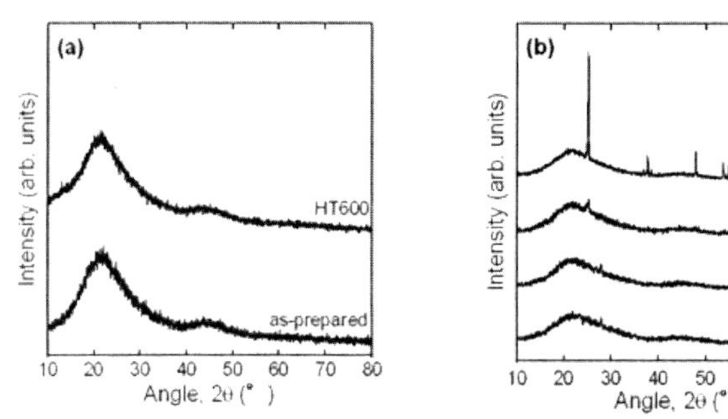

Figure 2. XRD patterns of the glasses, (a) No. 5 and (b) No. 6 as-prepared and after heat-treatment.

Chemical compositions of the glass phases after phase separation were estimated from ICP and XRF analyses. As shown in Table II, phosphorus in the TiO_2-free glass No. 2 is preferentially distributed into B_2O_3-rich phase after the phase separation. In the case of TiO_2-containing glasses, however, phosphorus is distributed also into SiO_2-rich phase, and the amount of phosphorus distributed into SiO_2-rich phase increases with the increase in TiO_2 content. About 90% of phosphorus in the glass No. 6 is distributed into SiO_2-rich phase. At the same time, titanium is also preferentially distributed into SiO_2-rich phase, which is observed in the slag glass of municipal waste.[2] However, in the specimens in which crystalline TiO_2 is precipitated (for example, No. 5' in Table II), the amount of phosphorus distributed into SiO_2-rich phase decreases, indicating that the distribution of phosphorus is strongly influenced by the existing state of titanium atom.

Table II. Analytical compositions of glass phases after phase separation.

(a) SiO_2-rich phase / mol% (XRF)

No.	Na_2O	B_2O_3	SiO_2	P_2O_5	TiO_2
2	0.1	-	99.8	0.1	
4	0.1	-	89.0	2.8	8.1
5	0.2	-	83.4	3.9	12.5
6	0.1	-	70.3	5.5	24.1
5'	0.2	-	84.3	0.5	15.0

5': TiO_2-precipitated sample heated at 660°C.

(b) B_2O_3-rich phase / mol% (ICP)

No.	Na_2O	B_2O_3	SiO_2	P_2O_5	Ti
2	18.1	73.1	2.8	6.0	
4	16.4	76.4	3.0	3.5	0
5	19.6	73.5	4.2	1.7	1
6	18.7	72.9	5.0	0.7	2
5'	19.0	72.5	3.6	4.8	0

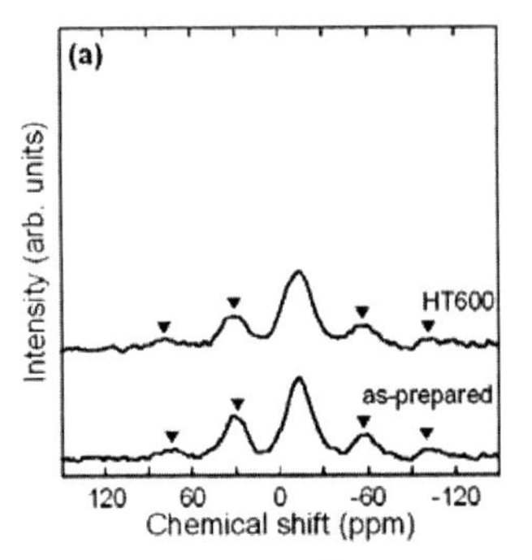

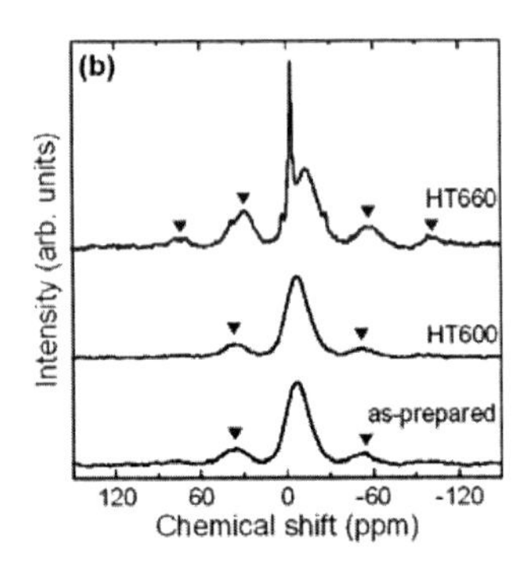

Figure 3. ^{31}P MAS NMR spectra of the glasses, (a) No. 2, and (b) No. 5. ▼: spinning side band

^{31}P MAS NMR spectra are shown in Fig. 3. In the TiO_2-free glass No. 2 (Fig. 3a), a peak at $\delta \approx -14$ ppm is observed, which is assigned to pure $Q^2{}_P$ unit and $Q^2{}_P$ unit connected to four-coordinated boron atoms,[5,6] where n in $Q^n{}_X$ means the number bridging oxygen (BO) connected to the X atoms. In the TiO_2-added glass No. 5 (Fig. 3b), the peak shifts to a lower magnetic field side due to TiO_2 addition, suggesting the formation of $Q^1{}_P$ or $Q^0{}_P$ units containing P-O-Ti bonds.[4] Comparing the spectra before and after the heat-treatment, significant change is not observed in the glass No. 2 and in the glass No. 5 until 600°C, indicating that local structure around phosphorus remains the same, and significant change is observed in the glass No. 5 heated at 660°C. A sharp peak is observed at $\delta \approx -3$ ppm, and the remaining broad peak is similar to that in glass No. 2, suggesting that the sharp peak relates to the precipitated TiO_2 crystal.

Fig. 4 shows IR spectra of the insoluble solids after acid treatment, that is, SiO_2-rich phase. In the IR spectra of TiO_2 containing glass No. 5, additional bands are observed at 650, 750 and 980 cm^{-1}. The bands at 650 and 750 cm^{-1} are due to TiO_6 and TiO_4 units, respectively,[7] and the band at 980 cm^{-1} is assigned to Si-O-Ti stretching mode.[8]

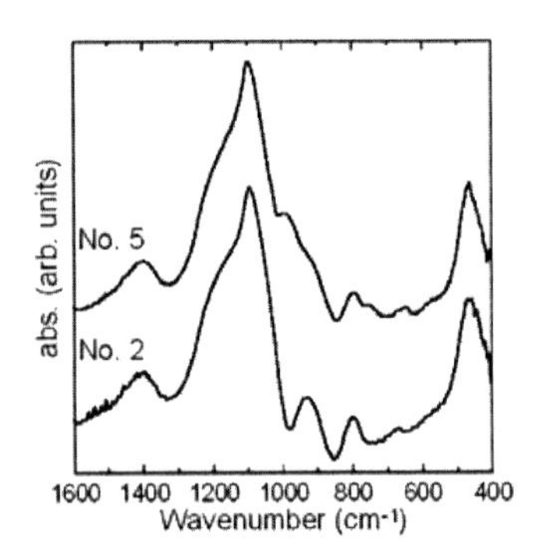

Figure 4. IR spectra of the insoluble solids after acid treatment.

It is revealed from the compositional analyses that phosphorus is preferentially distributed to B_2O_3-rich phase after the phase separation of TiO_2-free glass, and TiO_2 addition initiates the phosphorus distribution into SiO_2-rich phase. According to ^{31}P MAS NMR, phosphorus in TiO_2-free glass is mainly present in Q^2_P unit, and in the TiO_2 containing glasses, Q^1_P or Q^0_P units including P-O-Ti, bonds are formed. Since Q^n_P ($n = 0 \sim 2$) units are negatively charged, they must be coordinated by positively-charged species such as Na^+ to maintain electrical neutrality. In the TiO_2-free glass, Q^2_P units are compensated by Na^+ ion, and hence phosphorus is distributed to Na^+-abundant B_2O_3-rich phase. In the TiO_2-containing glass, however, Q^1_P or Q^0_P units distributed to SiO_2-rich phase are not compensated by Na^+ ions because Na^+ ions are absent in SiO_2-rich phase. It is therefore expected that Q^1_P or Q^0_P units are compensated by Ti^{4+} ions acting as network modifiers with the coordination numbers > 4. In the case of Al_2O_3 addition,[3] phosphorus distribution into SiO_2-rich phase is also observed, in which the formation of positively-charged Q^4_P units including P-O-Al bonds is confirmed. The effect of Al_2O_3 and TiO_2 on the phosphorus distribution seems to be the same, and the distribution mechanism, however, seems to be quite different, because Q^n_P units formed are different between the cases of Al_2O_3 and TiO_2 addition. It is revealed that the distribution of phosphorus is dependent on the states of PO_4 units, and furthermore, the state of PO_4 units is dependent on the neighboring ions of Na^+, Al^{3+} and Ti^{4+}. To clarify more detailed mechanism of phosphorus distribution, distribution mechanism of Al^{3+} and Ti^{4+} is also required.

CONCLUSION

Sodium borosilicate glasses containing P_2O_5 and TiO_2 were prepared by a conventional melt-quenching method and the effect of TiO_2 addition on the behavior of phosphorus associated with the phase separation were investigated. After the heat treatment, phase separation by spinodal decomposition was confirmed by SEM observation, and at TiO_2 content ≥ 5 mol%, precipitation of TiO_2 crystal was also found when heated near T_x. ^{31}P MAS NMR spectra suggested that in TiO_2-free glasses, phosphorus was essentially present in Q^2_P species, and Q^1_P or Q^0_P species formed after adding TiO_2. From ICP analysis, in TiO_2-free glasses, phosphorus was preferentially distributed into B_2O_3-rich phase after the phase separation. TiO_2 addition initiated the phosphorus distribution into SiO_2-rich phase, and with increasing TiO_2 content, the amount of phosphorus distributed into SiO_2-rich phase increased. It was revealed that the distribution of phosphorus depended on the existing state of PO_4 units.

ACKNOWLEDGMENTS

The authors gratefully acknowledge the financial support of Grant-in-Aid for Scientific Research in Establishing a Sound Material-Cycle Society (K2132, K22052), Ministry of Environment, Japan.

REFERENCES

[1]T. Nanba, S. Mikami, T. Imaoka, S. Sakida, Y. Miura, J. Ceram Soc. Jpn., 116(2) 220-223 (2008).
[2]T. Nanba, Y. Kuroda, S. Sakida, Y. Miura, J. Ceram Soc. Jpn., 117(11) 1195-1198 (2009).
[3]Y. Otsuki, S. Sakida, Y. Benino, T. Nanba, Proceedings of the 3rd International Congress on Ceramics (ICC3), S8-P033, (2010) (http://iopscience.iop.org/1757-899X/18/11/112022).
[4]B. Tiwari, M. Pandey, V. Sudarsan, S. K. Deb, G. P. Kothiyal, Physica B 404 (2009) 47.
[5]S. Elbers, W. Strojek, L. Koudelka, H. Eckert, Solid State Nucl. Magn. Reson. 27 (2005) 65.
[6]M. Zeyer-Dusterer, L. Montagne, G. Palavit, C. Jager, Solid State Nucl. Magn. Reson. 27 (2005) 50.
[7]G.M. Krishna, N. Veeraiah, N. Vekatramaiah, R. Venkatessan, J. Alloys and Compounds, 450 (2008) 477.
[8]H.C. Vasconcelos, J. Sol-Gel Technol., 55 (2010) 126.

*Corresponding author: gev422352@s.okayama-u.ac.jp

GLASS-CERAMICS FROM KINESCOPE GLASS CULLET

M. Reben,[1] J. Wasylak,[1] M. Kosmal[2]

[1]AGH University of Science and Technology, Faculty of Materials Science and Ceramics, al. Mickiewicza 30, 30-059 Cracow, Poland
[2] Institute of Ceramics and Building Materials, Division of Glass in Cracow

ABSTRACT

A series of glass-ceramics forming by bulk crystallization of CRT glass cullet mixed with calsiglass, fluidized bed combustions fly ash, and ground granulated blast furnace slag was investigated. Glass-ceramics, materials prepared by the control of crystallization of glasses, have a variety of established uses depending on their uniform reproducible fine-grain microstructure, absence of porosity and wide-ranging properties which can be tailored by changes in composition and heat treatment. The crystallization experiments were carried out on the basis of differential thermal analysis (DTA), X-ray diffraction (XRD), and scanning electron microscopy (SEM). A compressive strength of obtained glass-ceramic was sufficiently large for practical use.

INTRODUCTION

In Poland, on the projected date of 31 June 2013, analogue signal will be completely replaced with digital television; all old TV sets will become useless. The problem with the waste from electric and electronic equipment (WEEE), consisting of ~80% television sets and computers containing a cathode ray tube (CRT), is just now beginning to be dealt with. Cathode ray tubes (CRTs) are the video display components of televisions and computer monitors (EPA, 1995). CRT products represented by Standard Industrial Classification (SIC) code 3671, include tube glass, color picture tubes, monochrome picture tubes, and rebuilt tubes. The recycling of glass from CRT is quite problematic because a CRT is made up of four different chemical types of glass (screen or panel, cone or funnel, neck and frit junction), (Fig.1).[1-3] Glasses such as cone and neck contain principally different lead contents and other dangerous elements. Panel glass has other heavy metals (Ba, Sr, etc.) that forbid their recycling in the glass industry for the production of containers, domestic glassware and glass fiber.[4-6] Small amounts of contaminant may result in the contaminating of a large batch and may put production off for many hours. It is difficult to really proceed with glass remanufacturing for two reasons: a) composition issues: difficult to know what is in the glass, including foreign materials as well as other forms of cullet. b) reliability issues: because of the inconsistent nature of CRT recycling, there is not yet a way to guarantee industries a constant year-round supply of cullet.

There is an increasing urgency to develop new applications for CRT glass in agreement with Directives 2002/95/EC on the restriction of the use of certain hazardous substances in electrical and electronic equipment (RoHS), and 2002/96/EC and 2003/108/EC on waste electrical and electronic equipment (WEEE).[7] For a new environmentally friendly sustainable economy, wastes are to be considered as a real opportunity to produce clean high-quality secondary raw material. Arising problems related with recovery and recycling of CRT glass stimulates examinations aimed at making cullet-based glassy-ceramic materials of reinforced mechanical strength.[8,9] Used glassy-ceramic materials are manufactured on the basis of aluminosilicate and silicate glasses.[10] Properties of such materials depend on crystalline phase type, being their main component, as well as on size and shape of the crystalline grains. Suitable selection of the glass chemical composition, nucleators, and crystallization conditions allow production of the materials having miscellaneous properties; in general, glassy-ceramic materials are characterized with high hardness, thermal resistance, abrasion resistance, as well as resistance to actions of acids and lye, including good electro-insulation properties.[11-15] Our objective is to improve methods for the recycling of used electronics. We have chosen to focus on CRT glass recycling, and will attempt to find new options for dealing with this material. These will include new technology for recycling as well as new markets for the glass as currently processed.

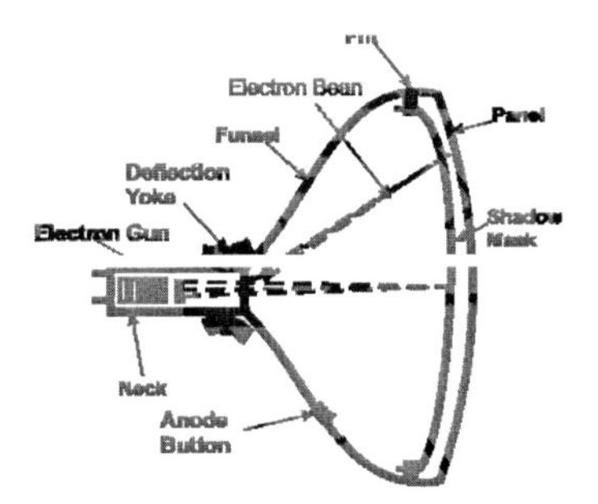

Figure 1. Basic elements of CRT.[16]

METHODS AND PROCEDURES

Cleaned panel glass from dismantling plants of TV kinescopes was used. The as-received panel glass was dry milled for 3h in a laboratory ball mill and sieved under 80 mm. Three glass compositions were prepared by dry mixing of CRT panel glass with different amounts of calsiglass, fluidized bed combustions fly ash, and ground granulated blast furnace slag. All the chemicals were mixed properly to ensure the homogeneity. The glasses were obtained by melting 50g batches in platinum crucibles in an electric furnace at the temperature 1450°C in air atmosphere. The melts were poured out onto a steel plate forming a layer about 2 to 5 mm thick. The chemical analysis of raw materials was performed by inductively coupled plasma. The composition of the investigated glasses is listed in Table I.

Table I. Glass compositions (wt%).

	Mixture (wt%)			
ID	CRT panel glass	FBC fly ash	calsiglass	GGBF slag
CF	60	40		
CC	60		40	
CS	60			40

The ability of the obtained glasses to crystallize was determined by DSC measurements conducted on the NETZSCH 5 System operating in heat flux DSC mode. The samples (60 mg) were heated in platinum crucibles at a rate 10°C/min^{-1} in dry nitrogen atmosphere to the temperature 1100°C. The glass transition temperature T_g was determined from the inflection point on the enthalpy curve; the jump-like changes of the specific heat ΔCp accompanying the glass transition, enthalpy ($\Delta H_{cryst.}$) of crystallization were calculated using the NETZSCH 5 Thermal Analysis Software Library. The ability of glasses for crystallization was measured by the values of the thermal stability parameter of glasses ($\Delta T = T_{cryst.} - T_g$). Glasses revealing the crystallizations events were selected for further thermal treatment. To obtain glass-ceramics they were subjected to heating for 3h at the temperature of the maximum crystallization events, respectively. The type and size of the formed crystallites were examined by XRD and SEM methods, respectively.

RESULTS AND DISCUSSION

The chemical compositions of the CRT panel glass, calsiglass, fluidized bed combustions fly ash, ground granulated blast furnace slag used are presented in Table II. Chemical analysis of CRT panel glass showed that panel glass is characterized by high levels of BaO (7.27 wt%) and SrO (10.80 wt%) alkaline earth oxides. For the PbO content a leaching test performed on the as received glass showed lead release (0.003 ppm). Barium oxide (7.27wt.%) is present in order to avoid radiation exposure. The X-ray absorption is maintained by adding higher amounts of barium, strontium and zirconium. The browning occurs even in leadless glass because alkali ions such as sodium are reduced and generate the brown color. Therefore, it is common in the panel glass composition to find more than two kinds of alkali oxides in order to prevent the coloring. The ratio of sodium and potassium is optimized and the amounts of both are almost equal. The presence in.the composition of small amounts of cerium has the effect of preventing the X-ray browning; the antimony is present (0.40) as a fining agent to refine the melt and thus avoiding the bubbles in the glass. Besides, the addition of titanium to glass ingredients has the effect of preventing solarization (effect of sunlight).

Table II. The chemical composition of used raw materials (wt%).

Oxides	CRT panel glass	calsiglass	FBC fly ash	GG blast furnace slag
SiO_2	61.82	39.60	36.27	33.54
CaO	-	43.20	18.73	41.30
MgO	0.01	7.30	1.79	5.00
Al_2O_3	2.03	6.73	22.87	12.59
Fe_2O_3	0.07	0.15	5.63	0.78
TiO_2	0.06	0. 20	0.40	0.43
Na_2O	7.95	0.60	1.31	1.01
K_2O	7.35	0.37	1.66	1.34
BaO	7.27	-	-	-
SrO	10.80	-	-	-
ZrO_2	1.70	-	-	-
SO_3	-	-	6.63	1.53
Others	0.94	1.85	4.71	2.48

By DTA curves performed on the quenched glasses, two kinds of thermal events were carried out: an infection point (T_g) corresponding to glass transition temperature, and an exothermic peak ($T_{cryst.}$) indicating the crystallization (Fig.2). In the case of CRT glass panel, the thermal behavior can be explained by the relevant quantities of alkaline oxides (~15 wt%) (Table II). The DTA curve of CRT glass panel did not present exothermal events, indicating that the glass does not crystallize in the temperature range studied, as confirmed by the low percentages of CaO, MgO and Fe_2O_3, typical modifier oxides that induce crystallization (Fig.2). The increase of Al_2O_3 content in the group of glass (CC, CS, CF) causes an increase of the transformation temperature T_g and increase of the specific heat

(ΔCp) accompanying the glass transition region (Table III). The increase of Al_2O_3 content causes the increase in the number of broken bonds during the course of the transformation of the glassy state. Due to the increase of CaO and MgO content (glass CF, CC, CS) the temperature of the maximum events of crystallization is shifted towards a lower temperature. This is evidence of the increasing ability of the glass for crystallization, manifested by decreasing value of the index of thermal stability of the glass ΔT (Table III).

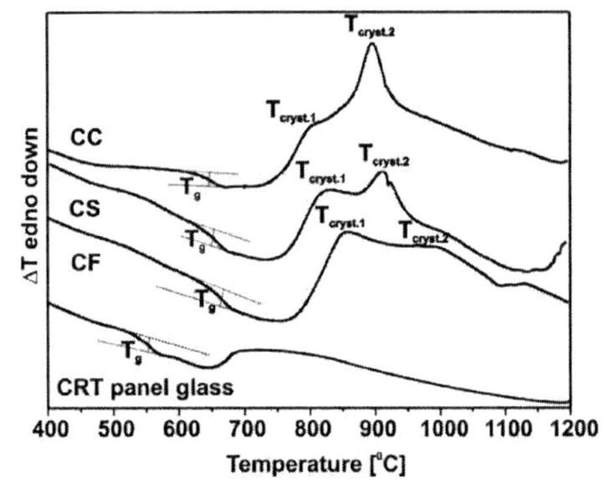

Fig.2. DTA curves of glasses: CRT panel glass, CF-CRT+fly ash, CS-CRT+slag, CC-CRT+calsiglass.

Table III. Thermal characteristic of glasses.

ID	T_g [^{0}C]	ΔCp [$J \cdot g^{-1} \cdot {}^0C^{-1}$]	$T_{cryst.1}$ [^{0}C]	ΔH_1 [$J \cdot g^{-1}$]	ΔT_1 [^{0}C]	$T_{cryst.2}$ [^{0}C]	ΔH_2 [$J \cdot g^{-1}$]	ΔT_2 [^{0}C]
CRT	549	0.227	-	-	-	-	-	
CC	650	0.353	820	-	170	903	143	253
CS	655	0.429	818	67	163	913	53	258
CF	662	0.577	850	135	188	1003	51	341

Using the Kissinger method, the activation energy (Ea) of crystallization was determined according to:[17]

$$\ln(T_p^2/\phi)=E_a/RT_p+\text{constant} \quad (1)$$

where T_p is the temperature of the maximum crystallization peak, ϕ is the heating rate and R is the universal gas constant. Crystallization kinetics was studied using the DTA/DSC method at three heating rates (5, 10, and 15°C/min). The activation energy was calculated from the slopes of the linear fits to the experimental data. The activation energy was calculated based on first and second crystallization events. The Kissinger plot is illustrated in Fig. 3 for the CF, CS, CC glass. Table IV summarizes the E_a value determined from the slope of each line as obtained from linear least-square regression analysis.

The glass transition temperature T_g and T_p increase as ϕ increases. For all glasses the activation energies of the first crystallization event are greater than activation energies of the second crystallization events. It is due to high viscosity in the lower range of the temperature. Moreover, the glass CC is characterized by the lowest value of E_a, what favors the crystallization. Crystalline phases in the heat treated glasses at the max. of the crystallization events were selectively analyzed by using a powder X-ray diffraction (XRD) (Fig.5-7). XRD measurements for the glass-ceramic CC (CRT+calsiglass) obtained by heat treating the glass at its first and second crystallization effect temperature showed several sharp diffraction peaks overlapped on the amorphous hump, which indicates that crystals are precipitated from the precursor glasses. It is easy to identify that the crystal phase is wollastonite ($CaSiO_3$).

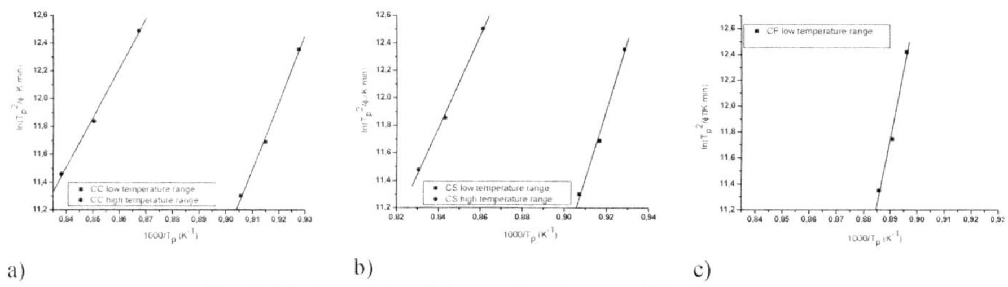

Fig. 4. Kissinger plot of the exothermic crystallization events:
a) for glass CC, b) for glass CS, C) for glass CF.

Table IV. Calculated activation energies.

ID	ϕ	T_g [°C]	T_{p1} [°C]	E_a [kJ/mol]	T_{p2} [°C]	E_a [kJ/mol]
CC	5	647	805	403.07	880	296.05
	10	650	820		903	
	15	651	831		920	
CS	5	653	804	417.29	888	279.30
	10	655	818		913	
	15	656	829		931	
CF	5	658	843	872.06	-	-

In glass–ceramics, the glass–ceramics with 3h crystallization times at the 820°C and 903°C, $CaSiO_3$ was the only crystal phase separated from the vitreous matrix (Fig.5). The glass CC is characterized by the lowest Na_2O and K_2O content. Consequently, in order for glass-ceramic materials to have good chemical resistance, it is necessary that their residual glass phase contain low concentrations of alkali metal oxides in particular. Generally, the chemical stability of glass-ceramic materials is affected by the composition of the crystalline phase and by the composition and amount of residual glass phase and its morphology. Wollastonite ($CaSiO_3$) is an important substance in the ceramic and cement industries.[18] A host of favorable properties such as low shrinkage, good strength, lack of volatile constituents, body permeability, fluxing characteristics, whiteness and acicular shape renders wollastonite useful in several ceramic and other applications.[19] Furthermore, in glass sample CS (CRT+slag) after heat treatment at 913^{0}C the appearance of a small peak at $\theta = 27.5^0$ and the broadening of the main reflection of wollastonite at $\theta = 30^0$ can be attributed to the emergence of the new crystal phase identified as a diopside-like structure ($CaMgSi_2O_6$) (Fig.6). Due to higher iron content in glass CS in comparison to glass CC, the heat treatment at the 818^{0}C causes the appearance of crystal phase identified as an augite-like structure $Ca(Fe,Mg)Si_2O_6$.

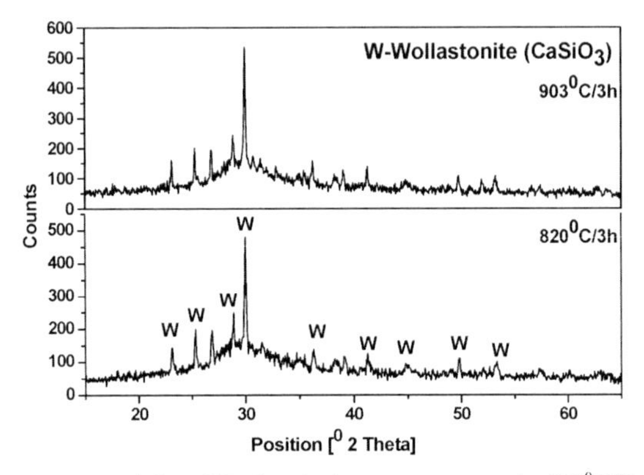

Figure 5. XRD patterns of glass CC after the heat treatment at the 820^{0}C/3h, and 903^{0}C/3h.

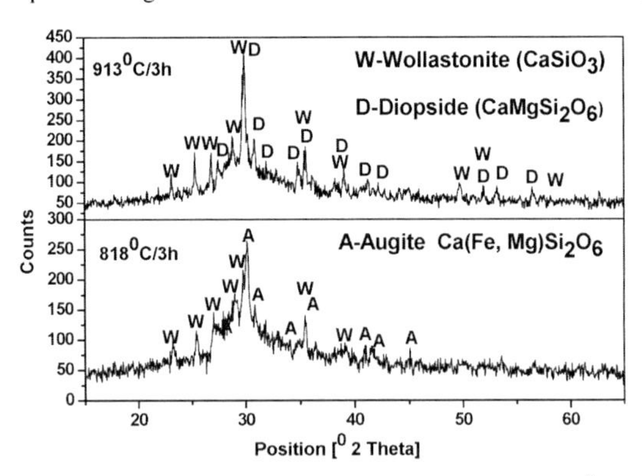

Figure 6. XRD patterns of glass CS after the heat treatment at the 818^{0}C/3h, and 913^{0}C/3h.

In the mixture containing CRT glass and fly ash after the heat treatment at the 850^{0}C orthoclase $K(Al,Fe)Si_2O_8$ is formed (Fig.7). This phenomenon is probably due to higher potassium/sodium oxides content ratio (K_2O/Na_2O:0.97 wt%) and the highest aluminum oxide content (22.87 wt%).

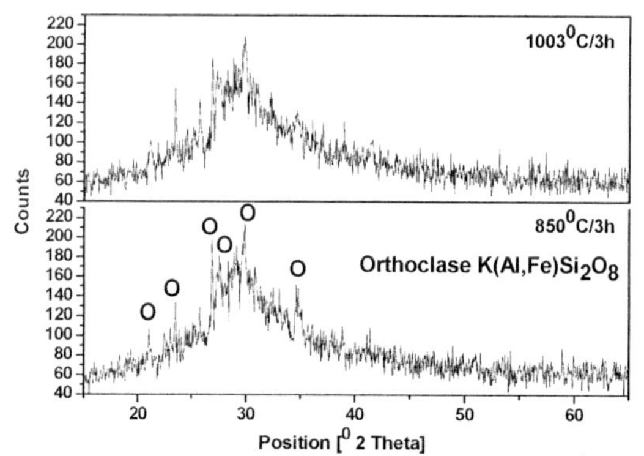

Figure 7. XRD patterns of glass CF after the heat treatment at the 850^{0}C/3h, and 1033^{0}C/3h.

Table V summarizes the complete study of the crystallization and XRD phase analysis of three glasses. The morphology of the crystals formed after the heat treatment of glass CC and CS was studied by SEM microscopy. The microstructures of glass CC after the heat treatment at 820^{0}C, and 903^{0}C, in agreement with the XRD results, showed well formed crystals of wollastonite ($CaSiO_3$) (Fig.8). It was observed that the crystallization time of 3h seems to be enough to form crystals. In the case of glass CF after the heat treatment at the temperature 850^{0}C, the crystallization time of 3h led to the formation of smaller orthoclase $K(Al,Fe)Si_2O_8$ crystals (Fig.9).

Table V. XRD phase analysis.

ID	Thermal history		Crystallizing phases
CC	820^{0}C	3h	Wollastonite ($CaSiO_3$)
	903^{0}C	3h	Wollastonite ($CaSiO_3$)
CS	818^{0}C	3h	Augite $Ca(Fe,Mg)Si_2O_6$
	913^{0}C	3h	Wollastonite ($CaSiO_3$) Diopside $(CaMg)Si_2O_6$
CF	850^{0}C	3h	Orthoclase $K(Al,Fe)Si_2O_8$
	1003^{0}C	3h	Amorphous

Figure 8. SEM images of the glass CC heat-treated at 820°C/3h.

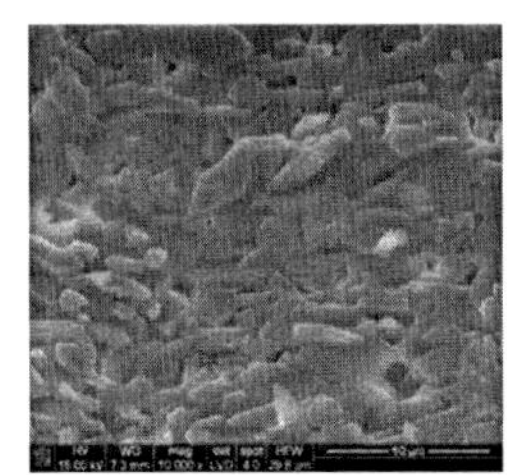

Figure 9. SEM image of the glass CF heat-treated at $850^0C/3h$.

CONCLUSIONS

Based on the results obtained from this study, the following conclusions can be reached:

1. The CRT panel glass with different waste materials such as: FBCFA, GGBBS, calsiglass during heating demonstrates two crystallization events, while in case of CRT panel glass crystallization effect does not appear.
2. The CC (CRT+calsiglass) and CS (CRT+slag) glasses are characterized by greater crystallization ability, due to the addition of CaO and MgO. Glass CF tends to have greater resistance to crystallization due to high content of Al_2O_3.
3. The lowest activation energy (Ea) value of crystallization in case of glass CC, favors the crystallization.
4. The mixture of 60 (wt%) of CRT panel glass with the wastes: calsiglass and GGBFS can be used to obtain wollastonite glass-ceramics in sintering process. The main applications of glass-ceramic are in the field of abrasion- resistant materials, i.e., industrial floor coverings, wall facings, abrasion-resistant linings, and high-temperature insulators. Moreover, low cost and availability of the raw materials make them very attractive from the economic point of view.

ACKNOWLEDGEMENTS

The work was supported by Grant No. N R08 0025 10, of the Ministry of Science and Informatisation of Poland.

REFERENCES

[1]F. Andreola, L. Barbieri, F. Bondioli, I. Lancellotti, P. Miselli and A.M. Ferrari, International Journal of Applied Ceramic Technology, 7 (6), 909-917 (2010).

[2]M. Marshall, J. Henderson, New approaches to the challenge of CRT recycling. In: Dhir K, Limbachiya MC, Dyer TD (eds), Recycling and reuse of glass cullet. Thomas Telford Publishing, London, pp75–83 (2001).

[3]F. Andreola, L.Barbieri, A. Corradi, I. Lancellotti, CRT glass state of the art: A case study: Recycling in ceramic glazes, Journal of the European Ceramic Society, **Vol. 27**, Issues 2-3, p.1623-1629 (2007).
[4]F. Andreola, Cathode ray tube glass recycling: an example of clean technology, Waste Management & Research, **Vol. 23**, No. 4, 314-321 (2005).
[5]F. Andreola, L. Barbieri, A. Corradi, A. M. Ferrari, I. Lancellotti and P. Neri, Recycling of EOL CRT glass into ceramic glaze formulations and its environmental impact by LCA approach, The International Journal of Life Cycle Assessment, **vol.12**, No 6, (2007).
[6]F.Andreola, L. Barbieri, A. Corradi and I. Lancellotti, Cathode ray tube glass recycling: an example of clean technology, *Waste Manage. Res.*, **23**, 1–8 (2005).
[7]Dir.2002/95/CE. Art.3, All.I B.
[8]F. Andreola, L. Barbieri, E. Karamanova, I. Lancellotti, M. Pelino, Journal of the European Ceramic Society **23**, 827–832 (2003).
[9]E. Karamanova, A. Karamanov, Waste Managment&Research **27**, 87-92 (2009).
[10]A. Karamanov, M. Pelino, A. Hreglich, Journal of the European Ceramic Society **23**, 827–832 (2003).
[11]N. Menzler, H. Mortel, R. Weibmann, and V. Balec, Examination of two glass compositions for the vitrification of toxic products from waste incineration. Glass Technology, **40**, 65–70 (1999).
[12]A. Boccaccini, A., Petitmeret, M. and Wintermantel, E., Glass-ceramics from municipal incinerator fly ash. The American Ceramic Society Bulletin, **11**, 75–78 (1997).
[13]M. Romero, R. Rawlings, and J. Rincon, Development of a new glass-ceramics by means of control vitrification and crystallization of inorganic wastes from urban incineration. J. Eur. Ceram. Soc., **19**, 2049–2058 (1999).
[14]A. Karamanov, and M. Pelino, Evaluation of the degree of crystallization in glass-ceramics by density measurements. J. European Cer. Soc., **19**, 649–654 (1999).
[15]A. Karamanov, G. Taglieri, and M. Pelino, Iron-rich sintered glass-ceramics from industrial wastes. J. Am. Ceram. Soc., **82**, 3012–3016 (1999).
[16]Y. Miwa, H. Yamazaki, and S. Yamamoto, The history of CRT glass material. In Proceedings of the XX International Congress of Glass (ICG), pp. 1–8 (2004).
[17]H.E. Kissinger, J. Res. Natl. Bur. Stand., **57**, 217-221 (1956).
[18]Byung-Hoon Kim, Bo-An Kang, Yeon-Hum Yun, Kyu-Seog Hwang, Chemical durability of β-wollastonite-reinforced glass-ceramics prepared from waste fluorescent glass and calcium carbonate, Materials Science-Poland, **Vol. 22**, No. 2, (2004).
[19]I. Tsilika, I. Eleftheriadis, Th. Kehagias, E. Pavlidou, Th. Karakostas, Ph. Komninou, Crystallization process of thermally treated vitrified EAFD waste, Journal of the European Ceramic Society **30** 2009–2015 (2010).

*Corresponding author: manuelar@agh.edu.pl

FUNCTIONAL AND STRUCTURAL CHARACTERIZATIONS OF FRESNOITE GLASS-CERAMICS ORIENTED WITH UST TECHNIQUE

Y. Benino,[1] A. Endo,[1] S. Sakida[2] and T. Nanba[1]
[1] Graduate School of Environmental Science, Okayama University
[2] Environmental Management Center, Okayama University
3-1-1 Tsushima-naka, Okayama 700-8530, Japan

ABSTRACT

Transparent glass ceramics with highly oriented surface crystalline layers of fresnoite $Ba_2TiSi_2O_8$ were prepared by heat treatment of precursor glasses in $BaO\text{-}TiO_2\text{-}SiO_2$ system. The orientation along *c*-axis of fresnoite crystals normal to the glass surface was evaluated by the surface XRD patterns, and it was remarkably enhanced by ultrasonic surface treatment (UST) before the crystallization using fresnoite powder suspension in water. The effect of UST disappeared after heat treatment in the glass transition region, and the mechanism of UST was discussed on the basis of the persistence at elevated temperatures.

INTRODUCTION

Glass ceramic materials, including fresnoite $Ba_2TiSi_2O_8$ phase, are of interest because of their excellent nonlinear optical (NLO) functions[1] induced by the crystalline phase formation. There has been other research on the preparation and NLO evaluations of the glass ceramic materials in $BaO\text{-}TiO_2\text{-}SiO_2$ and related glass systems that has been reported.[2-6] Among them a transparent glass ceramic with oriented fresnoite crystalline layer was an outstanding approach to realize a NLO device application of second harmonic conversion by transparent glass ceramics in a bulk form. Fresnoite glass ceramics with optical transparency and crystallographic orientation were only fabricated in a narrow region near a specific composition of $30BaO\text{-}20TiO_2\text{-}50SiO_2$ together with a well-controlled heat treatment.[7] In this case, two mechanisms of crystallization, the inhomogeneous nucleation on the glass surface and the competitive growth of needle-like crystals along the selected axis, are important, and therefore the resulting morphology or texture of glass ceramics is expected to be controlled for the refinement of the transparency and optical functions of the surface-crystallized glass ceramics. Here, ultrasonic surface treatment (UST) is a novel technique for controlling crystallization behavior, which enables the phase selection, size and orientation control of precipitates in various glass ceramic systems.[4,8-10] It can also be applied to the fresnoite system for the transparent glass ceramics with NLO function. In the present study the effect and mechanism of UST were investigated in the fresnoite glass ceramic system by evaluating the orientation of the surface crystalline layer, especially focusing on the enhancement by UST effect and the degradation after heat treatment in the glass transition region.

EXPERIMENTAL PROCEDURE

The glass with a composition of $30BaO\text{-}20TiO_2\text{-}50SiO_2$ (mol%) was prepared by a conventional melt-quenching method. Commercial powders of reagent grade $BaCO_3$, TiO_2 and SiO_2 were mixed thoroughly and melted in a platinum crucible in an electric furnace at 1500°C for 1h. The melt was quenched onto a preheated steel plate and then annealed at 730°C for 1h. The glass thus formed was cut into plates with ~1 mm in thickness, and they were mechanically polished with alumina slurry to obtain mirror surfaces.

UST was carried out by exposing the polished glass samples to ultrasonic energy transmitted through fresnoite/water suspension held in an ultrasonic cleaning bath (VS-100 II SUNPAR 100W Velvo Clear Co.) with a constant condition, that is, 28 kHz for ultrasonic frequency and 15 min for exposure time. The weight ratio of solid particles of fresnoite to water was also fixed to 1 wt% throughout the present experiments. The fresnoite powder in the suspension was prepared by firing and pulverizing the stoichiometric mixture of reagent grade $BaCO_3$, TiO_2 and SiO_2 at 1100°C for 120h and 1200°C for 20 h in an electric furnace, and it was confirmed to be a single phase of fresnoite (JCPDS 22-513) by the XRD measurement at room temperature.

The effect of UST induced on the glass surface was confirmed by the surface XRD patterns after crystallization heat treatment (HT-2) at 840°C for 48h, where the formation of oriented crystalline layer of fresnoite is a criterion of the nucleation on the glass surface activated by UST. In order to evaluate the persistence and degradation of the UST effect in the glass transition region, heat treatments (HT-1) at elevated temperatures, T_1 = 200 – 750°C for various durations, t_1 = 0 – 600 min were carried out and then crystallized in the same manner as HT-2. The polar orientation of fresnoite phase on the surface was evaluated by Lotgering factor (L.F.) for (00*l*) diffraction lines derived from the surface and powder XRD patterns with Cu Kα radiation at room temperature.

RESULTS AND DISCUSSION

Disappearance of UST Effect after HT-1

Figure 1 shows the surface XRD patterns of the full-crystallized glass ceramics after HT-2, where the glass ceramic samples whose XRD patterns denoted by temperatures were conducted by the preceding HT-1 for 1h after UST. The orientation of fresnoite surface layer along *c*-axis was remarkable compared with the normal crystallization without UST, and such behavior continues up to T_1 = 650°C. At higher temperatures than 650°C, on the other hand, the effect of UST on the surface orientation disappeared drastically after 1h heat treatment.

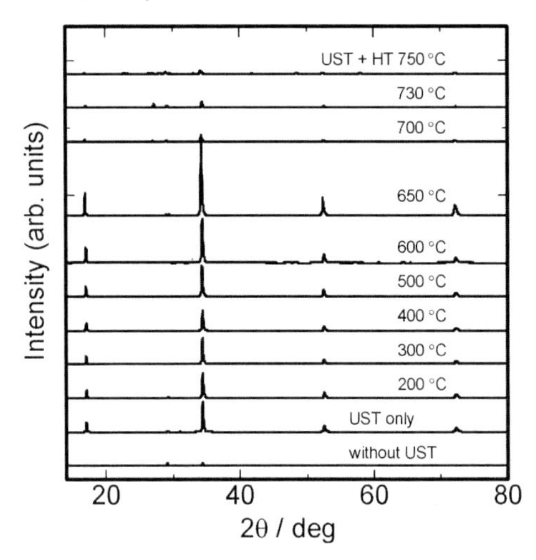

Figure 1. Surface XRD patterns of full crystallized (after HT-2) glass ceramics with various temperatures T1 of preceding heat treatment HT-1 for 1h.

The Lotgering factors (L.F.) for the c-axis shown in Fig. 2 as a function of HT-1 temperatures also demonstrated such a drastic change for $T_1 \geq 700°C$. This fact suggests that the effect of UST should be attributed to some physical trace, in other words glass-origin defects on the glass surface imprinted by fresnoite particles in the suspension and ultrasonic cavitation power, and that they can disappear in the glass transition region ($T_g = 710°C$) through structural relaxation processes like crack healing or fire polishing. If crystalline traces or nuclei were formed on the glass surface by UST, they could have an effect on the crystallization behavior after heat treatment at T_1 less than the melting point of fresnoite. In fact, L.F. for $T_1 = 750°C$ did not arrive at the intrinsic (without UST) value of ~0.3, which indicates crystalline traces, instead of pure physical damages, were partially contributed to the surface crystallization. Otherwise it is possible that another mechanism for the crystalline orientation was induced during HT-1 at 750°C. For detailed examination, however, further investigation should be done by quantitative treatment of nucleation sites enhanced by UST.

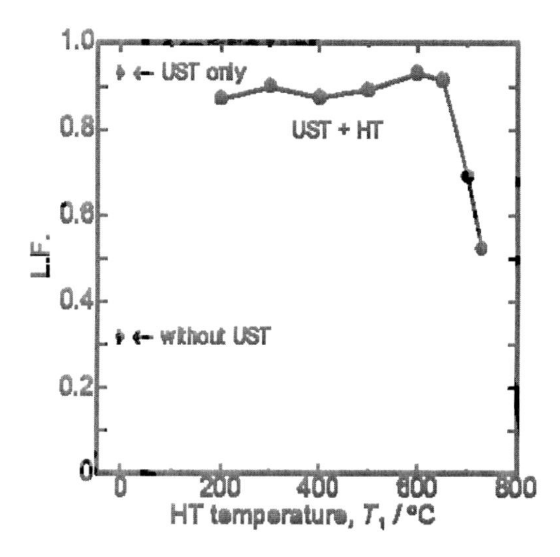

Figure 2. Change in orientation of surface crystals represented by Lotgering factors for surface XRD patterns in Fig. 1.

Change in UST Effect during Long Term Heat Treatment

In order to verify the time scale of the persistence of UST effect, L.F. values were plotted as a function of HT-1 duration t_1 for typical temperatures around the glass transition. For the case of T_1 = 650°C low enough for the structural relaxation of the corresponding glass the L.F. values kept high and so did UST effects, while these values for T_1 = 700°C have steeply decayed as shown in Fig. 3. This difference is consistent with the results shown in Fig. 2, and it is concluded that UST effect disappears readily in the glass transition region, implying the formation of glass-origin nucleation sites.

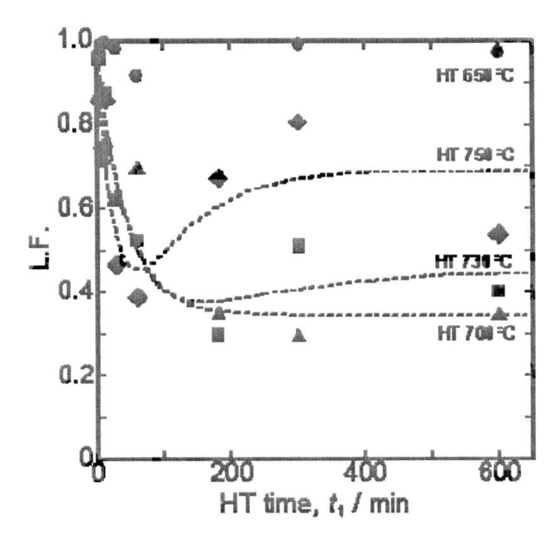

Fig. 3. Change in orientation of surface crystals represented by Lotgering factors, characterized by degradation and regeneration of surface crystallization during long term heat treatment HT-1 up to 600 min.

The change in L.F. values were rather complicated at higher temperatures, T_1 = 730 and 750°C, where they were characterized by faster decay and gradual increase. This is because the degradation of UST effects and the inhomogeneous nucleation on the glass surface have simultaneously occurred with different time constants. The latter progress dominated especially for longer heat treatment at higher temperature. In order to evaluate the UST effects, consequently, it is important to carefully separate several factors, (1) intrinsic orientation, (2) UST effect (glass surface defect origin), (3) permanent UST effect (stuck/embedded crystal origin), and (4) inhomogeneous nucleation during heat treatment. Instead of indirect evaluation by L.F. values it would be an effective approach to perform direct counting of surface nucleation sites using microscopic observations, which have been planned as the next work in future.

CONCLUSIONS

For the transparent glass ceramics with *c*-axis oriented surface crystalline layers of fresnoite, UST effects on the crystallization behavior were investigated in terms of the disappearance and degradation during heat treatment in the glass transition region. Lotgering factor derived from the surface XRD patterns steeply decreased by heat treatment at 700°C before the surface crystallization, indicating that some extent of UST effects on the glass surface can be lost by the relaxation process of glass. The existence of permanent effect of UST, stuck/embedded crystals on the glass surface, could not be confirmed in the present results. However, the formation of surface nucleation sites during UST and heat treatment above the glass transition should be clarified by continuous investigations, such as the evaluation of surface orientation and the microscopic observation of surface crystals.

REFERENCES

[1]P.S. Bechthold, S. Haussühl, E'. Michael, J. Eckstein, K. Recker and F. Wallrafen, Phys. Lett. 65A, 453-454 (1978).

[2]A. Halliyal, A.S. Bhalla and R.E. Newnham, Mat. Res. Bull. 18, 1007-1019 (1983).

[3]R. Kedding, C. Rüssel, J. Non-Cryst. Solids 219, 136-141 (1997).

[4]Y. Ding, A. Osaka and Y. Miura, J. Am. Ceram. Soc. 77, 2905-2910 (1994).

[5]Y. Takahashi, K. Kitamura, Y. Benino, T. Fujiwara and T. Komatsu, Appl. Phys. Lett. 86, 091110-1-3 (2005).

[6]N. Maruyama, T. Honma and T. Komatsu, Opt. Mater. 32, 35-41 (2009).

[7]H. Masai, S. Tsuji, T. Fujiwara, Y. Benino, and T. Komatsu, J. Non-Cryst. Solids 353, 2258-2262 (2007).

[8]Y. Ding, A. Osaka and Y. Miura, J. Am. Ceram. Soc. 77, 749-752 (1994).

[9]Y. Ding, A. Osaka and Y. Miura, J. Non-Cryst. Solids 176, 200-207 (1994).

[10]Y. Ding, N. Masuda, A. Osaka and Y. Miura, J. Non-Cryst. Solids 203, 88-95 (1996).

*Corresponding author: benino@cc.okayama-u.ac.jp

SURFACE TENSION OF Bi_2O_3-B_2O_3-SiO_2 GLASS MELTS

Chawon Hwang,[1] Bong Ki Ryu,[1,*] and Shigeru Fujino[2]
[1]Department of Materials Science and Engineering, Pusan National University
San 30, Jangjeon-dong, Geumjeong-gu, Busan 609-735, South Korea
[2]Department of Chemical Engineering, Kyushu University
744, Motooka, Nishi-ku, Fukuoka-shi, Fukuoka 819-0395, Japan
*Corresponding author: chawon.hwang@pusan.ac.kr

ABSTRACT

The surface tensions of Bi_2O_3-B_2O_3-SiO_2 melts were measured using the ring method over the temperature range 973 to 1573 K. The compositional and temperature dependences of surface tension were investigated. The surface tension decreased with decreasing SiO_2/B_2O_3 ratio and showed a maximum near 50-60 mol% Bi_2O_3 at 1273 K. Based on the linear relationship between the surface tension and the Bi_2O_3 content in binary silicate melts, the surface tension factor of Bi_2O_3 was calculated and had the value of 206 $mN{\cdot}m^{-1}$ at 1673 K. The surface tension values calculated using the surface tension factor of each single component showed a largest deviation near 50~60 mol% Bi_2O_3 from the measured surface tension values. It was inferred from this result that the deviation may be attributed to the coordination number conversion of BO_3 to BO_4.

INTRODUCTION

The surface tension is an important property of glass melt since it is related to some working conditions of glass production and its application.[1-3] For example, in the process of glass melting, the surface tension plays a great role in controlling the homogeneity of glass melt,[1] the bubble elimination from glass melt,[2] and refractory corrosion.[3] In addition, the compositional dependence of surface tension provides us with important information on the glass melt structure 4. From these points of view, many reports on the surface tension of glass melt have been carried out.[5-7] However, there are few reports on the surface tension of Bi_2O_3-B_2O_3-SiO_2 melts. Based on the surface tension and its compositional dependence, many writers attempted to determine the surface tension factor that is needed in order to predict the surface tension from its composition.[8-10]

The objectives of this study were i) to measure the surface tension of Bi_2O_3-B_2O_3-SiO_2 melts, ii) to determine the surface tension factor of Bi_2O_3, and iii) to investigate the compositional dependence of surface tension.

EXPERIMENTAL PROCEDURE

Sample Preparation

The compositions of samples are given by the symbols in Fig 1. Compositions were selected based on the glass forming region of the Bi_2O_3-B_2O_3-SiO_2 system. There are five series of compositions; Bi_2O_3-SiO_2 and Bi_2O_3-B_2O_3 binary systems, and three ternary Bi_2O_3-xB_2O_3-$(100-x)SiO_2$ series for which the SiO_2/B_2O_3 ratio is 3/1 (x=25), 1/1 (x=50), and 1/3 (x=75). The batch of each sample was prepared by mixing appropriate amounts of reagent grade Bi_2O_3, B_2O_3, and SiO_2. The batch of each composition was melted in Pt crucibles for 30 minutes in air at 1073 ~ 1473K, and was poured on the graphite mold. The samples were pulverized until powder size was reduced to less than 5 mm with an alumina mortar and pestle. Following the above procedure, the preparation of sample for measurement was repeated until the total amount of each sample reached 0.3 ~ 0.4 kg.

Measurement Method

The measurement method of surface tension was the ring method.[11, 12] Details of the apparatus and the measurement procedure have been given elsewhere.[13] After the measurements, the compositions of the samples were confirmed using X-ray fluorescence. The differences between analyzed and batch compositions were within about 1.0 mol% Bi_2O_3. The overall error of this measurement was estimated to be within 1.0% of the surface tension value.

RESULTS AND DISCUSSION

Compositional Dependence of Surface Tension

Surface tensions of Bi_2O_3-B_2O_3-SiO_2 melts at 1273K are shown in Fig. 1. The surface tensions of Bi_2O_3-SiO_2 melts decreased linearly with increasing Bi_2O_3 content. The surface tensions of Bi_2O_3-B_2O_3-SiO_2 melts decreased with decreasing SiO_2/B_2O_3 ratio and showed a maximum near 50-60 mol% Bi_2O_3. The compositional dependence of surface tension can be seen obviously in the iso-surface tension lines in Fig. 2. The surface tensions of Bi_2O_3-B_2O_3- SiO_2 melts are dependent on the content of B_2O_3 at low Bi_2O_3 content (less than 60 mol%) and on the content of Bi_2O_3 at high Bi_2O_3 content at 1273 K. If it compares the surface tensions of single component melts, they lie in the order of B_2O_3 (88.3 $mN{\cdot}m^{-1}$) < Bi_2O_3 (214.8 $mN{\cdot}m^{-1}$) < SiO_2 (280 $mN{\cdot}m^{-1}$) [14] at 1273 K, and the surface tension of B_2O_3 is particularly low. For the low surface tension of B_2O_3, the explanation by Scholze [15] is generally accepted. It states that the low surface tension of the pure B_2O_3 melt is attributed to the parallel arrangement of the triangular planes of three-oxygen-coordinated BO_3 at the surface of the melt. This leads to the weak binding force perpendicular to surface.

The surface tension in the range of 0 to 20 mol% Bi_2O_3 has almost same value to that of pure B_2O_3 melt and this fact can be explained by the two-liquid region in the phase equilibrium diagram of Bi_2O_3-B_2O_3 system [16]. From this fact, it is expected that the upper layer of melt in the two-liquid region consists almost of pure B_2O_3 melt.

Surface Tension Factor of Bi_2O_3 in Silicate Melt

From the linear relationship between surface tension and metal oxide content in silicate melts there are some attempts to calculate the surface tension of glass melt from its composition.[8-10] The basic principle of the surface tension calculation from its composition is to solve simultaneous equations as follows, where γ is surface tension, M_i is molar fraction of the ith component, and F_i is its surface tension factor.

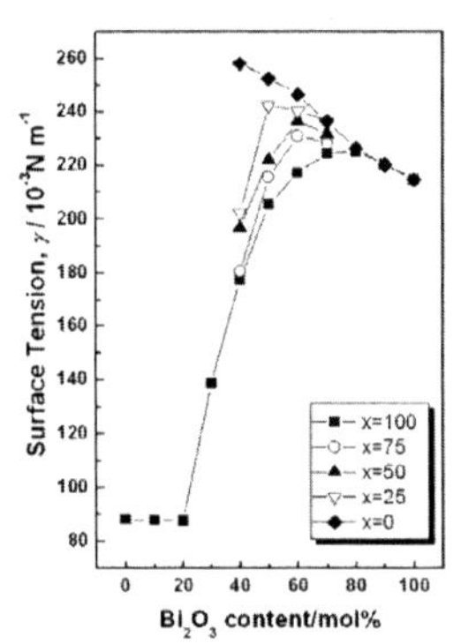

Figure 1. Surface tensions of Bi_2O_3-xB_2O_3-$(100-x)SiO_2$ melts at 1273 K.

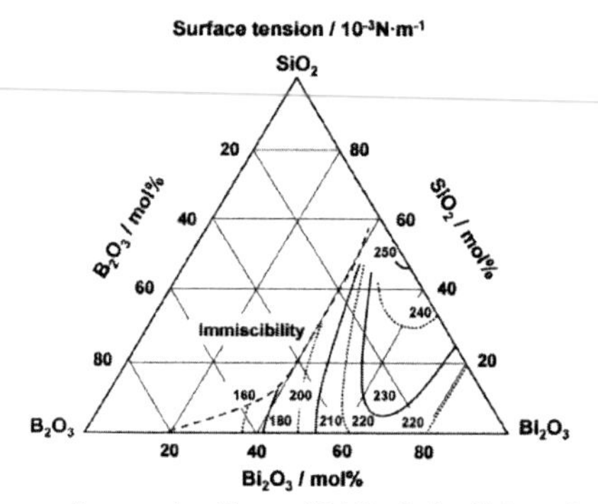

Figure 2. Iso-surface tension lines of Bi_2O_3-B_2O_3-SiO_2 melts at 1273 K.

$$\gamma = M_1F_1 + M_2F_2 + - - - - - + M_iF_i \qquad (1)$$

Boni[10] determined the surface tension factors of general oxides by solving two simultaneous equations similar to Eq. 1, using the surface tension data of their binary silicates (in some cases, ternary silicates or binary borates). The results for other oxides at 1673 K are shown in Fig. 3 and Table I along with the result of this study for Bi_2O_3. The radius of each cation is referred from the report by R. D. Shannon et al.[17] Based on the linear correlation between surface tension and Bi_2O_3 content in Bi_2O_3-SiO_2 binary system, the surface tension factor of Bi_2O_3 was calculated using the method of Boni[10] and had the value of 206 $mN \cdot m^{-1}$ at 1673 K.

Boni[10] classified the surface tension factor of each oxide into two branches. Oxides on one branch of the curve (left part in Fig. 3) are N.W.M. (Network modifier) and the surface tension factor of one oxide is proportional to the ionic potential of its cation. Then, oxides on the other branch of the curve (right part in Fig. 3) show the tendency of N.W.F. (Network former) and the surface tension factor of one oxide is inverse proportional to the ionic potential of its cation. The surface tension factor of Bi_2O_3 deviates from the branch of N.W.M. and shows a smaller value than would be expected from the relatively high ionic potential (2.94 $Å^{-1}$) of Bi^{3+} ion. Based on the explanation of Boni, the small surface tension factor of Bi_2O_3 (and PbO) may be attributed to the possibility of network-forming that is due to its high electronegativity.

The iso-surface tension lines of calculated values (using surface tension factors of Bi_2O_3, B_2O_3, and SiO_2 single component melts) are shown in Fig. 4 (a) and the differences between the calculated and measured values are shown (b), respectively. The differences (deviations from ideality) show a maximum near 50~60 mol% Bi_2O_3 in Bi_2O_3-B_2O_3 binary melts and are dependent on the B_2O_3 and Bi_2O_3 content. These results suggest some relationships between the deviation of surface tension from ideality and the coordination number (C. N.) of boron. In the previous report on the density of Bi_2O_3-B_2O_3 glass melts,[18] it was inferred that the BO_4 fraction goes through a maximum near 40 mol% Bi_2O_3 at 1273 K. Considering the above inference, the deviation of surface tension from ideality seems to be proportional to the BO_4 fraction. Consequently, it was inferred that the deviation may be attributed to the coordination number conversion of BO_3 to BO_4.

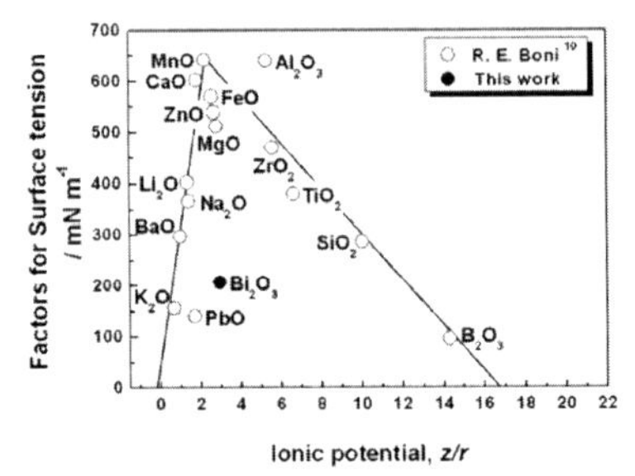

Figure 3. Surface tension factors of general oxides in silicate melts at 1673 K.

Table I. Surface tension factors of general oxides in silicate melts at 1673 K.

Oxide	Al_2O_3	B_2O_3	BaO	CaO	K_2O	Li_2O	MgO	Na_2O
Ionic potential, (z/r)/Å^{-1}	5.26	14.29	1.41	1.79	0.66	1.35	2.78	0.98
Factor /10^{-3}N·m^{-1}	640	96	366	602	156	403	512	297
Oxide	**MnO**	**FeO**	**PbO**	**SiO_2**	**TiO_2**	**ZnO**	**ZrO_2**	**Bi_2O_3**
Ionic potential, (z/r)/Å^{-1}	2.23	2.56	1.69	10.00	6.61	2.67	5.56	2.94
Factor /10^{-3}N·m^{-1}	641	570	140	286	380	540	470	206.8

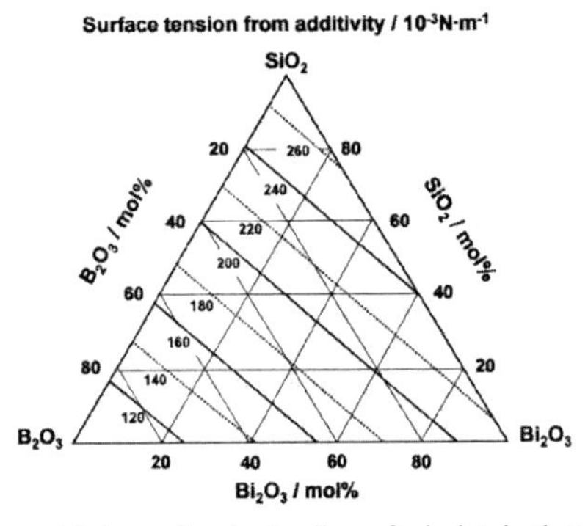

(a) Iso-surface tension lines of calculated values (b) using surface tension factors at 1273 K.

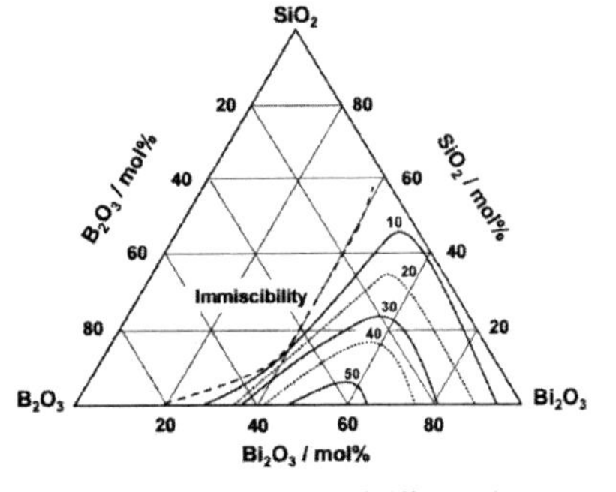

(b) Iso-surface tension lines of difference between calculated values and measured values at 1273 K.

Figure 4. Iso-surface tension lines of Bi_2O_3-B_2O_3-SiO_2 melts at 1273 K.

CONCLUSION

The surface tensions of Bi_2O_3-B_2O_3-SiO_2 melts were measured systematically over a 973 to 1573 K temperature range using the ring method. The composition dependence of the surface tension in Bi_2O_3-B_2O_3-SiO_2 melts was discussed. The following conclusions can be drawn from this work.

- Compositional dependence of surface tension

The surface tension decreased with decreasing SiO_2/B_2O_3 ratio and showed a maximum near 50-60 mol% Bi_2O_3. It was found from the iso-surface tension lines that the surface tensions are dependent on the content of B_2O_3 at low Bi_2O_3 content (less than 60 mol%) and on the content of Bi_2O_3 at high Bi_2O_3 content (over 60 mol%) at 1273 K.

- Surface tension factor of Bi_2O_3 in silicate melt

Based on the linear relationship between the surface tension and the Bi_2O_3 content in binary silicate melts, the surface tension factor of Bi_2O_3 was calculated and had the value of 206 $mN{\cdot}m^{-1}$ at 1673 K. The small surface tension factor of Bi_2O_3 seems to be attributed to the possibility of network-forming by Bi_2O_3. The surface tension values calculated using the surface tension factor of each single component showed the largest deviation near 50~60 mol% Bi_2O_3 from the measured surface tension values. It was inferred from this result that the deviation may be attributed to the BO_4 fraction and the C. N. conversion of BO_3 unit to BO_4 unit increases the surface tension of its melt.

ACKNOWLEDGEMENTS

Supported by Japanese Government (*Ministry of Education, Culture, Sports, Science and Technology, Japan*) Scholarship in 2001-2004 and the Brain Korea 21 Program in 2011.

REFERENCES

[1]W.B. Silverman, Surface Tension of Glass and Its Effect on Cords, *J. Am. Ceram. Soc.*, **25** 168-173 (1942).

[2]M.C. Weinberg, Surface Tension Effects in Gas Bubble Dissolution and Growth, *Chem. Eng. Sci.*, **36** 137-141 (1981).

[3]Z. Yu et al, Relation between Corrosion Rate of Magnesia Refractories by Molten Slag and Penetration Rate of Slag into Refractories, *J. Ceram. Soc., Jpn.*, **101** 533-539 (1993).

[4]T.B. King, XVI.-The Surface Tension and Structure of Silicate Slags, *Journal of the Society of Glass Technology*, **35** 241-259 (1951).

[5]L. Shartsis and A.W. Smock, Surface Tensions of Some Optical Glasses, *J. Am. Ceram. Soc.*, **30** 130-136 (1947).

[6]H. Ito, T. Yanagase, and Y. Suginohara, On the Surface Tension of Molten Lead Silicate, *Journal of the Mining and Metallurgical Institute of Japan*, **77** 895-898 (1961).

[7]S. Toyoda, S. Fujino, and K. Morinaga, Density, Viscosity and Surface Tension of $50RO$-$50P_2O_5$ (R: Mg, Ca, Sr, Ba, and Zn) Glass Melts, *J. Non-Cryst. Solids*, **321** 169-174 (2003).

[8]A. Dietzel, Relation between the Surface Tension and the Structure of Molten Glass, *Kolloid-Z.*, **100** 368-80 (1942).

[9]K.C. Lyon, Calculation of Surface Tensions of Glasses, *J. Am. Ceram. Soc.*, **27** 186-189 (1944).

[10]R.E. Boni and G. Derge, Surface Tensions of Silicates, *Transactions AIME*, **206** 53-59 (1956).

[11]H. Sasaki et al., Surface Tension Variation of Molten Silicon Measured by the Ring Method, *Jpn. J. Appl. Phys.* **34**, 414-418 (1955).

[12]W.D. Harkins and H.F. Jordan, A Method for the Determination of Surface and Interfacial Tension from the Maximum Pull on a Ring, *J. Am. Chem. Soc.*, **52**, 1751-1772 (1930).

[13]C. Hwang, S. Fujino, and K. Morinaga, Surface tension of Bi_2O_3-B_2O_3 glass melts, *J. Chem Soc. Jap an*, **112**, 1200-1205 (2004).

[14]W.D. Kingery, Surface Tension of Some Liquid Oxides and Their Temperature Coefficients, *J. Am. Ceram. Soc.*, **42** 6-10 (1959).

[15]H. Scholze, Nature and Structure of Glass-3.7 Surface Tension, pp. 322 in *Glass: Nature, Structure, and Properties*, Translated by M. J. Lakin. Springer-Verlag, New York, 1991.

[16]E.M. Levin and C.L. McDaniel, The System Bi_2O_3-B_2O_3, *J. Am. Ceram. Soc.*, **45** 355-360 (1962).

[17]R.D. Shannon and C.T. Prewitt, Effective ionic radii in oxides and fluorides, *Acta Crystallogr.*, **B25** 925-946 (1969).

[18]C. Hwang, S. Fujino, and K. Morinaga, Density of Bi_2O_3-B_2O_3 glass melts, *J. Am. Ceram. Soc.*, **87** **[9]**, 1677-1682 (2004).

OXIDATION BEHAVIOR OF NITROGEN RICH *AE*-Si-O-N GLASSES (*AE* = Ca, Sr, Ba)

Sharafat Ali and Bo Jonson
School of Engineering, Glass Group
Linnæus University
SE-351 95 Växjö, Sweden

ABSTRACT

AE-based silicon oxynitride glasses (AE = Ca, Sr, Ba) with high nitrogen content have been synthesized using AE hydrides as primary precursors. The oxidation behavior of AE-Si-O-N glasses in ordinary atmosphere at different temperature has been investigated. These glasses react with air oxygen when heated just above the glass transition temperatures. The oxidation starts with bubble formation on the surface and continued oxidation leads to formation of a white layer on the surface. The oxidation of AE-Si-O-N glasses involves concurrently ongoing inward diffusion of oxygen and outward diffusion of AE elements and nitrogen, resulting in compositional gradient. EDX analysis showed substantial enrichment in AE content at the surfaces of the oxidized layer.

INTRODUCTION

Oxynitride silicate glasses are a branch of high performance glasses, obtained by incorporating nitrogen atoms into oxide glass networks of *e.g.* silicates, borates and phosphates. The majority of studied oxynitride glasses are silicate based, often also containing Al or, less frequently, B. The addition of nitrogen to silicate glass networks causes profound changes in a number of physical properties. Density, glass transition and crystallization temperatures, micro-hardness, Young's modulus, shear modulus, fracture toughness, viscosity, dielectric constant and refractive index increase with increasing nitrogen content, while the thermal expansion coefficient decreases with increasing nitrogen content.[1-19] Most of these changes are attributed to an increased cross-linking of the network due to the presence of three-coordinated nitrogen atoms.

The unique properties of silicon oxynitride and Sialon glasses have led to a search for areas of application. Commercial applications are at present held back by high cost, relatively difficult preparative procedures, lack of transparency and poor oxidation resistance. Potential areas of application that have been proposed include high elastic modulus glasses for computer hard discs, ceramic seals, coatings on metals, encapsulation of nuclear waste for long term storage, joining of structural ceramics, high electrical resistivity coatings for use at high temperatures and glass fibers.[11-20] Oxynitride glasses can potentially be used in special windows where their higher elastic moduli would allow them to remain stiff in thinner sections, thus allowing weight and energy savings. Oxynitride glasses may find applications due to their high refractivity indices, for the design of new miniaturized lenses and prisms, as hosts for luminescent ions or, when containing large amounts of rare earth elements, as Faraday rotators.[20]

Oxynitride glasses have mostly been obtained by conventional melting of mixtures of glass modifier metal oxides, SiO_2 and Si_3N_4. Recent studies have shown that La-Si-O-N,[21] Ca-Si-O-N[8] and Sr-Si-O-N[9] glasses can be formed over wider compositional ranges, by using the electropositive elements in their metallic state or as metal hydrides as batch materials. The present oxynitride glasses in AE-Si-O-N systems were prepared by using AEH_2 as the primary AE source. These glasses retain high amounts of the electropositive elements and nitrogen and exhibit very high values of glass transition temperature, micro-hardness and refractive index.

One disadvantage of oxynitride glasses is their tendency to oxidize at temperatures above T_g, e.g. an Y-Si-Al-O-N glass with 20 e/o N (N content in equivalents; e/o =3[N]/(3[N]+2[O]), where [N] and [O] are the atomic concentrations of N and O respectively), yielded upon heat treatment in air at

1200°C for 20 min, a foamy white product with a soft surface.[22] The susceptibility towards oxidation is dependent on the composition of the glass. Wusirika[23] surveyed the oxidation behaviors for different M-Si-Al-O-N glasses. For the modifiers M = Li, Mg and Mn, coherent, crystalline oxide layers of ca. 0.06 – 0.1mm thickness had formed after 24 h at 1000°C. The susceptibility to oxidation is affected by the presence of water vapor in the air, which reacts with bridging Si-O-Si bonds at the glass surface to produce ≡Si-OH groups, thus lowering the glass viscosity and facilitating the diffusion of oxygen into the glass. The aim of this work is to investigate the oxidation behavior of the nitrogen rich glasses in AE-Si-O-N systems.

EXPERIMENTAL

Oxynitride glasses in the *AE*-Si-O-N systems were prepared from mixtures of AEH_2 (≥ 96 % metal basis, Alfa Aesar GmbH & Co), Si_3N_4 (ChemPur GmbH), and SiO_2 (99.9%, ABCR GmbH & Co) powders. All chemicals were stored in a glove box compartment under Ar atmosphere to avoid exposure to air. The powders were weighed inside the glove box and then pressed into pellets. Six gram batches for each composition were placed in niobium crucibles, covered with parafilm to avoid air contact during the transport between the glove box and the furnace. The mixtures were melted in niobium crucibles in nitrogen atmosphere at temperatures between 1500 to 1700°C. Details concerning synthesis and additional characterizations are given in[8] and in.[9]

The oxidation studies of nitrogen rich glasses in the AE-Si-O-N systems were carried out in a radio frequency furnace, with using sample size of (3×3×3 mm^3). Samples were heat treated up to 1400°C in air in a Pt crucible. Specimens were examined by light microscope (Olympus PMG3, Japan) and scanning electron microscopy (SEM; JEOL JSM - 7000F, Tokyo Japan) equipped with an energy dispersive spectrometry (EDS; Link Isis, Oxford Instruments, Buckinghamshire, UK).

RESULTS AND DISCUSSION

Not unexpected, the reduced oxynitride glasses are thermally less stable in the presence of oxygen, i.e. more prone to oxidize, than their oxygen-rich counterparts. Their oxidation reaction limits high temperature uses of oxynitride glasses in air. In order to explore the oxidation behavior for the present high-nitrogen AE-based glasses, 5 selected compositions (three in Ca, one in Sr, and one in Ba systems) having varying amount of nitrogen content, were heat treated up to 1400°C in air in a Pt crucible. Measured physical properties for the glasses are given in Table I.[8-10]

Table I. Glass designation, determined glass composition, contents of AE and N in equivalent % (e/o), density (ρ), glass transition temperatures (T_g) and crystallization temperatures (T_c).

Glass ID	Glass composition	*AE*/ e/o	N/ e/o	ρ/ g/cm^3	T_g / ^{0}C	T_C/ ^{0}C
Ca	$Ca_{9.94}Si_{10}O_{27.73}N_{8.14}$	33.2	40.0	3.02	948	1054
Ca 1	$Ca_{12.90}Si_{10}O_{20.93}N_{7.98}$	39.2	36.4	3.02	912	1040
Ca 2	$Ca_{11.04}Si_{10}O_{13.21}N_{11.89}$	35.6	58.0	3.24	1010	1170
Sr	$Sr_{7.66}Si_{10}O_{20.27}N_{4.93}$	27.6	26.3	3.59	883	1020
Ba	$Ba_{9.5}Si_{10}O_{21.25}N_{5.56}$	32.2	28.2	3.92	845	1017

No noticeable oxidation was observed below 910°C for the Ca-based oxynitride glasses, since they have glass transition temperatures > 910°C. For all of the investigated glasses, severe oxidation occurred at temperatures above the glass transition temperature (Fig. 1).

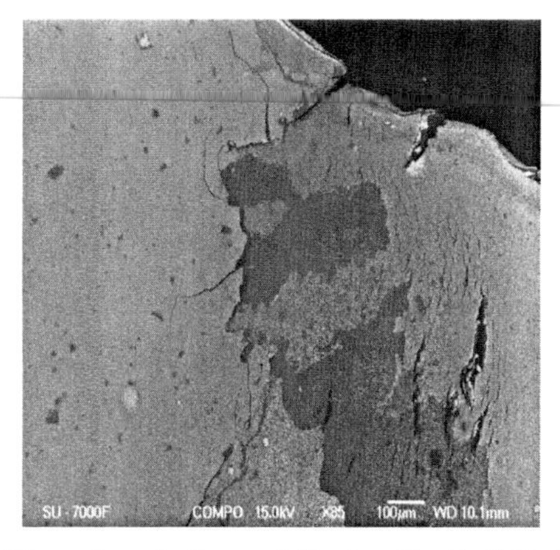

Figure 1: SEM image of the surface of a Ca oxynitride glass (Ca 1 in Table I) after oxidation in air.

After 30 minutes at 1050°C, the surfaces of Ca and Ca1 and at 1100°C the surfaces of Ca 1, became porous and contained bubbles, evidently from gas evolution. Similar observations were reported by Wusirika and others.[23-27] The presence of the bubbles led to the formation of an irregular interface and was followed by the formation of a white oxide layer. Continued heating resulted in a collapse of the bubbles and the formation of a thick foamy oxide. An increase in temperature favored the formation of bubbles and increased the porosity of the oxidized layer, as shown in Fig. 2, which facilitates further oxygen access. For longer times of treatment, the oxidized layer appears either as big porous bulges or a smoother scale with severe cracks and pores.

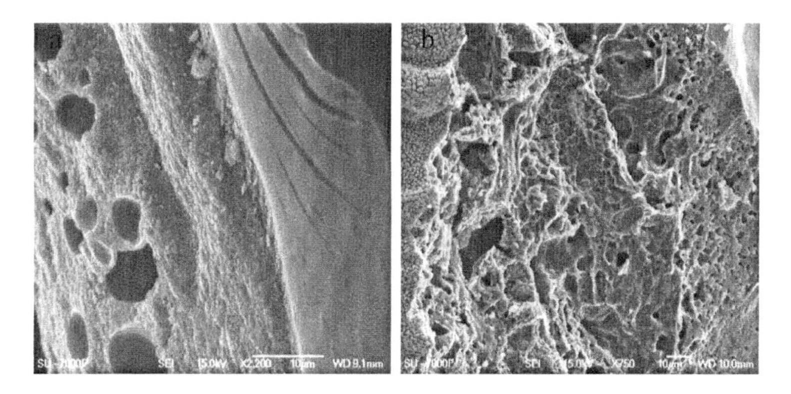

Figure 2: SEM images showing the surface morphology for a Ca oxynitride glass (Ca 1 in Table I) after oxidation in air at (a) 1100°C and (b) 1200°C.

During high temperature treatment, around 1400°C, the platinum crucible reacted with the glass and formed platinum crystals on the glass surface. The SEM observations suggest that the glass is very soft at this temperature and is squeezed out in some parts of the sample to form the bulges. The oxidation of Ca-based oxynitride glasses involves concurrently ongoing inward diffusion of oxygen and outward diffusion of Ca cations and nitrogen products, resulting in compositional gradient. EDX analysis showed substantial enrichment in Ca content at the surfaces of the oxidized layer. For example sample Ca 1 (Table I), having a Ca:Si ratio of 82:28 at%, in comparison with 50:50 at% within the bulk. The oxidation products, for Ca 1, identified via their X-ray powder diffraction patterns indicates the presence of cristobalite (SiO_2) and wollastoniote ($CaSiO_3$) are the main crystalline phases detected in the surface of oxide scale.

Similar oxidation behavior was found for the Sr and Ba containing oxynitride glasses. The behavior of the glasses during heat treatment in air was strongly temperature dependent. Below 830°C, no noticeable oxidation was noticed for both Sr and Ba-based oxynitride glasses, since both glasses have glass transition temperature > 840°C. Oxidations were noticed at temperatures above the glass transition temperature at 860 and 900°C for Sr and Ba containing glasses respectively. After 30 minutes at 1000°C, the surfaces of Sr and Ba containing glasses became porous and bubbles appeared. With increasing temperatures a disorder in the glass structure is induced. This disorder is followed by a greater average number of non-bridging anions (oxygen and nitrogen) per polyhedron which decreases the glass viscosity. EDX analysis showed substantial enrichment in Sr/Ba content at the surfaces of the oxidized layer. The results for glasses of varying nitrogen content on the oxidation resistance of AE-Si-O-N systems show that increasing the incorporation of nitrogen into the glass composition increases the oxidation resistance for a given temperature.

CONCLUSIONS

Nitrogen rich AE-Si-O-N glasses oxidize in air when they are heated just above their glass transition temperatures. The oxidation starts with bubble formation on the surface and involves concurrently ongoing inward diffusion of oxygen and outward diffusion of AE and nitrogen, resulting in compositional gradient. EDX analysis showed substantial enrichment in AE content at the surfaces of the oxidized layer.

REFERENCES

[1]R.A.L. Drew, S. Hampshire, and K.H. Jack, "Nitrogen glasses". Proc. Br. Ceram. Soc., 1981(31): p. 119-132.

[2]S. Hampshire, R.A.L. Drew, and K.H. Jack, "Oxynitride Glasses". Phys. Chem. Glasses., 1985. **26**(5): p. 182-186.

[3]R.E. Loeham, "Oxynitride Glasses". J. Non-Cryst. Solids., 1980. **42**(1-3): p. 433-445.

[4]R.E. Loeham, "Preparation and Properties of Oxynitride Glasses". J. Non-Cryst. Solids., 1983. **56**(1-3): p. 123-134.

[5]S. Sakka, "Oxynitride Glasses". Annu. Rev. Mater. Sci., 1986. **16**: p. 29-46.

[6]S. Sakka, "Structure, Properties and Application of Oxynitride Glasses". J. Non-Cryst. Solids., 1995. **181**(3): p. 215-224.

[7]S. Hampshire, "Oxynitride glasses and their properties - an overview". Key Eng Mat, 2003. **247**: p. 155-160.

[8]A. Sharafat, J. Grins, and S. Esmaeilzadeh, "Glass-forming region in the Ca-Si-O-N system using CaH_2 as Ca source". J. Eur. Ceram. Soc., 2008. **28**(14): p. 2659-2664.

[9]A. Sharafat, B. Forslund, J. Grins, and S. Esmaeilzadeh, "Formation and properties of nitrogen-rich strontium silicon oxynitride glasses". J. Mater. Sci., 2009. **44**(2): p. 664-670.

[10]A. Sharafat, J. Grins, and S. Esmaeilzadeh, "Properties of high nitrogen content mixed alkali earth oxynitride glasses $(AE_{(x)}Ca_{(1-x)})_{1.2(1)}SiO_{1.9(1)}N_{0.86(6)}$, AE = Mg, Sr, Ba". J. Non-Cryst. Solids., 2009. **355**(22-23): p. 1259-1263.

[11]A. Sharafat, J. Grins, and S. Esmaeilzadeh, "Hardness and refractive index of Ca-Si-O-N glasses". J. Non-Cryst. Solids., 2009. **355**(4-5): p. 301-304.
[12]D.R. Messier and A. Broz, "Microhardness and Elastic-Moduli of Si-Y-Al-O-N Glasses". J. Am. Cer. Soc., 1982. **65**(8): p. C123-C123.
[13]D.N. Coon, J.G. Rapp, R.C. Bradt, and C.G. Pantano, "Mechanical-Properties of Silicon-Oxynitride Glasses". J. Non-Cryst. Solids., 1983. **56**(1-3): p. 161-166.
[14]J. Homeny and D.L. Mcgarry, "Preparation and Mechanical-Properties of Mg-Al-Si-O-N Glasses". J. Am. Cer. Soc., 1984. **67**(11): p. C225-C227.
[15]D.N. Coon and J.R. Weidner, "Elastic-Moduli of Y-Al-Si-O-N Glasses". J. Non-Cryst. Solids., 1990. **116**(2-3): p. 201-205.
[16]S. Hampshire, E. Nestor, R. Flynn, J.L. Besson, T. Rouxel, H. Lemercier, P. Goursat, M. Sebai, D.P. Thompson, and K. Liddell, "Yttrium Oxynitride Glasses - Properties and Potential for Crystallization to Glass-Ceramics". J. Eur. Ceram. Soc., 1994. **14**(3): p. 261-273.
[17]I.M. Peterson and T.Y. Tien, "Thermal-Expansion and Glass-Transition Temperatures of Y-Mg-Si-Al-O-N Classes". J. Am. Cer. Soc., 1995. **78**(7): p. 1977-1979.
[18]F. Lofaj, R. Satet, M.J. Hoffmann, and A.R. de Arellano Lopez, "Thermal expansion and glass transition temperature of the rare-earth doped oxynitride glasses". J. Eur. Ceram. Soc., 2004. **24**(12): p. 3377-3385.
[19]Hakeem, A.S., J. Grins, and S. Esmaeilzadeh, "La-Si-O-N glasses - Part II: Vickers hardness and refractive index". J. Eur. Ceram. Soc., 2007. **27**(16): p. 4783-4787.
[20]Sharafat. A, "Preparation, characterization and properties of nitrogen rich glasses in alkaline earth-Si-O-N systems". Stockholm University, Ph.D thesis, 2009.
[21]A.S. Hakeem, J. Grins, and S. Esmaeilzadeh, "La-Si-O-N glasses - Part I. Extension of the glass forming region". J. Eur. Ceram. Soc., 2007. **27**(16): p. 4773-4781.
[22]T. Das, "Oxynitride glasses - An overview". Bull. Mater. Sci., 2000. **23**(6): p. 499-507.
[23]R.R. Wusirika, "Oxidation Behavior of Oxynitride Glasses". J. Am. Cer. Soc., 1985. **68**(11): p. C294-C297.
[24]C. O'Meara, G.L. Dunlop, and R. Pompe, "Formation, crystallisation and oxidation of selected glasses in the Y-Si-Al-O-N system". J. Eur. Ceram. Soc., 1991. **8**(3): p. 161-170.
[25]M. Sebaï, C. Penot, P. Goursat, K. Liddell, D.P. Thompson, E. Nestor, R. Ramesh, and S. Hampshire, "Oxidation resistance of Nd-Si-Al-O-N glasses and glass-ceramics". J. Eur. Ceram. Soc., 1998. **18**(2): p. 169-182.
[26]M. Desmaisonbrut and J. Desmaison, "The Reactivity in Oxygen of 2 Ca-Si-Al-O-N, Mn-Si-Al-O-N Oxynitride Glasses". Solid State Ionics, 1988. **26**(2): p. 153-153.
[27]M. Desmaisonbrut, J. Desmaison, and P. Verdier, "Oxidation Behavior of an Oxynitride Glass in the System Ca-Si-Al-O-N". J. Non-Cryst. Solids., 1988. **105**(3): p. 323-329.

*Corresponding author: sharafat.ali@lnu.se

PART B: STRUCTURE, PROPERTIES, AND PHOTONIC APPLICATIONS OF GLASS

TELLURIUM OXIDE THIN FILM WAVEGUIDES FOR INTEGRATED PHOTONICS

Khu T. Vu and Steve J. Madden
Laser Physics Centre, Research School of Physics and Engineering
The Australian National University
Canberra, Australia

ABSTRACT

We report the fabrication of Tellurium oxide films with deposited propagation losses that are sufficiently low to allow the fabrication of high quality planar integrated optical devices. Stoichiometric low loss films were produced by reactive magnetron RF sputtering and the effect of a wide range of sputtering parameters on stoichiometry, refractive index, Raman spectra and optical loss of thin films were investigated. As deposited TeO_2 films with propagation losses below 0.1dB/cm at 1550nm have been achieved which are therefore well suited for integrated optical applications.

INTRODUCTION

Tellurite glasses, with Tellurium dioxide (TeO_2) as the main component, have been intensively studied on account of their attractive properties, in particular for applications in optical amplification.[1] These glasses are transparent from around 300nm in the UV to ~6μm in the Mid infra-red; have high acousto-optic figures of merit (20-27x10^{-15} s^3kg^{-1} or ~3 times that of quartz crystals); the largest optical nonlinearity of any oxide glass at up to 100 times that of silica;[2] high linear refractive indices (up to 2.2 at 1.55μm) and low material dispersion. Pure TeO_2 glass has zero dispersion near 1.7μm; more complex tellurite glasses have zero dispersion beyond 2μm,[3] and the zero dispersion region in a waveguide device can therefore easily be tuned to the telecommunication window.[3] Tellurite glasses have large Raman shifts (up to 1200cm^{-1} in some compositions)[4] and large Raman gain coefficients of order of 50 times that of silica.[5] Tellurite glasses are also very efficient hosts for many rare earth ions such as Er^{3+}, Tm^{3+} and Ho^{3+} which are crucial for active components in integrated optics.[6] Furthermore, Tellurite glasses are thermally and mechanically stable. The glass transition temperature is typically above 300°C and the melting point above 600°C making them highly suitable for device fabrication and processing. This retinue of desirable properties has made tellurites of considerable interest for integrated optical applications, but there has not until now been a demonstration of the fabrication of suitably low loss films to enable planar devices which can take advantage of these attractive characteristics.

Thin films of amorphous Tellurium oxide have been produced by several methods including chemical and sol-gel;[7,8] large dose ion implantation;[9] thermal evaporation;[10,11] reactive radio frequency (RF) sputtering using pure Tellurium;[12,13] and pulsed laser deposition.[14,15] The previous works on Tellurite thin films have not reported low propagation losses at 1.55μm for useful functional planar applications.

In this paper, we report for the use of reactive RF sputtering to fabricate as-deposited films with the required stoichiometry and optical propagation losses below 0.1dB/cm, low enough to enable high quality integrated optics devices. In doing so, we explored a wide range of sputtering parameters using the design of experiment (DOE) method to obtain the desired stoichiometry, high index, and low loss planar waveguides. The obtained films from these series of runs had O/Te ratios ranging from 1 to 3.5. The refractive index variation, band gaps, Raman spectra, and planar waveguide propagation losses at 1550nm against stoichiometry are reported.

FABRICATION AND CHARACTERIZATION METHODS

TeO_x thin films were produced by reactive magnetron RF sputtering from a pure (99.95%) sintered Tellurium target in O_2/Ar plasma. The RF magnetron sputtering chamber had a load lock and was usually pumped to base pressures below $5x10^{-7}$ Torr. The RF gun was cooled with water at 20°C and the rotating substrate holder was placed at a distance of ~10cm from the target and was not actively cooled.

The effects of several sputtering parameters were investigated to obtain suitable thin films. The sputtering parameters that were investigated were: chamber pressure (2.0mTorr to 20mTorr), RF power (120W to 360W), and percentage of oxygen in the gas mix (20% to 80%) - the total flow of O_2 and Ar was kept at 15sccm. This flow was chosen as it was the maximum possible while allowing chamber pressures across the entire desired range (limited by the turbomolecular pump's pumping speed). The thin films produced from this investigation had thickness in the range 500nm to 2μm. Since there were 3 varying parameters and several optimization targets, design of experiment (DOE) software (D.O.E Fusion from S-Matrix), was used to screen the optimum experimental conditions and find the best sputtering conditions under which the films were stoichiometric, dense and low loss.

Once the conditions close to optimum were identified from the DOE, a series of TeO_x films was deposited at various O_2 flows with all other parameters fixed at the optimum conditions. A number of different substrates were used in each run including 4" thermally oxidised silicon wafers (2 microns of oxide), Silicon pieces, silica slides and 200nm thick silicon nitride membranes on silicon (back etched to make a very thin window for Raman measurements).

The films obtained were characterized using EDXA (Jeol 6400 SEM) and RBS (alpha particle at 2MeV) for composition; by Raman scattering and X-ray diffraction; transmissive spectroscopy to determine the Tauc band gap; with a dual angle spectroscopic reflectometer (SCI Filmtek 4000) for thickness and refractive index, and by a Metricon prism coupler for optical propagation loss at 1.55μm. Using the m-line technique, thin film propagation loss was measured though prism coupling where EDFA amplified laser light was coupled to the propagation mode of the waveguides by a prism in contact with the films. The TE fundamental modes were chosen for the optical loss measurement for consistency between different wafers. The optical loss was measured by processing of the light streak images captured by a low noise InGaAs NIR camera. 100mm diameter wafers were used to produce a streak of light with lengths up to 80mm to provide the capability of measuring below 0.1dB/cm.

The Raman spectra were obtained using a backscatter configuration with a pump laser at 808nm. Since the Tellurium oxide films for this characterization were deposited on very thin silicon nitride membranes, the substrate background signal was effectively suppressed leaving only the unobscured Raman signal from the tellurium oxide film.

TELLURIUM OXIDE THIN FILM OPTICAL PROPERTIES

Refractive Index, Band Gap and Composition

Figure 1a summarizes the dependence of the refractive indices at 1.55μm of the sputtered films against the ratio of oxygen to tellurium in the films as determined by EDXA. Each data point corresponds to one sputtering run. As a wide range of sputtering parameters was used during film deposition the obtained films had a range of optical and physical properties. The as-deposited amorphous TeO_x films were classified into three types: tellurium rich ($x<2$); stoichiometric ($x\sim2$); and oxygen rich ($x>2$). Tellurium rich films were produced when there was an oxygen deficiency in the chamber, or a high flux of tellurium from the target. This condition occurred at high RF power, low oxygen flow or low chamber pressure. Tellurium rich films had excessive levels of metallic Te atoms and were therefore highly absorbing. These films had an effective band edge high up in the red or infrared and refractive indices at 1.55μm greater than 2.1. Oxygen rich films on the other hand were highly transmissive down to ~350nm, had low absorption in the infrared as well as lower density and refractive index at 1.55μm below 2.1. These films were produced when the sputtering power was low, or the oxygen flow or chamber pressure were high. Stoichiometric films were obtained when the balance of all three parameters was maintained which corresponded to 5mTorr pressure; 150W RF power; 6.75sccm oxygen flow (45%) and 8.25sccm argon flow (55%).

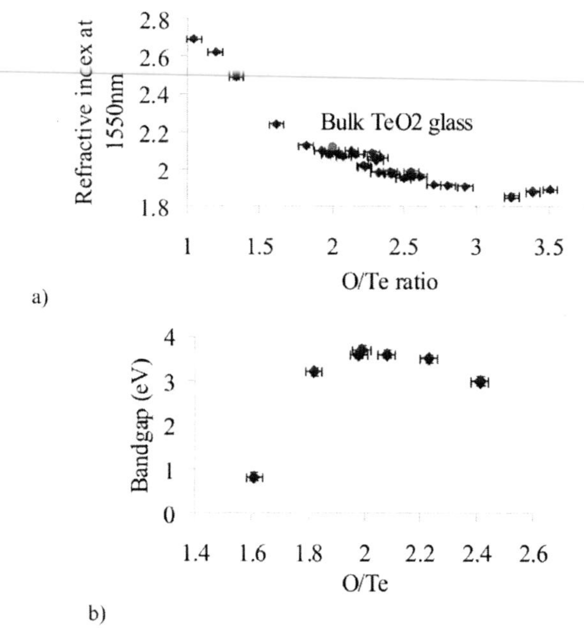

Figure 1. a) Dependence of refractive indices at 1550nm on composition. The index also depends on the sputtering conditions. The dots with error bars are data points; the isolated dot is represents bulk TeO_2 glass (taken from Kim[16] and Gosh[17]); b) Variation of optical band gaps with composition.

The results of Figure 1a represent a much wider range of compositions and refractive indices than reported by previous work.[13] The graph shows that the refractive indices of films around TeO_2 composition are very close to bulk TeO_2. At an O/Te ratio of 2.0, the film refractive indices are as high as 2.08 compared to bulk amorphous TeO_2 glass of 2.12.[1,16,17] Furthermore, the slope of the composition/index relationship appears to flatten in the vicinity of the stoichiometric point. The refractive indices remain almost constant for O/Te higher than 2.6. This is slightly different to the results from[13] which reported a constant index of around 1.98 (at 1500nm) for O/Te ratios between 2.3 and 2.6. The thin films obtained here have higher refractive indices and are therefore denser than previously reported according to the Lorentz-Lorenz relation.[18]

In addition, the refractive index does not only depend on the stoichiometric coordination but also the sputtering conditions. The cluster around the O/Te ratio of 1.9-2.4 on Figure 1a clearly supports this case. Higher index films were usually obtained with lower oxygen flow, therefore, higher deposition rate. This indicates that higher deposition rate increases the density of the films leading to higher refractive indices.

The stoichiometric film, which has the refractive index close to the value of the bulk TeO_2 glass of 2.12,[16,17] of TeO_2 were obtained when the oxygen flow percentage was around 45% of the total flow. The index difference from bulk was around 0.04 at 1550nm. One would expect from these results that the optical properties of the films should be very similar to that of the bulk glass. Figure 2b shows the variation of optical band gaps inferred from Tauc plots of $(\alpha E)^{1/2}$ vs E, where α is absorption and E is photon energy. The band gaps reach a maximum of 3.7eV at stoichiometric composition. This is closely approaching the value of bulk TeO_2 glass of 3.79eV.[1] With Te rich films, the optical band gaps decrease very rapidly due to metallic absorption. On the other hand Oxygen rich films retain relatively high optical band gaps.

Raman Spectra

The Raman spectra of the films are shown in Figure 2. There are some significant trends in the result. For thin films with O/Te ≥ 2, there are three broadened peaks at: 770cm^{-1} (stretching vibration of Te=O), 660cm^{-1} (coupled symmetric vibration along Te-O-Te axes), and 450cm^{-1} (symmetric and bending vibration of Te-O-Te linkage at corner sharing sites). The spectra for this case are in agreement with those reported in literature.[1,19,20] The spectra for films with O/Te ≤ 2 are remarkably similar to that of the α-TeO2 crystal with relatively narrow peaks at 660cm^{-1} and 400cm^{-1}. However, X-ray diffraction measurements showed no indication of crystallization in any of the as deposited films. In the crystal case, the two peaks at 150 and 120cm^{-1} are due to vibrational modes of trigonal bi-pyramid TeO_4 units.[20] However, in Tellurium rich films, the peaks between 100 and 200cm^{-1} are dominated by the vibrations of Te-Te bond.[21]

In a pure TeO_2 crystal and in Te rich films, TeO_4 bi-pyramid structures and Te-Te bonds dominate. TeO_4 units are linked by corner or edge sharing oxygen atoms resulting in very similar Raman spectra. With the addition of oxygen, the TeO_4 trigonal bi-pyramid units evolve into TeO_3 trigonal pyramid resulting in non-bridging oxygen in Te=O bonds with strong stretching vibration at 770cm^{-1}.

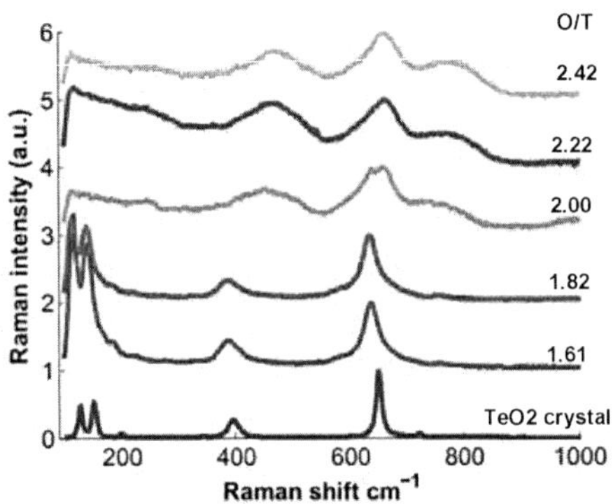

Figure 2. Raman spectra of TeOx thin films, x value is on the right side of the graph. The curves have been normalized based on the main peaks between 600-700cm^{-1} and shifted upward by 1 unit consecutively.

Thin Film Propagation Loss

The propagation losses at 1550nm were determined from observation of the light streaks from a set of as-deposited TeO_x thin films on thermally oxidized silicon substrates with film thickness of around 1.5μm and composition O/Te ratios of 1.6 to 2.4. The results are shown in Figure 3. The minimum loss observed was less than 0.1dB/cm at 1550nm, which is the first time such low losses have been reported in this material and is even more remarkable given this is an as-deposited tellurium oxide film. As mentioned above, an excessive level of Te in the films, $x<2$, is expected to produce high losses and this is indeed observed. However, the loss curve dips to a minimum right at the stoichiometric point before gradually increasing as the oxygen content increases. The oxygen rich thin films are relatively more porous and inhomogeneous than TeO_2. This leads to an increase in loss in oxygen rich films.[12,18]

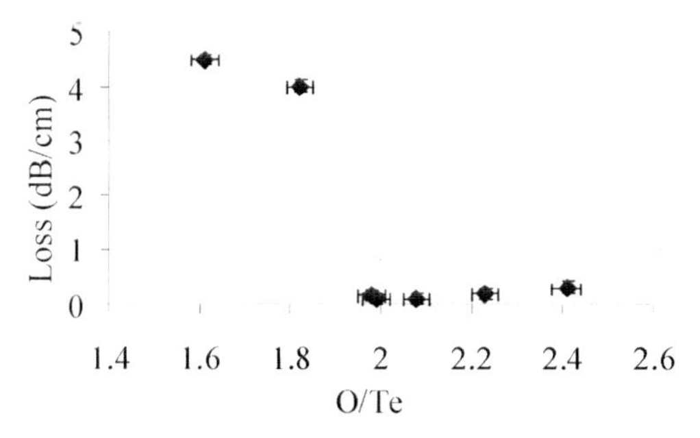

Figure 3. 1550nm propagation loss at various O/Te ratios.

CONCLUSION

We have shown for the first time that reactive RF sputtering of tellurium in O_2 and Ar gases can produce TeO_2 films with optical propagation losses less than 0.1dB/cm at 1550nm in as-deposited films. This level of performance is a requirement to obtain high quality integrated optics waveguide devices in tellurite glasses, a goal that is being pursued by a number of groups over the last 5 years. A wide range of sputtering conditions was investigated to achieve stoichiometric films with high refractive indices and low optical propagation loss.

ACKNOWLEDGEMENT

The authors acknowledge Prof. Rob Elliman at the Department of Electronic Materials Engineering at the School of Physics and Engineering at the Australian National University for performing RBS measurement, and Dr. Frank Blink at Electronic Microscopic Unit at the Australian National University for helping with EDX analysis. This work is financially supported by the Australian Research Council under Discovery Project Grant DP070333.

REFERENCES

[1]R. A. H. El-Mallawany, *Tellurite glasses handbook: physical properties and data.* (CRC Press, US, 2002).

[2]B. Jeansannetas, S. Blanchandin, P. Thomas, P. Marchet, J. C. Champarnaud-Mesjard, T. Merle-Mejean, B. Frit, V. Nazabal, E. Fargin, G. Le Flem, M. O. Martin, B. Bousquet, L. Canioni, S. Le Boiteux, P. Segonds, and L. Sarger, Glass structure and optical nonlinearities in thallium(I) Tellurium(IV) oxide glasses, J. Solid State Chem., **146**, 335 (1999).

[3]D. A. Gaponov and A. S. Biryukov, Optical properties of microstructure tellurite glass fibres, Quantum Electron., **36**, 343 (2006).

[4]R. Jose and Y. Ohishi, Enhanced Raman gain coefficients and bandwidths in P2O5 and WO added tellurite glasses for Raman gain media, Appl. Phys. Lett., **89** (2006).

[5]R. Jose and Y. Ohishi, Ultra-broadband Raman gain media for photonics device applications, Proceedings of Optical Components and Materials IV, San Jose, CA, United States (SPIE, 2007), p. 64690.

[6]E. B. Intyushin and V. A. Novikov, Tungsten-tellurite glasses and thin films doped with rare-earth elements produced by radio frequency magnetron deposition, Thin Solid Films, **516**, 4194 (2008).

[7]S. N. B. Hodgson and L. Weng, Chemical and sol-gel processing of tellurite glasses for optoelectronics, J. Mater. Sci. - Mater. Electron., **17**, 723 (2006).

[8]S. Coste, A. Lecomte, P. Thomas, T. Merle-Mejean, and J. C. Champarnaud-Mesjard, Sol-gel synthesis of TeO_2-based materials using citric acid as hydrolysis modifier, J. Sol-Gel Sci. Technol., **41**, 79 (2007).
[9]A. Singh, E. J. Knystautas, and R. Lapointe, Synthesis of Tellurium oxide by high dose Oxygen ion-implantation, Appl. Surf. Sci., **22-23**, 2 (1984).
[10]M. F. Al-Kuhaili, S. M. A. Durrani, E. E. Khawaja, and J. Shirokoff, Effects of preparation conditions on the optical properties of thin films of Tellurium oxide, J. Phys. D: Appl. Phys., **35**, 910 (2002).
[11]K. Arshak and O. Korostynska, Gamma radiation-induced changes in the electrical and optical properties of Tellurium dioxide thin films, IEEE Sens. J., **3**, 717 (2003).
[12]R. Nayak, V. Gupta, A. L. Dawar, and K. Sreenivas, Optical waveguiding in amorphous Tellurium oxide thin films, Thin Solid Films, **445**, 118 (2003).
[13]S. M. Pietralunga, F. D'Amore, and A. Zappettini, Optical properties of reactively sputtered TeO_x amorphous films, Appl. Opt., **44**, 534 (2005).
[14]A. P. Caricato, M. Fernandez, M. Ferrari, G. Leggieri, M. Martino, M. Mattarelli, M. Montagna, V. Resta, L. Zampedri, R. M. Almeida, M. C. Concalves, L. Fortes, and L. F. Santos, Er3+ doped tellurite waveguides deposited by excimer laser ablation, Mater. Sci. Eng., B, **105**, 65 (2003).
[15]M. Bouazaoui, B. Capoen, A. P. Caricato, A. Chiasera, A. Fazzi, M. Ferrari, G. Leggieri, M. Martino, M. Mattarelli, M. Montagna, F. Romano, T. Tunno, S. Turrel, and K. Vishnubhatla, Pulsed Laser Deposition of Er doped tellurite films on large area, J. Phys. Conf. Ser., **59**, 475 (2007).
[16]S.-H. Kim, T. Yoko, and S. Sakka, Linear and nonlinear optical properties of TeO2 glass, J. Amer. Cer. Soc., **76**, 2490 (1993).
[17]G. Ghosh, Sellmeier coefficients and chromatic dispersions for some tellurite glasses, J. Amer. Cer. Soc., **78**, 2828 (1995).
[18]H. K. Pulker, Characterization of optical thin films, Appl. Opt., **18**, 1969 (1979).
[19]M. D. O'Donnell, K. Richardson, R. Stolen, C. Rivero, T. Cardinal, M. Couzi, D. Furniss, and A. B. Seddon, Raman gain of selected tellurite glasses for IR fibre lasers calculated from spontaneous scattering spectra, Opt. Mater., **30**, 946 (2008).
[20]M. Ceriotti, F. Pietrucci, and M. Bernasconi, Ab initio study of the vibrational properties of crystalline TeO_2: the alpha, beta, and gamma phases, Phys. Rev. B: Condens. Matter, **73**, 17 (2006).
[21]M. H. Brodsky, R. J. Gambino, J. E. Smith, Jr., and Y. Yacoby, Raman spectrum of amorphous Tellurium, Phys. Status Solidi B, **52**, 609 (1972).

*Corresponding author: khu.vu@anu.edu.au

FABRICATION OF MICRO STRUCTURES COMPOSED OF METALLIC GLASSES DISPERSED OXIDE GLASSES BY USING MICRO STEREOLITHOGRAPHY

Maasa Nakano
Graduate School of Engineering, Osaka University
Osaka, Japan
Satoko Tasaki and Soshu Kirihara
Joining and Welding Research Institute, Osaka University
Osaka, Japan

ABSTRACT

Magnetophotonic crystals with periodic arranged magnetic materials can reflect the terahertz waves through Bragg diffraction. The micrometer order magnetophotonic crystals were fabricated by stereolithographic methods. In this process, the photo sensitive acrylic resin paste mixed with micrometer sized metallic glass, a kind of amorphous alloy, and oxide glass particles were spread on a glass substrate with 10 μm in layer thickness by using a mechanical knife edge, and two dimensional images of UV ray were exposed with 2 μm in part accuracy. Through the layer by layer stacking, micrometer order three dimensional structures were formed. The metallic glass and oxide glass composite structures could be obtained through the dewaxing and sintering process with the lower temperature under the transition point of metallic glass. The amorphous structure formation after the heat treatment was verified by a X-ray diffraction analysis. A transmission spectrum of electromagnetic wave in terahertz frequency ranges for the formed magnetophotonic crystals with a diamond lattice structure was measured by using a terahertz time domain spectroscopy.

INTRODUCTION

Periodically arranged structures of magnetic materials are called magnetophotonic crystals. These artificial materials can form photonic band gaps to totally reflect electromagnetic waves with wavelengths comparable to the lattice constant though Bragg diffraction.[1-2] Electromagnetic waves in terahertz frequency ranges have lately attracted considerable attention as novel analytical light sources. Because the terahertz wave in terahertz frequency can be synchronized with collective vibration modes of various harmful substances, the spectroscopic technologies are expected to be applied to novel imaging sensors for real time detecting toxic materials in aqueous phase environments.[3,4] However, the terahertz waves are difficult to transmit into the water solvents due to electromagnetic absorptions.[5] In our investigation group, micrometer order alumina or zirconia lattices with diamond structures were fabricated successfully by using a micro stereolithography system to resonate the terahertz waves.[6] In this study, the micrometer order magnetophotonic crystals with the diamond structure composed of the metallic glass dispersed oxide glass were fabricated by the micro stereolithography. Metallic glass has random atomic arrangement and is changed to crystalline structured material by heat treatment at over the glass transition temperature. Crystallized metallic glass is inferior or loses its magnetic, mechanical and other properties. This matter causes difficulty in secondary processing of metallic glass. We suggest the new metallic glass forming process. The fabricated artificial crystals can be applied to the terahertz wave sensor devices. These devices are composed of the magnetophotonic crystals with through holes and micro tube as shown in the schematic illustrated in figure 1. The artificial crystal, having a cylindrical structural defect, can resonate the specific terahertz wave with wavelength comparable to the diameter of cylinder, and a localized mode of a transmission peak appears in the band gap. Water solvents flow into the micro tube and the terahertz wave spectra will be measured. If the resonated terahertz wave can harmonize with the collective vibration mode of the introduced water solvents, the transmission peak will disappear through the electromagnetic wave absorptions. The magnetophotonic crystal sensors will be applied to various scientific and engineering fields by utilizing the terahertz wave spectroscopic database.

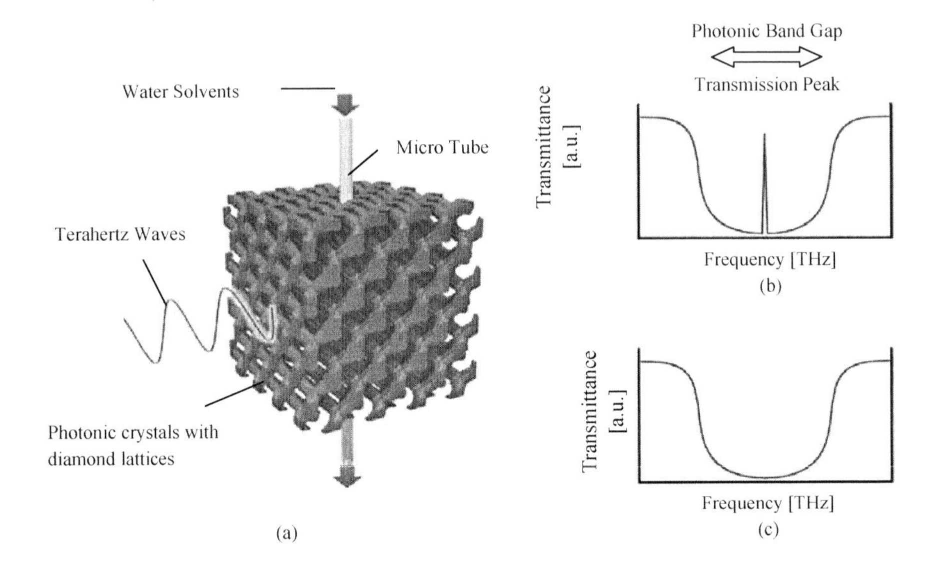

Figure 1. A schematic illustration of a magnetophotonic crystal sensor device (a). Transmission spectra of terahertz waves for the device including water solvents without (b) or with (c) harmful substances.

EXPERIMENTAL

The magnetophotonic crystal was fabricated by using the micro stereolithography. Figure 2 shows a schematic illustration of the method. A three dimensional model was designed by using computer graphic software (Materialise Japan Co. Ltd., Japan, Magics Ver. 14). The designed model was sliced into a series of two dimensional cross sectional data of 10 μm in layer thickness. These data were transferred into the micro stereolithography equipment (D-MEC Co. Ltd., Japan, SI-C 1000). In this machine, photo sensitive acrylic resins dispersed with metallic glass ($Fe_{72}B_{14.4}Si_{9.6}Nb_4$) and oxide glass ($B_2O_3 \cdot Bi_2O_3$) particles of 2.6 μm and 1.0 μm diameter were supplied on a substrate and spread uniformly by a mechanical knife edge. The thickness of each layer was controlled to 10 μm. The two dimensional pattern was formed clearly by illuminating ultra violet laser of 405 nm in wavelength on the slurry surface. The high resolution was achieved by using a digital micro-mirror device (DMD) and an objective lens. The DMD is composed of micro mirrors of 14 μm in edge length with 1024×768 in numbers. Each tiny mirror can be tilted independently according to the two dimensional cross sectional data by a computer operation. The three dimensional structures were built by layer stacking.

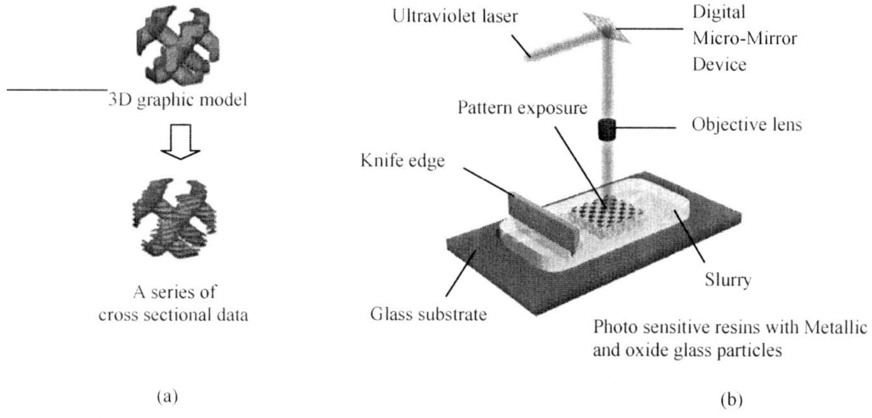

Figure 2. Schematic illustration of a stereolithography method. A three dimensional model is sliced into two dimensional cross sections (a). The two dimensional layers are stacked up to create the three dimensional object in micrometer order (b).

Process parameters of the micro stereolithography were optimized to fabricate the magnetophotonic crystals exactly. Three kinds of slurry were mixed to create the three dimensional objects. The metallic glass and oxide glass particles were dispersed acrylic resin at 40 vol. % in total, and these two kinds of particles were mixed with each other at 16:24, 17:23 and 18:22 volume ratios. The model shown in figure 3 was fabricated to measure the thickness of one layer and the size tolerance by the laser scattering. The process parameters to expose and stack the layers were optimized and imputed into the stereolithic system. The results of the optimization are discussed in the next section. The magnetophotonic crystals with diamond lattice structures were fabricated along the optimized condition. The rod diameter and length were 144 and 217 μm, and the lattice constant of the diamond structure was 500 μm, respectively. The whole structure was $5 \times 5 \times 0.5$ mm in size consisting of $10 \times 10 \times 1$ unit cells. Under the glass transition temperature at 552°C of the selected metallic glass is, the formed precursor was dewaxed to remove the acrylic resin by thermal decomposition at 420°C for 8.0 hs with the heating rate of 1.0°C /min, and sintered at 460°C for 0.5 hs with the heating rate of 2.0°C /min in an Ar atmosphere.[7] The diamond lattice model was corrected and re-designed according to the linear shrinkage ratios. Microstructures of the lattices were observed by an optical microscope and a scanning electron microscopy (SEM), and the microstructural stabilities of the amorphous phase alloy were analyzed by using a X-ray diffraction (XRD). The transmittance of the terahertz waves were analyzed by using a terahertz time domain spectroscopy (Aispec Co. Ltd., Japan, J-spec 2001 spc/ou).

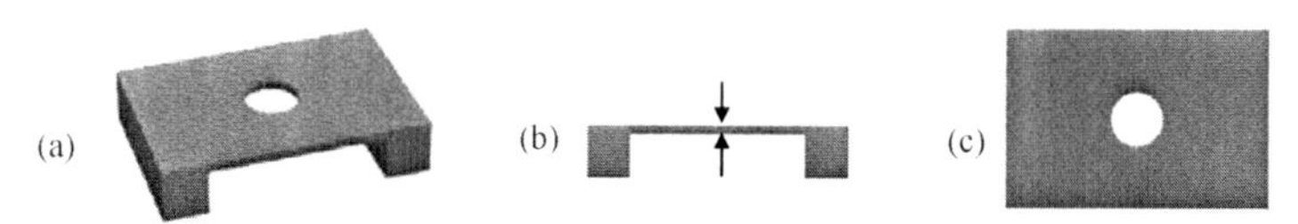

Figure 3. A graphic model of solid specimen (a) to measure layer thicknesses (b) and scale tolerances (c) at ultra violet lay exposing for the photo sensitive resin slurry with fine particles in the micro stereolithographic process.

RESULTS AND DISCUSSION

The Influence of laser power for the layer thickness and size tolerances of the solid objects were compared and investigated. The compositions of metallic glass and oxide glass were varied systematically. The thickness of one layer was required over 14.5 μm to stack layers every 10 μm successfully. The slurry including metallic glass: oxide glass at 16:24 formed too thick layer, and the slurry at 18:22 formed too thin a layer. The slurry compositions of the acrylic resin, metallic glass and oxide glass were 60%, 17% and 23% in volume ratios was adapted to this method. The laser power was optimized at 700 mJ/cm^2 to create the fine cross sectional micro patterns. Figure 5(a) shows the magnetophotonic crystals with the diamond lattices fabricated by the micro stereolithography. Tolerance between the designed model and formed sample was converged within ± 3μm. The sintered diamond lattice structure with 500 μm in lattice constant is shown in figure 5 (b). The micrometer order three dimensional structure was formed successfully. The linear shrinkage ratios of horizontal and vertical axis were 10.2% and 12.5%, respectively. Figure 5(c) shows the microstructure of the sintered metallic glass and oxide glass composite lattice. The metallic glass particles dispersed into the oxide glass matrix. The XRD patterns of metallic glass particles before and after heat treatments were analyzed as shown in figure 6. The metallic glass was not crystallized through the dewaxing and sintering heat treatment. This event shows that metallic glass has kept magnetic property. The magnetophotonic crystals were formed by periodically arranged rods composed of amorphous metallic and oxide glasses. Figure 7 shows the measured terahertz wave transmission spectrum for the fabricated magnetophotonic crystal with the diamond structure. The electromagnetic band gap can be formed at the frequency range from 0.19 to 1.02 THz. Because added metallic glass had the effect of decreasing the transmittance of decay area, the artificial crystals could form the band gap clearly.

(a) (b) (c)

Figure 5. Magnetophotonic crystals with diamond lattices composed of metallic glass and oxide glass particles dispersed acrylic resin fabricated by using the micro stereolithography (a) sintered metallic glass and oxide glass composite lattices (b). SEM image of the oxide glass lattice with metallic glass particles dispersion (c).

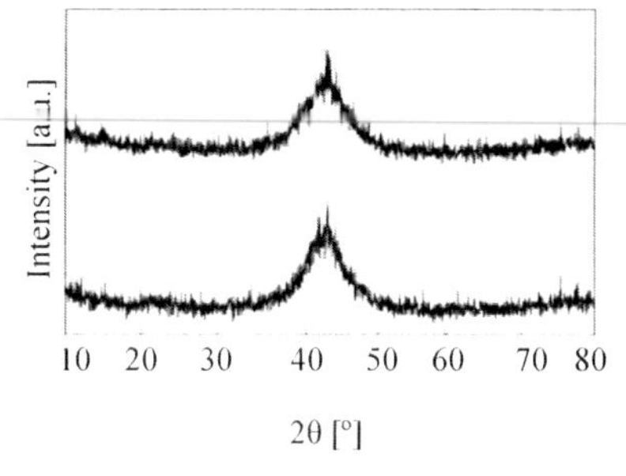

Figure 6. XRD pattern of the metallic glass particle before and after the dewaxing and sintering treatments.

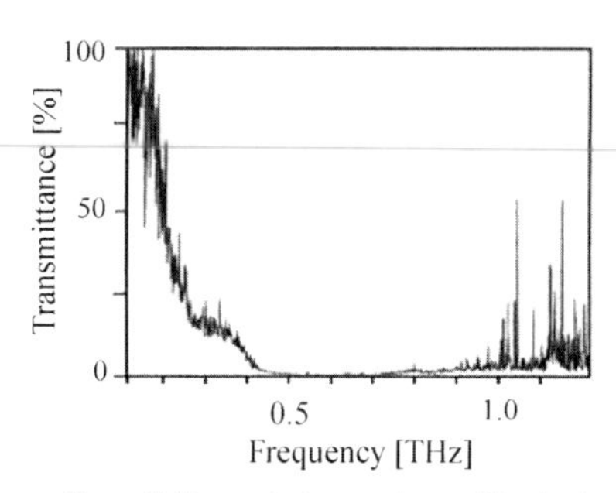

Figure 7. Transmission spectrum of terahertz waves through the fabricated magnetophotonic crystal with the micro diamond structure.

CONCLUSION

Through careful optimizations of process parameters regarding a stereolithography method, three dimensional micrometer order diamond lattices composed of acrylic resin including metallic glass and oxide glass particles at 17% and 23% in volume ratios could be fabricated using the micro stereolithography. The metallic glass and oxide glass composite crystals were obtained through the dewaxing and sintering process. The metallic glass was not crystallized through the heat treatment. We can say that a new way of metallic glass processing technology was established. The sintered magnetophotonic crystals formed electromagnetic band gap in a terahertz frequency range from 0.19 to 1.02 THz. The artificial crystals with micro through holes are expected to resonate the terahertz waves in wavelengths comparable to the hole diameters, and to be applied to real time sensing of harmful substances in aqueous phase environments.

ACKNOWLEDGEMENT

The authors would like to thank Prof. Hidemi Kato (Institute for Materials Research, Tohoku University) for providing metallic glass particle samples.

REFERENCES

[1]E. Yablonobitch, Inhibited Spontaneous Emission in Solid-State Physics and Electronics, *Phys. Rev. Lett.*, **58**, 2059-62 (1987).

[2]M. Inoue, Magnetophotonic crystals, *J. Phys D: Appl. Phys.*, **39**, R151-R161 (2006).

[3]S.M. Angel, Thomas J. Kulp, and Thomas M. Vess, Remote-Raman Spectroscopy at Intermediate Ranges Using Low-Power cw Lasers, *Appl. Spectrosc.*, **46**, 1085-91 (1992).

[4]H. Wang, Q.Wang, Spectrum characteristics of nitrofen by terahertz time-domai spectroscopy, *Phys. Conf. Ser.*, **276**, 012209-14 (2011).

[5]M.R. Kutteruf, C.M. Brown, L.K. Iwaki, M.B. Campbell, T.M. Korter, E.J. Heilwail, Terahertz spectroscopy of short-chain polypeptides, *Chem. Phys. Lett.*, **375**, 337-43 (2003).

[6]W. Chen, S. Kirihara, Y. Miyamoto, Three-dimensional Microphotonic Crystals of ZrO_2 Toughened Al_2O_3 for Terahertz wave applications, *Appl. Phys. Lett.*, **91**, 153507-09 (2007).

[7]A. Makino, New Functional Materials, Fundamentals of Metallic Glasses and their Applications to Industry, Edited by A. Inoue, *Technosystem*, Japan, 322 (2009).

*Corresponding author: nakano@jwri.osaka-u.ac.jp

CRYSTALLIZATION AND OPTICAL LOSS STUDIES OF Dy^{3+}-DOPED, LOW Ga CONTENT, SELENIDE CHALCOGENIDE BULK GLASSES AND OPTICAL FIBERS

Zhuoqi Tang,[1-3] David Furniss,[1-3] Michael Fay,[4] Nigel C. Neate,[2] Slawomir Sujecki,[3] Trevor M. Benson[3] and Angela B. Seddon[1-3*]

[1] Novel Photonic Glasses Research Group; [2]Centre for Advanced Materials; [3]George Green Institute of Electromagnetics Research Electrical Systems and Optics Research Division, Faculty of Engineering; [4]Nottingham Nanotechnology and Nanoscience Centre, University of Nottingham, University Park, Nottingham, NG7 2RD, UK.

ABSTRACT

Dysprosium (Dy^{3+}) doped GeAsGaSe chalcogenide glasses are a significant candidate for novel mid-infrared fiber lasers beyond 4 μm wavelength. This work is a crystallization and optical loss study of Dy^{3+}-doped GeAsGaSe chalcogenide bulk, as-prepared glasses and Dy^{3+}-doped optical fiber with low Ga content. In the bulk glasses of 0-3000 ppm Dy^{3+}-doped $Ge_{16.5}As_{16}Ga_3Se_{64.5}$ (at% (atomic%)), despite none of their X-ray diffraction patterns showing any crystallization peaks, a sub-micron crystal is imaged in the 3000 ppm Dy^{3+}-doped $Ge_{16.5}As_{16}Ga_3Se_{64.5}$ glass and analyzed using high resolution transmission electron microscopy to be a modified Ga_2Se_3 incorporating a small amount of Ge. Extra background optical scattering loss of the bulk glasses is found in Fourier transform infrared spectra. Finally, this work presents the optical loss spectrum of a 500 ppm Dy^{3+}-doped $Ge_{16.5}As_{16}Ga_3Se_{64.5}$ optical fiber, exhibiting 1.16 dB/m loss at 6.56 μm. An X-ray diffraction pattern of the powdered glass of the fiber exhibited small crystallisation peaks despite the low doping level of Dy^{3+}.

INTRODUCTION

The rare earth doped chalcogenide glasses are an important candidate material for mid-infrared fiber lasers due to the low phonon energy of these glasses and their suitable optical transparency at both the pumping and lasing wavelengths. As yet, there is no mid-infrared fiber laser reported beyond 4 μm wavelength in the world.[1-3] It is suspected that crystallization problems are found, although we have not found any studies to date on the phase-crystallization behaviour of rare earth doped chalcogenide glasses, other than our prior work,[8] especially regarding the procedure of drawing low loss and high rare earth concentration fiber, which is a key issue for the fabrication of mid-infrared fiber lasers. Ga has been reported to stabilise the solubilisation of rare earth dopants in sulfide glasses[4] and the rare earth doped GeAsGaSe glass system has been suggested as stable during fiber drawing as well as a competitive candidate for mid-infrared fiber lasers.[5-7]

In our previous work,[8] crystallization of Dy^{3+}-doped $Ge_{16.5}As_{16}Ga_{10}Se_{64.5}$ glasses was reported and Ga_2Se_3 crystals were found to be prevalent in the as-made bulk glasses. Here, for similar GeAsGaSe based glasses, the Ga content was decreased to 3 at% (atomic%)[1,9] and the crystallization behaviour of the 0-3000 ppm Dy^{3+}-doped $Ge_{16.5}As_{16}Ga_3Se_{64.5}$ glasses is presented. Finally, in order to understand better the potential for crystal growth during fiber drawing, the loss spectrum of a 500 ppm Dy^{3+}-doped Ga_3 glass fiber and the fiber glass XRD (X-ray diffraction) pattern are discussed.

EXPERIMENTAL PROCEDURE

In this work, the Dy^{3+}($DyCl_3$) precursor dopant concentration was from 0 ppm to 3000 ppm and the host glass was $Ge_{16.5}As_{16}Ga_3Se_{64.5}$. The glasses were prepared by the conventional melt-quenching method. High purity raw materials (arsenic and selenium were treated to distil off oxides by chemical heat treatment under vacuum (10^{-3} Pa) that had a 1 hour dwell at 300°C and 0.5 hour dwell at 250°C for arsenic and selenium, respectively. But no further particular purification was carried out) were batched inside a nitrogen-circulated glove box into a silica glass ampoule (purified by heating in air and then under vacuum at 1000°C) which was then sealed under vacuum. The ampoule was set in a rocking furnace for 96 hours at 930°C, following by liquid metal quenching and annealing around T_g for one hour in an annealing furnace. In addition, a 500 ppm Dy^{3+}-doped glass fiber-optic rod preform (8 mm diameter) was prepared under the same conditions then drawn to fiber on a customized Heathway tower with winder.

For bulk and fiber glass characterization, a Siemens Krystalloflex 810 machine was used for powder XRD pattern collection at 0.48°2θ/minute over 2 hours. For TEM (transmission electron microscopy), only bulk glass was studied and care was taken not to include the surface glass which had been in direct contact with the silica-glass melt containment ampoule during the melting, quenching and annealing. A JEOL 2100F FEG-TEM with a Gatan Orius camera was used for TEM. TEM-EDX (energy dispersive X-ray spectroscopy) results were obtained with an Oxford Instruments INCA TEM 250 system with ~±1 at% accuracy for elements heavier than oxygen. Bulk glass FTIR (Fourier transform infrared) spectra and optical fiber spectral loss were collected using an IFS 66/S (Bruker) FTIR system. Fiber cut-back was used in the fiber loss measurement.

RESULTS AND DISCUSSION

Bulk Crystallization StudiesFigure 1 shows the XRD patterns of 0 ppm, 500 ppm, 1000 ppm, 2000 ppm and 3000 Dy^{3+}-doped $Ge_{16.5}As_{16}Ga_3Se_{64.5}$ as-prepared glasses, after Kα2 striping and normalization of the background intensity. In figure 1, no crystallization peak could be found in any of the XRD patterns, suggesting amorphicity and a glassy structure. Thus, for GeAsGaSe glasses with 3 at% Ga, notwithstanding the presence of potential Dy_2O_3 nucleation agents coming from the high concentrations of Dy^{3+} up to 3000 ppm, the Ga tended to remain in the glass matrix instead of phase separating.

Although there were no crystallisation peaks observed in the XRD patterns (see figure 1), a very few nano- and sub-micron crystals were indeed found (very rarely) inside a 3000 ppm Dy^{3+}-doped $Ge_{16.5}As_{16}Ga_3Se_{64.5}$ bulk glass sample using TEM (see bright features in figure 2). In figure 2, the dash-line-highlighted-area shows a large crystal ~ 80 nm x 80 nm and the SAED (selected area electron diffraction) image at the upper right corner of figure 2 proved that this structure was crystalline.

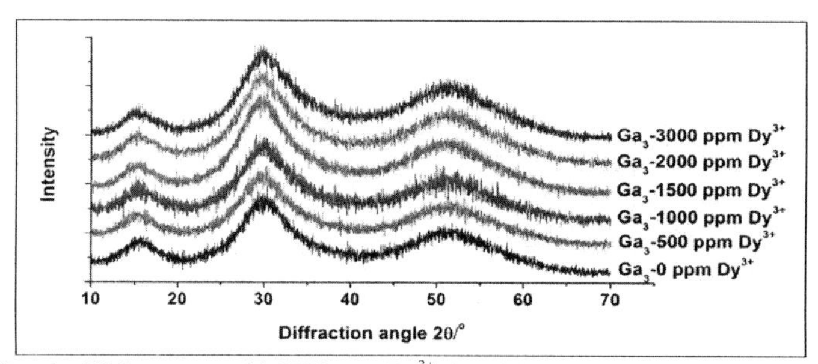

Figure 1. Powder XRD patterns of 0-3000 ppm Dy^{3+}-doped $Ge_{16.5}As_{16}Ga_3Se_{64.5}$ as-prepared glasses.

The TEM-EDX spectrum of the highlighted area is at the lower right corner of figure 2 (the Cu peak in the spectrum was from the sample holder). The corresponding quantification results (Table I), were Ga (41±1 at%) and Se (57±1 at%) and close to theoretical Ga_2Se_3. Also, there was a small amount of Ge (2±1 at%) in the crystal. Thus, the TEM-EDX results suggested that crystals were modified Ga_2Se_3 with a small amount of Ge in the Dy^{3+}-doped $Ge_{16.5}As_{16}Ga_3Se_{64.5}$ bulk glass.

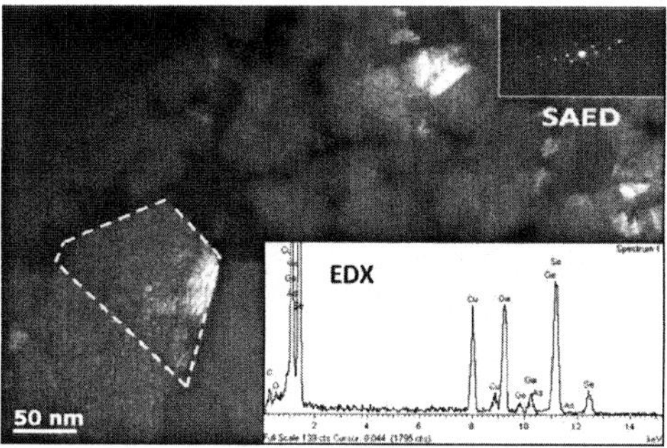

Figure 2. TEM image of sub-micron crystal which were rare inside the bulk of 3000 ppm Dy^{3+}-doped $Ge_{16.5}As_{16}Ga_3Se_{64.5}$ bulk glass. (SAED is selected area electron diffraction.)

Table I. TEM-EDX quantification results (±1) of the dash line highlighted 80 nm x 80 nm crystal grown inside the 3000 ppm Dy^{3+}-doped Ga_3 bulk glass (see figure 2).

	Ge	As	Ga	Se
Material batched/(at%)	16.5	16	3	64.5
Crystal EDX result/(at%)	2	0	41	57
Ga_2Se_3 (Theory)/(at%)	-	-	40	60

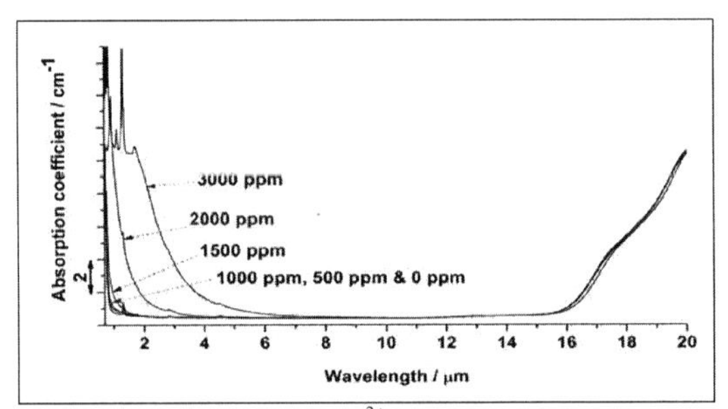

Figure 3. FTIR spectra of 0-3000 ppm Dy^{3+}-doped $Ge_{16.5}As_{16}Ga_3Se_{64.5}$ bulk glasses.

Figure 3 presents the FTIR spectra of 0-3000 ppm Dy^{3+}-doped $Ge_{16.5}As_{16}Ga_3Se_{64.5}$ bulk glasses. Extra background loss, assumed to be due to scattering in the bulk, was found to appear for ≥ 1500 ppm Dy^{3+}-doped Ga_3 glasses. As the Dy^{3+} concentration was increased, the scattering level became higher, as can be seen in spectra of the 2000 ppm, and 3000 ppm Dy^{3+}-doped, glasses where the scattering started from ~ 4μm, and 8μm, respectively. As no crystallisation peaks could be found in XRD patterns of the bulk glasses, it might be supposed that the modified-Ga_2Se_3 nano- and/or sub-micron crystals should not have contributed much to this scattering because the crystals are small, and few in number. Possibly Dy_2O_3 particles were present in the bulk glass,[8] because the Dy^{3+} concentration was ≤0.3 wt%, Dy_2O_3 particles would not have contributed much to the optical scattering shown in figure 3. As discussed in our previous work,[8] we believe that the $DyCl_3$ additive tended to etch the silica-glass melt containment ampoule, especially as the temperature was raised during chalcogenide glass melting, to produce [Si-O-Dy] particles in bulk glasses during melting. In summary, it is suggested the Dy_2O_3 particles, Ga_2Se_3 crystals and [Si-O-Dy] particles all contributed to the scattering loss manifested in Dy^{3+} doped Ga_3 bulk glasses. We suspect that [Si-O-Dy] particles play a main role because of the reservoir of silica in the melt-containment and evidence in our previous work.[8]

Optical Loss Spectrum and XRD of a 500 ppm Dy^{3+}-doped Fiber

The 500 ppm Dy^{3+}-doped $Ge_{16.5}As_{16}Ga_3Se_{64.5}$ glass optical fiber spectrum (figure 4) exhibits vibrational bands at 3.5 μm, 4.1 μm and 4.6 μm, due to H_2Se or Se-H contamination[10] and bands at 1.3 μm, 1.7 μm and 2.8 μm due to Dy^{3+} electronic absorption. In the fiber spectrum, the lowest loss was 1.16 dB/m at 6.56 μm. The background loss in both wavelength ranges 1.9 μm-2.2 μm and 3.3 μm-3.8 μm was around 2.4 dB/m. Also, no obvious scattering was found at wavelengths > ~ 1.5 μm.

XRD patterns in figure 5 are for the powder samples of the 500 ppm and 1000 ppm Dy^{3+}-doped Ga_3 bulk glasses as-prepared and the glass constituting the 500 ppm Dy^{3+}-doped Ga_3 fiber. Before fiber drawing, there were no crystallization peaks in the XRD pattern of the 500 ppm Dy^{3+}-doped Ga_3 bulk glass. After fiber drawing, small crystallisation peaks at 28.3 °2θ, 47.0 °2θ and 55.8 °2θ appeared in the XRD pattern of the 500 ppm Dy^{3+}-doped Ga_3 glass fiber. These three peak positions matched the main peak positions: 28.3 °2θ, 47.0 °2θ and 55.8 °2θ in the XRD pattern of 1000 ppm Dy^{3+}-doped Ga_{10} glass,[8] where the crystals were confirmed to be modified Ga_2Se_3 with a small amount of Ge. Also, from section 3.1, the crystals inside 3000 ppm Dy^{3+}-doped Ga_3 bulk glass have been confirmed as the modified Ga_2Se_3 with a small amount of Ge by TEM-EDX.

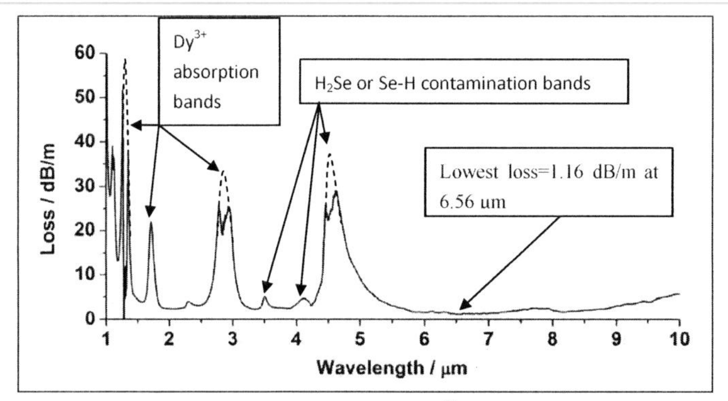

Figure 4. Optical fiber loss spectrum of 500 ppm Dy^{3+}-doped $Ge_{16.5}As_{16}Ga_3Se_{64.5}$ glass. (No particular purification was applied to remove the Se-H, H_2Se bands.)

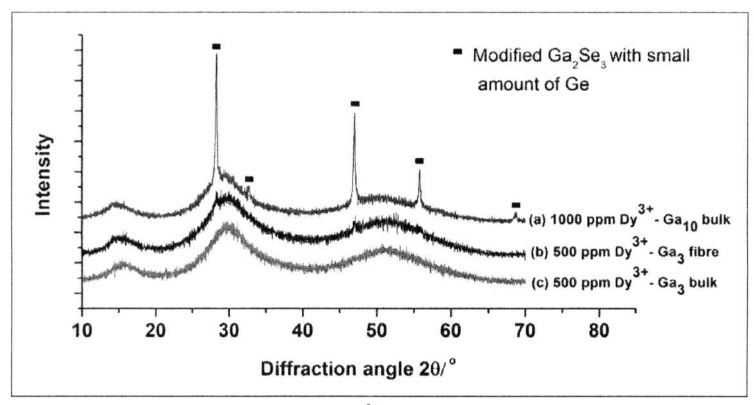

Figure 5. XRD patterns of: (a) 500 ppm Dy^{3+}-doped $Ge_{16.5}As_{16}Ga_3Se_{64.5}$ as-prepared, bulk glass and (b) fiber and (c) 1000 ppm Dy^{3+}-doped $Ge_{16.5}As_9Ga_{10}Se_{64.5}$ bulk glass.

Thus, the crystals formed in the 500 ppm Dy^{3+}-doped Ga_3 glass fiber are the same as the crystals in both Ga_3 and Ga_{10} as–prepared bulk glasses. Therefore crystals which grew during the reheating for fiber drawing were the same as the crystals grown after melt-quenching in the as-prepared Dy^{3+} ($DyCl_3$)-doped GeAsGaSe glasses. This indicates that, in order to get low loss rare earth doped chalcogenide glass fiber, it is important to reduce the level of small crystals and potential nucleation agents which may already be present in glass fiberoptic preforms formed by the melt-quenching method. This is because small crystals may grow larger and nucleation agents may nucleate new crystals during the fibre drawing process. However, although small crystallization peaks could be found by XRD for the fibre, the lowest loss in the 500 ppm Dy^{3+}-doped Ga_3 glass fiber still reached 1.16 dB/m with apparently no scattering beyond ~ 1.5 μm wavelength, which indicated the number and/or size of crystals grown during fiber drawing was small.

CONCLUSIONS

In this work, although XRD patterns did not show any crystallization peaks for 0-3000 ppm Dy^{3+}-doped $Ge_{16.5}As_{16}Ga_3Se_{64.5}$ bulk glasses, rare nano- and sub-micron crystals were indeed found in the as-prepared 3000 ppm Dy^{3+}-doped bulk glass using TEM and the TEM-EDX results suggested them to be a modified Ga_2Se_3 with some substitution of Ge. Because there were no crystallization peaks in the XRD patterns of the bulk glasses, optical scattering found in FTIR spectra of Dy^{3+}-doped $Ge_{16.5}As_{16}Ga_3Se_{64.5}$ glass was suggested to be mainly caused by silica glass and/or silica-dysprosia particles potentially produced during the etching of the silica-glass melt containment ampoule by the rare earth additive. Finally, a 500 ppm Dy^{3+}-doped $Ge_{16.5}As_{16}Ga_3Se_{64.5}$ fiber spectrum has been presented with no scattering beyond ~ 1.5 μm and lowest loss of 1.16 dB/m at 6.56 μm, although small crystallisation peaks were found in the XRD pattern of the powdered glass of the fiber.

ACKNOWLEDGEMENTS

Zhuoqi Tang gratefully thanks: (1) receipt of the prestigious Travel Award from the AFPG-9 conference and the selection committee from Alfred University, USA; (2) the Andrew Carnegie Research Fund from the Institute of Materials, Minerals and Mining (IOM3), UK, and (3) a travel grant from the Electrical Systems and Optics Research Division, Faculty of Engineering, University of Nottingham, UK enabling him to attend the AFPG-9 conference.

REFERENCES

[1] A.B. Seddon, Z. Tang, D. Furniss, S. Sujecki, and T.M. Benson, Progress in rare-earth-doped mid-infrared fiber lasers, *Opt. Express*, 18(25), 26704-26719 (2010).

[2] M. Ebrahim-Zadeh and I.T. Sorokina, Mid-Infrared coherent sources and applications, *Springer*, The Netherlands, 2005.

[3] J.S. Sanghera, L.B. Shaw and I.D. Aggarwal, Chalcogenide glass fiber based Mid-IR sources and applications, *IEEE J. Sel. Top. Quantum Electron*, Vol. 15. No.1, pp 114-119 (2009).

[4] B.G. Aitken, C.W. Ponader, and R.S. Quimby, Clustering of rare earths in GeAs sulfide glass, *C. R. Chimie*, 5, 865-872 (2002).

[5] L.B. Shaw, B. Cole, P.A. Thielen, J.S. Sanghera, and I.D. Aggarwal, Mid-wave IR and long-wave IR laser potential of rare-earth doped chalcogenide glass fiber, *IEEE J. Quantum Electron.* 48 (9) 1127-1137 (2001).

[6] R.S. Quimby, L.B. Shaw, J.S. Sanghera, and I.D. Aggarwal, Modeling of cascade lasing in Dy:chalcogenide glass fiber laser with efficient output at 4.5 μm, *IEEE Photon. Tech. Lett.* 20 (2) 123-125 (2008).

[7] S. Sujecki, L. Sójka, E. Bereś-Pawlik, Z. Tang, D. Furniss, A.B. Seddon, and T.M. Benson, Modeling of a simple Dy^{3+} doped chalcogenide glass fibre laser for mid-infrared light generation, *Opt. Quantum Electron*, 42, 69-79 (2010).

[8]Z. Tang, N.C. Neate, D. Furniss, T.M. Benson, and A.B. Seddon, Crystallization behavior of Dy^{3+}-doped selenide glasses, *J. Non-Cryst. Solids*, 357, 2453-2462 (2011).
[9]Z. Tang, D. Furniss, M. Fay, Y. Cheng, N.C. Neate, T.M. Benson, and A.B. Seddon, Comparison of the crystallization behavior of Dy^{3+} doped GeAsSe glasses with different levels of Ga addition, *J. Am. Ceram. Soc*, to be submitted in 2012.
[10]C.T. Moynihan, P.B. Macedo, M.S. Maklad, R.K. Mohr, and R.E. Howard, Intrinsic and impurity infrared-absorption in As_2Se_3 glass, *J. Non-Cryst. Solids*, 17, 369-385 (1975).

*Corresponding author: Angela.Seddon@nottingham.ac.uk

FABRICATION AND CHARACTERIZATION OF Er^{3+}-DOPED TELLURITE GLASS WAVEGUIDES BY Ag^{+}-Na^{+} ION-EXCHANGE METHOD USING A DRY ELECTROMIGRATION PROCESS

S. Sakida,[1] K. Kimura,[2] Y. Benino[2] and T. Nanba[2]

[1]Environmental Management Center, Okayama University, 3-1-1, Tsushima-Naka, Kita-ku, Okayama 700-8530, Japan; [2]Graduate School of Environmental Science, Okayama University, 3-1-1, Tsushima-Naka, Kita-ku, Okayama 700-8530, Japan

ABSTRACT

Planar waveguides were prepared on $12Na_2O \cdot 10NbO_{2.5} \cdot 25WO_3 \cdot 53TeO_2 \cdot 1Er_2O_3$ ([NbWEr]) substrate glass by Ag^{+}-Na^{+} ion-exchange using a dry electromigration process. The optical properties of the waveguides were characterized and compared with those of [NbWEr] glass waveguides prepared by Ag^{+}-Na^{+} ion-exchange by thermal diffusion. After ion-exchange by both a dry electromigration process and thermal diffusion, the transmittance of the glasses slightly decreased from that of the [NbWEr] substrate glass but frosting was not observed on these ion-exchanged glasses. The depth of the waveguide fabricated at 300°C by a dry electromigration process was about 8 μm and deeper than that at 380°C by thermal diffusion. On the [NbWEr] glass ion-exchanged at 300°C by a dry electromigration process, the stains seen in the [NbWEr] glass ion-exchanged at 380°C by thermal diffusion were hardly observed. Therefore, the dry electromigration process is promising due to the reduction of the process temperature and fewer stains.

INTRODUCTION

It has been requested that the speed and density of telecommunications be enhanced due to an increase in the demand for optical communication. As one of the solutions, the wavelength division multiplexing (WDM) method attracts attention. A wideband optical amplifier as one of the optical devices is indispensable in order to amplify communication light that attenuates by transmission. Er^{3+}-doped tellurite glass is an attractive material as a broadband amplifier at a 1.5 μm wide band in the WDM network. Application to a small wideband optical amplifier with high optical gain and low-cost as one of the optical devices can be expected by fabricating a waveguide on this glass.

An optical waveguide is the basis for an integrated optical device applicable to an amplifier and laser for high-speed signal processing in telecommunication. Optical waveguides have been fabricated on substrate glasses by means of various techniques such as ion-exchange, sol-gel, plasma enhanced chemical vapor deposition (PECVD), physical vapor deposition (PVD), flame hydrolysis deposition (FHD), pulsed laser deposition (PLD), rf-sputtering and laser writing. Among these techniques, the ion-exchange method is effective to make an optical waveguide on substrate glass owing to its simplicity, flexibility, reliability and low cost. So far, many studies about optical waveguides fabricated on silicate,[1] soda-lime,[2] borosilicate[3] and phosphate[4] substrate glasses using ion-exchange by thermal diffusion have been reported. However, studies about the characterization of tellurite glass waveguides fabricated using ion-exchange by thermal diffusion are few.[5-8] Ion-exchange at high temperature is necessary in order to enhance the diffusion rate of ions but ion-exchange at a rather higher temperature than the glass transition temperature (T_g) is difficult. Since tellurite glasses generally have low T_g (about 110 - 440°C),[9] the composition of a tellurite glass as a substrate glass for ion-exchange by thermal diffusion is considerably limited. If a tellurite glass waveguide by ion-exchange can be fabricated at a low temperature for a short time, it is expected to be a tellurite glass waveguide with extensive glass composition. Since the fabrication of a waveguide by ion-exchange using a dry electromigration process is performed by applying high voltage, ion-exchange using it is considered to permit fabrication of a waveguide at a lower temperature for a shorter time than ion-exchange by thermal diffusion. However, the fabrication of tellurite glass waveguides by Ag^+-Na^+ ion-exchange using a dry electromigration process has hardly been reported.[10] Hence, further information about the fabrication and characterization of tellurite glass waveguides by Ag^+-Na^+ ion-exchange using a dry electromigration process is necessary.

In the present study, the planar waveguides were prepared on $12Na_2O \cdot 10NbO_{2.5} \cdot 25WO_3 \cdot 53TeO_2 \cdot 1Er_2O_3$ ([NbWEr]) glass by Ag^+-Na^+ ion-exchange using a dry electromigration process. The optical properties of the waveguides are characterized and compared with those of [NbWEr] glass waveguides prepared by Ag^+-Na^+ ion-exchange by thermal diffusion.

METHODS AND PROCEDURES

Er^{3+}-doped tellurite glasses with $12Na_2O \cdot 10NbO_{2.5} \cdot 25WO_3 \cdot 53TeO_2 \cdot 1Er_2O_3$ composition in mol% ([NbWEr]) were prepared as the substrate glass. High-purity reagents of Na_2CO_3, Nb_2O_5, WO_3, TeO_2, and Er_2O_3 powders were used as starting materials. A 20 g batch of well-mixed reagents was melted in a gold crucible covered with a lid of alumina using an electric furnace at 900°C for 30 min in air. The melt was poured onto a brass plate and immediately pressed by a stainless plate. The glasses prepared were annealed near the T_g for 1.5h. After annealing, the glasses were cut into plates 50 × 15 × 2 mm in size and all faces mirror-polished for optical measurements and waveguide fabrication.

The planar waveguides were prepared on the [NbWEr] substrate glass by Ag^+-Na^+ ion-exchange. Ag^+-Na^+ ion-exchange by thermal diffusion was performed by immersing the glass samples in a $1.0AgNO_3 \cdot 49.5NaNO_3 \cdot 49.5KNO_3$ (mol%) molten salt at 320 - 380°C for 5h in a square alumina bath. Ag^+-Na^+ ion-exchange by a dry electromigration process was performed as follows. Electroconductive silver paste (Dotite, Fujikura Kasei Co., Ltd.) was applied to the glass samples. Carbon plates were used as electrodes. After drying the silver paste, Ag^+-Na^+ ion-exchange using a dry electromigration process was performed at an applied voltage of 440 V/mm and 290-300°C for 5h in an electric furnace.

The T_g of the [NbWEr] substrate glass was determined by TG-DTA (at a heating rate of 10 $K \cdot min^{-1}$). The refractive indices of the [NbWEr] substrate glass and the effective mode indices of the [NbWEr] waveguides at wavelengths of 473, 632.8, 983.1, and 1548 nm were measured by means of a prism coupler technique. The transmission spectra of the [NbWEr] substrate glass and waveguides were measured by UV/VIS/NIR spectrophotometer. Surface analysis by EDS was performed for the [NbWEr] waveguides.

RESULTS AND DISCUSSION

Glass transition temperature and refractive indices of [NbWEr] substrate glass (Table I.) lists the T_g and refractive indices at the wavelengths of 473, 632.8, 983.1, and 1548 nm (n_{473}, $n_{632.8}$, $n_{983.1}$, and n_{1548}, respectively) of the [NbWEr] substrate glass.

Table I. Glass transition temperature (T_g), and refractive indices at wavelengths of 473, 632.8, 983.1, and 1548 nm (n_{473}, $n_{632.8}$, $n_{983.1}$, and n_{1548}, respectively) of [NbWEr] substrate glass.

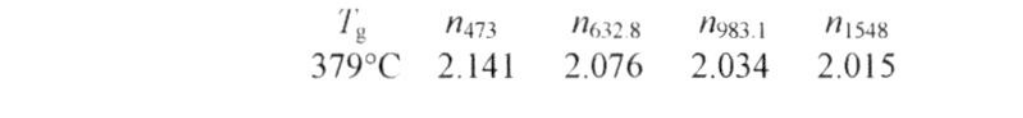

T_g	n_{473}	$n_{632.8}$	$n_{983.1}$	n_{1548}
379°C	2.141	2.076	2.034	2.015

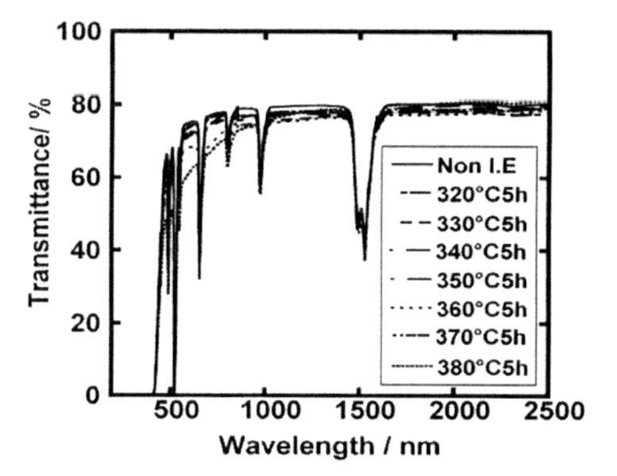

Figure 1. Transmission spectra of the substrate glass (Non I.E) and glasses ion-exchanged at 320-380°C for 5h by thermal diffusion for [NbWEr] glasses.

Ion-Exchange by Thermal Diffusion

Fig. 1 shows the transmission spectra of the substrate glass (Non I.E) and glasses ion-exchanged at 320-380°C for 5h by thermal diffusion for [NbWEr] glasses. After ion-exchange by thermal diffusion, the transmittance of the glasses slightly decreased from that of the [NbWEr] substrate glass. The transmittance of the glasses ion-exchanged at 360 and 380°C decreased compared to that at other temperatures at 500-1000 nm, suggesting that high ion-exchange temperature tended to decrease transmittance.

Fig. 2 shows photographs of [NbWEr] substrate glass (left), [NbWEr] glass ion-exchanged at 360°C for 5h by thermal diffusion (center), and [NbWEr] glass ion-exchanged at 380°C for 5h by thermal diffusion (right). The [NbWEr] substrate glass was obviously clear and had neither frosting nor stains. Although stains were observed on the [NbWEr] glasses ion-exchanged at 360 and 380°C, these ion-exchanged glasses had no frosting as evidenced by clear black lines under the glasses.

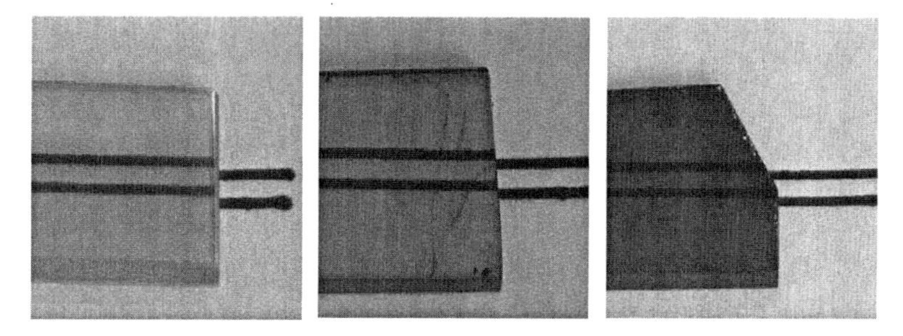

Figure 2. Photographs of [NbWEr] substrate glass (left), [NbWEr] glass ion-exchanged at 360°C for 5h by thermal diffusion (center), and [NbWEr] glass ion-exchanged at 380°C

In the prism coupling measurements, 3 to 10 waveguide modes were observed for all the glasses ion-exchanged by thermal diffusion, indicating the successful fabrication of the planar waveguides on the glass under all the ion-exchange conditions by thermal diffusion in this study. The number of modes increased by increasing the ion-exchange temperature. Refractive-index profiles were obtained from the measured mode indices using the inverse Wentzel-Kramers-Brillouin (WKB) method[11] without assuming any particular shape since it is well known that an ion-exchanged layer fabricated by thermal diffusion is generally gradient. This method is applicable when at least three modes are observed. Its reliability increases with an increase in the number of modes. The measured mode indices do not correspond to a glass substrate index or a glass surface refractive index. A glass surface refractive index can be obtained from a refractive-index profile.

Fig. 3 exhibits refractive-index profiles at 632.8 nm for [NbWEr] glass waveguides by ion-exchange at 320-380°C for 5h by thermal diffusion. The waveguide depths from the glass surface increased by increasing the ion-exchange temperature. The glass surface refractive indices were high at low ion-exchange temperatures. This is probably due to an error caused by a small number of waveguide modes.

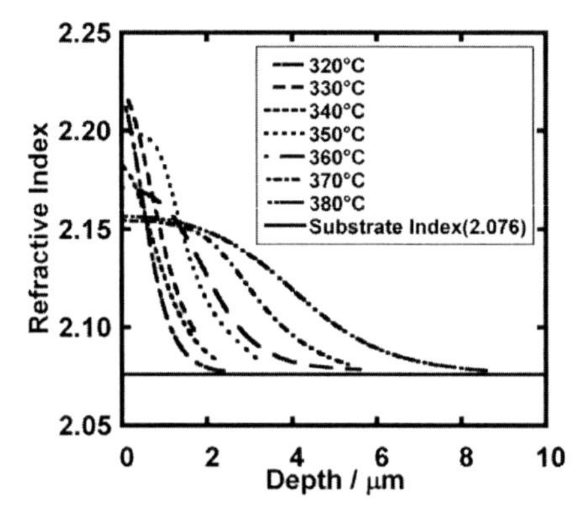

Figure 3. Refractive index profiles at 632.8 nm for [NbWEr] glass waveguides by ion-exchange at 320 - 380°C for 5h by thermal diffusion.

Ion-Exchange Using a Dry Electromigration Process

The relation between the fraction of $Ag^+/(Ag^++Na^+)$ and the depth of [NbWEr] waveguides fabricated by Ag^+-Na^+ ion-exchange at 290 and 300°C using a dry electromigration process by EDS is exhibited in Fig. 4. The depths of the [NbWEr] waveguides fabricated by Ag^+-Na^+ ion-exchange using a dry electromigration process were about 5 and 8 μm at 290 and 300°C, respectively, indicating the successful fabrication of the planar waveguides on the glass under all ion- exchange conditions by a dry electromigration process in this study. The depths of the waveguides fabricated at 290 and 300°C using a dry electromigration process were respectively deeper than those at 360 and 380°C by thermal diffusion. Hence, it can be said that ion-exchange using a dry electromigration process enables us to fabricate waveguides with a thicker waveguide layer at a lower ion-exchange temperature than ion-exchange by thermal diffusion.

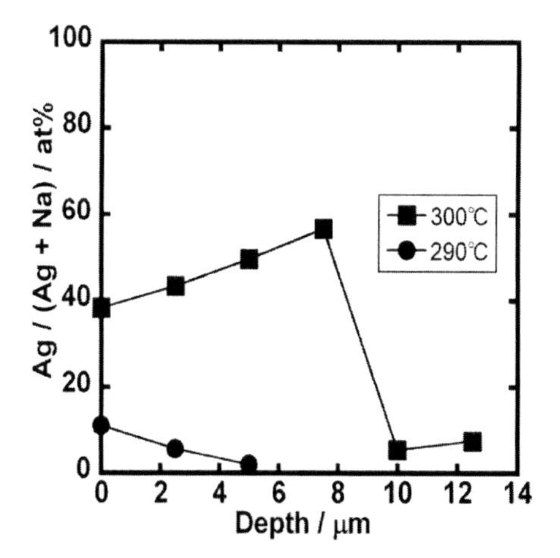

Figure 4. Relation between the fraction of $Ag^+/(Ag^++Na^+)$ and the depth of [NbWEr] waveguides fabricated by Ag^+-Na^+ ion-exchange at 290 and 300°C using a dry electromigration process by EDS.

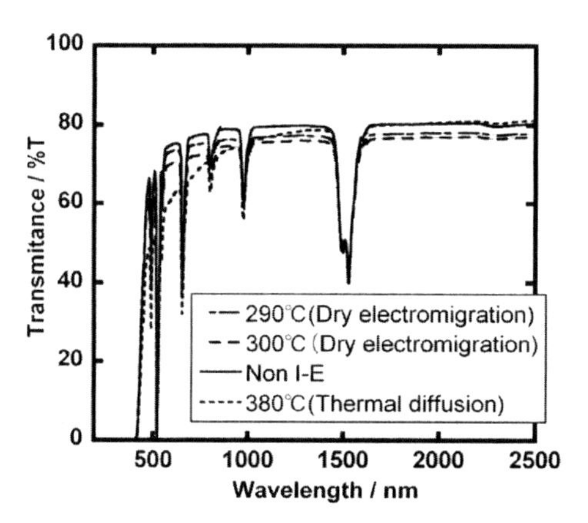

Figure 5. Transmission spectra of the substrate glass (Non I-E) and glasses ion-exchanged at 290-300°C for 5 h using a dry electromigration process and at 380°C for 5 h by thermal diffusion for [NbWEr].

Ag^+-Na^+ ion-exchange using a dry electromigration process were about 5 and 8 μm at 290 and 300°C, respectively, indicating the successful fabrication of the planar waveguides on the glass under all ion-exchange conditions by a dry electromigration process in this study. The depths of the waveguides fabricated at 290 and 300°C using a dry electromigration process were respectively deeper than those at 360 and 380°C by thermal diffusion. Hence, it can be said that ion-exchange using a dry electromigration process enables us to fabricate waveguides with a thicker waveguide layer at a lower ion-exchange temperature than ion-exchange by thermal diffusion.

Fig. 5 shows the transmission spectra of substrate glass (Non I-E) and glasses ion-exchanged at 290-300°C for 5h using a dry electromigration process and at 380°C for 5h by thermal diffusion for [NbWEr]. After ion-exchange at 290 and 300°C using a dry electromigration process, the transmittance of the glasses slightly decreased from that of the [NbWEr] substrate glass. The transmittance of the glass ion-exchanged at 380°C by thermal diffusion was lower than that at 290 and 300°C using a dry electromigration process at 500-1000 nm.

Fig. 6 shows photographs of [NbWEr] substrate glass (left) and [NbWEr] glass ion-exchanged at 300°C for 5h using a dry electromigration process (right). On the [NbWEr] glass ion-exchanged at 300°C using a dry electromigration process, stains as seen in the [NbWEr] glass ion-exchanged at 380°C by thermal diffusion were hardly observed. The [NbWEr] glass ion-exchanged at 300°C using a

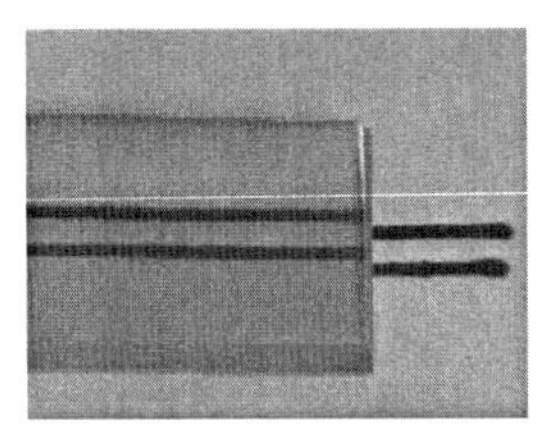

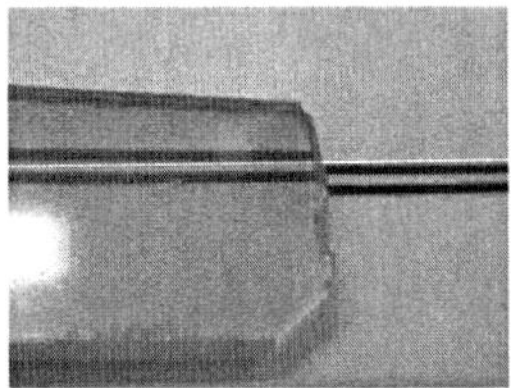

Figure 6. Photographs of [NbWEr] substrate glass (left) and [NbWEr] glass ion-exchanged at 300°C for 5 h using a dry electromigration process (right).

dry electromigration process was as clear as the [NbWEr] substrate glass. Therefore, Ag^+-Na^+ ion-exchange using a dry electromigration process for the fabrication of a tellurite glass waveguide is promising from the viewpoint of the reduction of process temperature and few stains.

CONCLUSION

(1) The planar waveguides of the [NbWEr] glasses could be fabricated under all ion-exchange conditions by both a dry electromigration process and thermal diffusion in this study.
(2) After ion-exchange by both a dry electromigration process and thermal diffusion, the transmittance of the glasses slightly decreased from that of the [NbWEr] substrate glass but frosting was not observed on these ion-exchanged glasses.
(3) The depths of the waveguides fabricated at 290 and 300°C by a dry electromigration process were deeper than those at 360 and 380°C by thermal diffusion, respectively.
(4) On the [NbWEr] glass ion-exchanged at 300°C by a dry electromigration process, the stains seen in the [NbWEr] glass ion-exchanged at 380°C by thermal diffusion were hardly observed.
(5) Ag^+-Na^+ ion-exchange using a dry electromigration process for the fabrication of a tellurite glass waveguide is promising from the viewpoint of the reduction of process temperature and few stains.

REFERENCES
[1]G.C. Righini, S. Pelli, M. Brenci, M. Ferrari, C. Duverger, M. Montagna and R. Dall'Igna, "Active optical waveguides based on Er- and Er/Yb-doped silicate glasses," J. Non-Cryst. Solids, Vol. [284], (2001), 223-229.
[2]C. De Bernardi, S. Morasca, D. Scarano, A. Carnera and M. Morra, "Compositional and stress-optical effects in glass waveguides: comparison between K-Na and Ag-Na ion exchange," J. Non-Cryst. Solids, Vol. [119], (1990), 195-204.
[3]T. Ohtsuki, S. Honkanen, N. Peyghambarian, M. Takahashi, Y. Kawamoto, J. Ingenhoff, A. Tervonen and K. Kadono, "Evanescent-field amplification in Nd^{3+}-doped fluoride planar waveguide," Appl. Phys. Lett., Vol. [69], (1996), 2012-2014.
[4]G. Sorbello, S. Taccheo, M. Marano, M. Marangoni, R. Osellame, R. Ramponi and P. Laporta, "Comparative study of Ag-Na thermal and field-assisted ion exchange on Er-doped phosphate glass," Opt. Mater., Vol. [17], (2001), 425-435.
[5]Y. Ding, S. Jiang, T. Luo, Y. Hu and N. Peyghambarian, "Optical waveguides prepared in Er^{3+}-doped tellurite glass by Ag^{+}-Na^{+} ion-exchange", Proc. SPIE-Int. Soc. Opt. Eng. (USA), Vol. [4282], (2001), 23-30.
[6]G.N. Conti, S. Berneschi, M. Bettineli, M. Brenci, B. Chen, S. Pelli, A. Speghini and G.C. Righini, "Rare-earth doped tungsten tellurite glasses and wavegides: fabrication and characterization," J. Non-Cryst. Solids, Vol. [345&346], (2004), 343-348.
[7]S. Sakida, T. Nanba and Y. Miura, "Refractive index profiles and propagation losses of Er^{3+}-doped tungsten tellurite glass waveguide by Ag^{+}-Na^{+} ion-exchange," Mater. Lett., Vol. [60], (2006), 3413-3415.
[8]S. Sakida, T. Nanba and Y. Miura, "Optical properties of Er^{3+}-doped tungsten tellurite glass waveguides by Ag^{+}-Na^{+} ion-exchange", Opt. Mater., Vol. [30], (2007), 586-593.
[9]N. Mochida, K. Takahashi, K. Nakata and S. Shibusawa, "Properties and structure of the binary tellurite glasses containing mono- and di-valent cations", Yogyo-Kyokai-Shi,Vol. [86], (1978) 316-326.
[10]B. Stepanov, J. Ren, T. Wagner, J. Lorincik, M. Frumar, M. Churbanov and Y. Chigirinsky, "Solid state field-assisted diffusion of silver in multi-component tellurite glasses," J. Non-Cryst. Solids, Vol. [357], (2011), 3022-3026.
[11]K.S. Chiang, "Construction of refractive-index profiles of planar dielectric waveguides from the distribution of effective indexes," J. Lightwave Technol. Vol. [LT-3], (1985) 385-391.

*Corresponding author: sakida@cc.okayama-u.ac.jp

SPECTRAL PROPERTIES OF Bi-Er-Yb TRIPLY DOPED BOROSILICATE GLASSES WITH 805nm EXCITATION

Dong Hoon Son,[1] Bok Hyeon Kim,[2] Seung Ho Lee,[1] Sang Youp Yim,[2] and Won-Taek Han[1,*]

[1]Department of Photonics and Applied Physics/Graduate School of Information and Mechatronics, Gwangju Institute of Science and Technology (GIST), Gwangju, 500-712, South Korea
[2]Advanced Photonics Research Institute (APRI), GIST, Gwangju, 500-712, South Korea
*Corresponding author: sdh@gist.ac.kr

ABSTRACT

Spectral properties of the Bi-Er-Yb triply doped borosilicate glasses with different concentrations of bismuth were investigated. Broad absorption bands appeared at 300-1700nm and the emission bands in near infrared and visible wavelength regions from Er^{3+} and Yb^{3+} ions showed efficient enhancement in peak intensity due to the incorporation of bismuth upon excitation at 805nm. Possible energy transfer mechanisms of Er^{3+}, Yb^{3+} and bismuth ions were proposed.

INTRODUCTION

Rare-earth ion doped glasses and crystals have drawn much attention due to their emission properties in the range of near infrared (NIR) and visible wavelengths.[1-5] Especially, Er^{3+} and Yb^{3+} codoped glasses have attracted great interest because of the emission in Er^{3+} ions near 1.5μm and the enhancement of the emission by Yb^{3+} ions codoped.[6-8]

Recently, various glasses, such as silicate, germanate, borate, phosphate, and chalcogenide glasses, doped with bismuth were extensively investigated due to their excellent ultra-broadband emission property covering 1100-1600nm upon excitation at 800 or 980nm.[9-13] Emission properties of Bi-Yb codoped glasses were intensively studied for the application of broadband amplifiers and their intensity enhancement was attributed to energy transition between Yb^{3+} and bismuth ions,[14,15] on the other hand, little attention has been paid to other glass systems such as Bi-Er codoped glasses. More recently, emission properties of a Bi-Er codoped zeolite which is a microporous crystalline solid were investigated and showed an efficient emission enhancement of Er^{3+} ions by incorporation of bismuth ions.[16] Ultra-broadband emission at 1000-1800nm was found in Bi-Er-Tm triply doped germanate glasses with 808nm excitation, being explained by emission overlapping originated from Er^{3+}, Tm^{3+} and bismuth ions. [17] Bismuth ions were known to emit a broad 1300nm band and partially transfer their energy to Er^{3+} and Tm^{3+} ions to enhance the emissions at 1534 and 1680 nm of the rare-earth ions. However, the effect of Bi_2O_3 addition on emission enhancement of Er^{3+} ions in erbium-doped bismuth silicate glasses was found to be significant only when the extent of Bi_2O_3 addition was less than 2mol% not over 20mol%. [15-18]

In this study, we have tried to fabricate Bi-Er-Yb triply doped alumino-borosilicate glasses with different concentrations of bismuth and to investigate the effect of the addition of bismuth on their spectroscopic properties. UV/VIS/NIR absorption and emission properties of the glasses upon 805nm excitation were measured and possible energy transfer process between bismuth ions and Er^{3+}/Yb^{3+} ions was studied.

EXPERIMENTAL PROCEDURES

Glass Sample Preparation

Glass samples were prepared by conventional glass melting method with the composition of $54SiO_2$-$10B_2O_3$-$15Al_2O_3$-$10Na_2O$-$8ZnO$-$1Ga_2O_3$-$2In_2O_3$ (mol% in batch composition). As doping materials, Er_2O_3, Yb_2O_3 and Bi_2O_3 were additionally incorporated to the host glass. The detailed glass compositions with dopant concentrations are listed in Table I. The SBA-A, B, C, D glasses were equally codoped with 1mol% of Er_2O_3 and 1mol% of Yb_2O_3 and different concentrations of bismuth were incorporated into the glasses in the range, 0-2mol%. As reference samples, several glasses (SBA-E, F, G) were also made with singly and doubly dopants as indicated in Table I.

For glass fabrication, SiO_2, B_2O_3, Al_2O_3, Na_2CO_3, ZnO, Ga_2O_3, and In_2O_3 (purity: 99.5-99.99%) in powder types were used as raw materials. Each 40g of a glass batch was well mixed and melted in an alumina crucible at 1550℃ for 60min in air. The glass melt was then casted onto a preheated brass mold at 500℃ and the glass was heat treated at 500℃ for 120min and cooled down to room temperature to relieve residual stress. Then, the glass was cut and polished to 10x10mm and a thickness of ~450μm.

Measurement of Optical Properties

Refractive indices of the glass samples were measured by a prism coupler (SPA-4000, Sairon Tech. Inc., Korea). Optical absorption spectra were obtained using a spectrophotometer (V-570, Jasco, Japan) in the range of 300-1700nm. Emission spectra were measured by optical pumping at 805nm with a cw-Ti:sapphire laser (Mira-900, Coherent, USA) at input power of 370mW. InGaAs and PMT detectors were used to collect the emission spectra of the glasses in the ranges of NIR and visible wavelengths, respectively.

SPECTRAL PROPERTIES OF THE GLASSES

Refractive Indices

The measured refractive indices of the glasses were summarized in Table I. By increasing bismuth concentration in the glasses, refractive indices were found to increase from 1.5380 to 1.5594 at 1550nm as expected from the fact that glass polarizability increased by incorporation of bismuth as a heavy metal element.

Table I. Glass compositions in mol% and measured refractive indices (RI)

Host glass	$54SiO_2$-$10B_2O_3$-$15Al_2O_3$-$10Na_2O$-18ZnO-$1Ga_2O_3$-$2In_2O_3$			RI
Dopant	Er_2O_3	Yb_2O_3	Bi_2O_3	
SBA-A	1	1	0	1.5380
SBA-B	1	1	0.5	1.5445
SBA-C	1	1	1	1.5498
SBA-D	1	1	2	1.5594
SBA-E	0	0	2	1.5590
SBA-F	1	0	2	1.5678
SBA-G	0	1	2	1.5664

Absorption Spectra

Fig. 1 (a) shows optical absorption spectra of the Er^{3+}-Yb^{3+} codoped glasses (SBA-A, B, C, D) with different bismuth concentrations. In the SBA-A glass without bismuth, the well known absorption peaks by Er^{3+} and Yb^{3+} ions appeared at 379, 407, 452, 487, 522, 544, 652, 795, and 1530nm and they are assigned to the glass transitions from $^4I_{15/2}$ energy level to $^4G_{11/2}$, $^2G_{9/2}$, $^4F_{5/2}$, $^4F_{7/2}$, $^2H_{11/2}$, $^4S_{3/2}$, $^4F_{9/2}$, $^4I_{9/2}$ and $^4I_{13/2}$ energy levels of Er^{3+} ions, respectively.[19] Note that a strong absorption peak appeared at 976nm and is due to the overlapped transitions, ${}^2F_{7/2}\rightarrow{}^2F_{5/2}$ of Yb^{3+} ions and ${}^4I_{15/2}\rightarrow{}^4I_{11/2}$ of Er^{3+} ions.

Fig. 1(b) shows optical absorption spectra purely by bismuth ions in the SBA-B, C, and D glasses and they were obtained by eliminating absorption from Er^{3+} and Yb^{3+} ions in the glasses. To obtain absorption spectra by bismuth ions discriminately, the measured absorption profiles of the SBA-B, C, and D glasses shown in Fig. 1(a) were subtracted by the absorption profile of the SBA-A glass only with Er^{3+} and Yb^{3+} ions. A clear absorption band centered at 470nm was found in the glasses and the absorption intensity increased with the increase of bismuth concentration. In previous studies, several absorptions by bismuth ions were additionally found at the wavelengths around 700, 800, and 1000nm as well as 470nm.[9-11] As shown in figure 1(b), weak absorption bands near 700 and 800nm were also found, on the other hand, absorption band around 1000nm was not clear in this study. The absorption baseline in the range from 300 to 1200nm increased slightly with the increase of bismuth concentration.

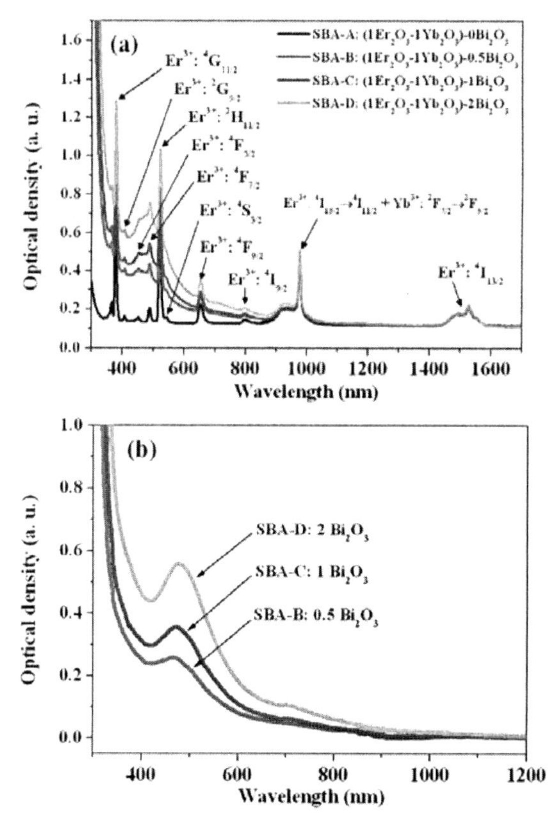

Figure 1. (a) Measured absorption spectra of the Bi-Er-Yb triply doped glasses (SBA-A, B, C, D) with different concentrations of bismuth, and (b) absorption spectra purely by bismuth ions in the SBA-B, C, and D glasses.

Near-Infrared (NIR) Emission Properties

Emission spectra of the reference glasses (SBA-E, F, G) in NIR wavelength upon pumping at 805nm by the cw-Ti:sapphire laser were represented in Fig. 2(a). A broad emission band centered at 1257nm was found to appear in the SBA-E glass only with bismuth. Similar emission bands were also reported in various glass compositions such as silicate, germanate, borate, phosphate glasses incorporated with bismuth.[9-15]

In the other glasses, similar broad emission bands were also found at slightly different center wavelengths of 1245nm (SBA-F) and 1260nm (SBA-G). But the emission intensity decreased considerably by the introduction of Er^{3+} and Yb^{3+} ions in the glasses. It is also noted that additional emission bands centered at 984 and 1020nm by Yb^{3+} ions (SBA-G) were also found and these bands were not made by Yb^{3+} ions themselves since Yb^{3+} ions don't have absorption at the excitation wavelength of 800nm. These results can be explained by energy transfer from bismuth ions to the Yb^{3+} ions. A part of the pump energy at 805nm absorbed by bismuth ions was transferred to Yb^{3+} ions and this energy transfer induced emission bands at 984 and 1020nm and in turn, decreased the emission intensity near 1245 and 1260nm.

Emission properties of the Bi-Er-Yb triply doped borosilicate glass system (SBA-A, B, C, D) upon pumping at 805nm were also investigated by varying Bi_2O_3 concentration in the range of 0-2mol% with the fixed concentrations of Er_2O_3(1mol%) of Yb_2O_3(1mol%). Fig. 2(b) shows the measured emission spectra of the glasses in NIR wavelength range by the 805nm excitation. In the SBA-A glass with Er^{3+}/Yb^{3+} ions and without bismuth ions, a strong 1540nm emission band was found and this band was assigned to the well known transition of $^4I_{13/2} \rightarrow {}^4I_{15/2}$ in Er^{3+} ions. Weak emission bands at 984 and 1020nm were also found and these bands were due to energy transfer from $^4I_{11/2}$ of Er^{3+} ions to $^2F_{5/2}$ of Yb^{3+} ions.[8]

In the case of the SBA-B, C, and D glasses, the broad emission bands near 1230nm by bismuth ions were observed together with the emission bands near 984, 1020, and 1540nm by Yb^{3+} and Er^{3+} ions. As shown in Fig. 2(b), the 1230nm emission band of bismuth ions increased by the increase of bismuth concentration. Interestingly, the 1540nm emission band of Er^{3+} ions also considerably increased by the increase of bismuth concentration due to energy transfer from bismuth ions to Er^{3+} ions. The peak intensity and the full width at half maximum (FWHM) of the 1540nm emission band were shown in the inset of Fig. 2(b). The peak intensity increased 1.5 times and the FWHM increased from 58 to 62nm by the incorporation of Bi_2O_3 up to 2.0mol%. Thus the broad emission appeared in the range of 1000-1600nm makes the Bi-Er-Yb triply doped glasses be regarded as a candidate material for broadband amplifiers and tunable lasers. Note that further investigation of quantum efficiency of Bi and Er^{3+} ions through lifetime measurement and upconversion efficiency of $^4I_{13/2}$ energy level of Er^{3+} ions, one of main loss for the NIR emission, are needed for practical application.

Fig. 2(c) compares the emission spectra of Yb^{3+} ions in the glasses with different bismuth concentrations in the range of 960-1060nm and these spectra were obtained by removing the baseline emission which originated from bismuth ions themselves. The emission profiles of the SBA-B, C, and D glasses with different bismuth concentrations shown in Fig. 2(b) were subtracted by the emission profile from bismuth ions of the SBA-B, C, and D glasses. Although the increase amount of the emission by bismuth incorporation is not so large, it is clear that the emission intensities at 984 and 1020nm gradually increased with the increase of the bismuth concentration. The inset of Fig. 2(c) shows that the peak intensity of the 984nm emission band linearly increased with bismuth concentration.

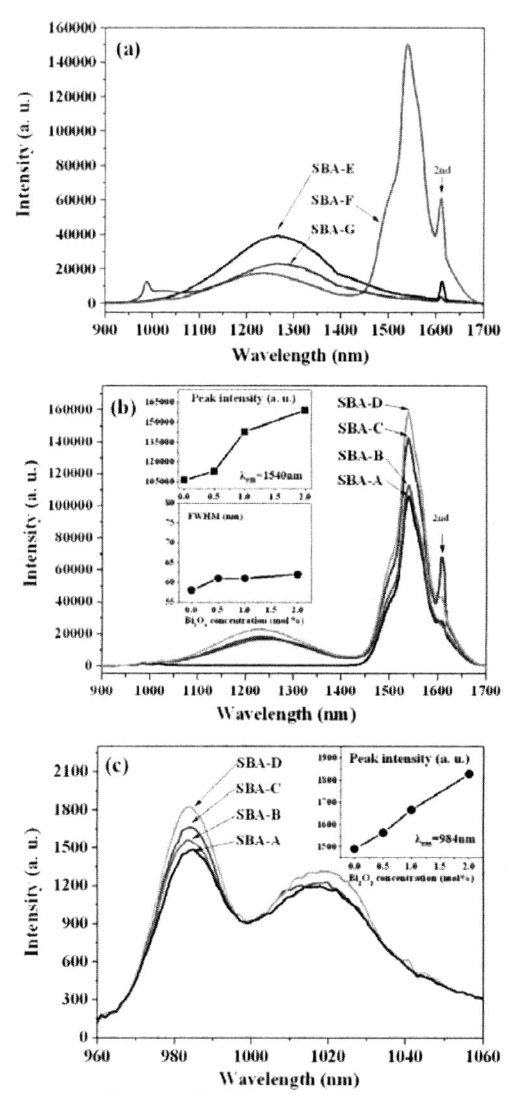

Figure 2. NIR emission spectra of (a) the SBA-E, F, and G glasses and (b) the Bi-Er-Yb triply doped glasses (SBA-A, B, C, D) with different concentrations of bismuth (Insets show the peak intensity and the FWHM at 1540nm emission band with respect to bismuth concentration.) and (c) emi (Inset shows the peak intensity at 984nm emission band.).

Visible Emission Properties

Visible emission spectra of the glasses (SBA-A, B, C, D) with Er^{3+}/Yb^{3+} ions and bismuth ions by the 805nm excitation are shown in Fig. 3(a). Two strong green emission bands centered at 524 and 548nm and a weak red emission band centered at 660nm were found and these bands are assigned to the transitions, $^2H_{11/2}\rightarrow{}^4I_{15/2}$, $^4S_{3/2}\rightarrow{}^4I_{15/2}$ and $^4F_{9/2}\rightarrow{}^4I_{15/2}$ of Er^{3+} ions, respectively.[18] These visible emission bands can be explained by an upconversion process including ground state absorption (GSA) and excited state absorption (ESA) between the energy levels of Er^{3+} ions as illustrated in Fig. 4.

It was also found that the visible emission bands of Er^{3+} ions were strongly enhanced by the introduction of bismuth ions. Fig. 3(b) shows the change in the peak intensities of the three emission bands with respect to the increase of bismuth concentration. As shown in figure 3(b), while intensities of two green emissions at 524 and 548nm increased about 2.4 and 1.8 times, respectively, the increase of red emission at 660nm was much larger (about 4.8 times) by the addition of 2mol% Bi_2O_3. These results indicate the effect of bismuth incorporation on enhancement of emission intensity was dominant on the red emission band and a possible energy transfer process will be given in the next.

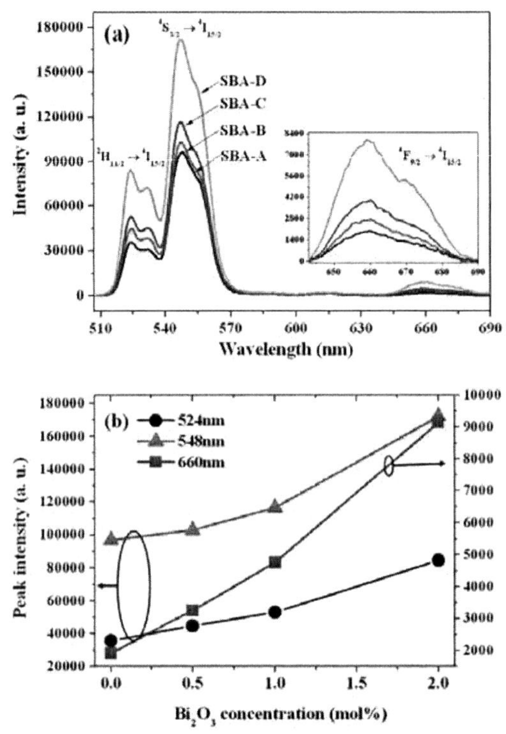

Figure 3. (a) Visible emission spectra of the Bi-Er-Yb triply doped glasses (SBA-A, B, C, D) with different concentrations of bismuth upon pumping at 805nm and (b) peak intensities of 524, 548, and 660nm emission bands with bismuth concentration.

Energy Transfer Mechanisms

To explain the enhancement of the NIR and visible emissions by bismuth ions in the Bi-Er-Yb triply doped glasses, energy transfer mechanism between Er^{3+}, Yb^{3+}, and bismuth ions was investigated with a possible energy band diagram shown in Fig. 4.

When the glass is pumped at 805nm, pump energy excites Er^{3+} ions from the ground state to $^4I_{9/2}$ energy level then moves to $^4I_{11/2}$ and $^4I_{13/2}$ by successive non-radiative multi-phonon relaxation (MPR) processes and finally returns to the ground state by emission at 1540nm. As for another path for the 1540nm emission, pump energy absorbed by bismuth ions excites the ions to ES_2 level and a part of the energy is transferred to $^4I_{9/2}$ energy level (ET_1) and thus the 1540nm emission intensity is enhanced by bismuth incorporation as shown in Fig. 2(b). The other part of the absorbed energy by bismuth ions moves from ES_2 energy level to lower ES_1 energy level through MPR and return to the ground state by emission near 1230nm.

As for the NIR emissions at 984 and 1020nm of Yb^{3+} ions, it can be explained by energy transfer from Er^{3+} and bismuth ions to Yb^{3+} ions as shown in figure 4. Pump energy in $^4I_{11/2}$ energy level of the ions after absorbed by Er^{3+} ions can transfer to $^2F_{5/2}$ energy level of Yb^{3+} ions (ET_2) and finally make emissions near 980 and 1020nm by returning to the ground state. Another path for the emissions, the pump energy excites bismuth ions and relaxes from ES_2 to ES_1 energy level of the ions by MPR then partially transfers to $^2F_{5/2}$ energy level of Yb^{3+} ions (ET_3) and makes emissions at 980 and 1020nm. Therefore, as shown in Fig. 2(a), the emission intensity near 1230nm of bismuth ions considerably decreased by Er^{3+} and Yb^{3+} ions in glasses since the absorbed energy by bismuth ions is transferred to Er^{3+} and Yb^{3+} ions by ET_1 and ET_3 processes, respectively.

Enhancement of the emission bands of Er^{3+} ions in visible range upon pumping at 805nm can be explained by cross relaxation process from bismuth ions to Er^{3+} ions. As a path of the emissions, the pump energy can be directly absorbed by Er^{3+} ions and excites the ions to $^4I_{9/2}$ energy level and moves to the higher energy levels of $^4F_{5/2}$ and $^2H_{11/2}$ in Er^{3+} ions by excited state absorption (ESA) then relaxes to $^2H_{11/2}$, $^4S_{3/2}$, and $^4F_{9/2}$ energy levels and finally returns to the ground state with the emissions near 524, 548, and 660nm, respectively. As for another path of the emissions, pump energy is absorbed by bismuth ions and moves to ES_1 energy level in the ions and the energy is transferred to $^2H_{11/2}$ and $^4F_{9/2}$ of Er^{3+} ions by cross relaxation processes (ET_4, ET_5) and added with the energy from the first path by Er^{3+} ions themselves. Therefore, this energy transfer process can explain emission enhancement by bismuth incorporation in the visible bands.

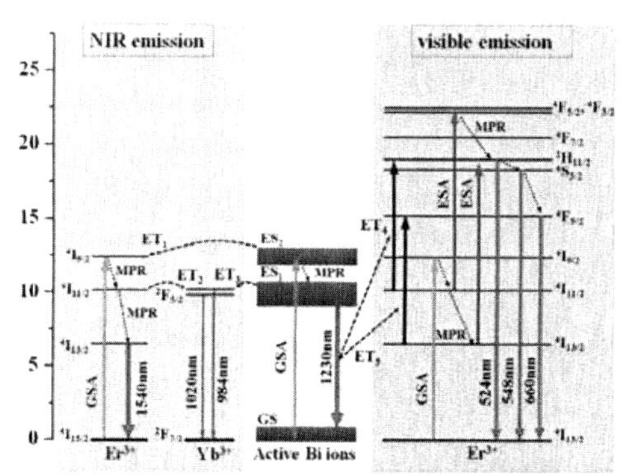

Figure 4. Possible energy band diagram to explain energy transfer process between Er^{3+}, Yb^{3+}, and bismuth ions in the Bi-Er-Yb triply doped glasses.

Fig. 5 shows enhancement rates of the peak intensities of the emission bands at 524, 548, and 660nm with bismuth concentration in the Bi-Er-Yb triply doped glasses (SBA-A, B, C, D). The increase rates were obtained from the peak intensity of the emissions with different bismuth concentrations divided by that of the SBA-A glass without bismuth using the results given in Fig. 3(b). As explained with Fig. 3(b), the enhancement rates at the green emission bands at 524 and 548nm increased about 2.4 and 1.8 times, respectively, by bismuth incorporation (2mol% of Bi_2O_3). On the other hand, the rate (4.8 times) for the red emission band at 660nm was much larger than those of the green bands. This result indicates that the energy transfer process (ET_5) is more dominant than the other process, ET_4. The slightly larger enhancement rate of the 524nm emission band than that of the 548nm band can be explained by the added energy from bismuth ions firstly populates $^2H_{11/2}$ energy level of Er^{3+} ions by ET_4 and is mostly exhausted by the emission at 524nm, then the small amount of the remained energy is transferred to $^4S_{3/2}$ energy level of Er^{3+} ions for the emission at 548nm.

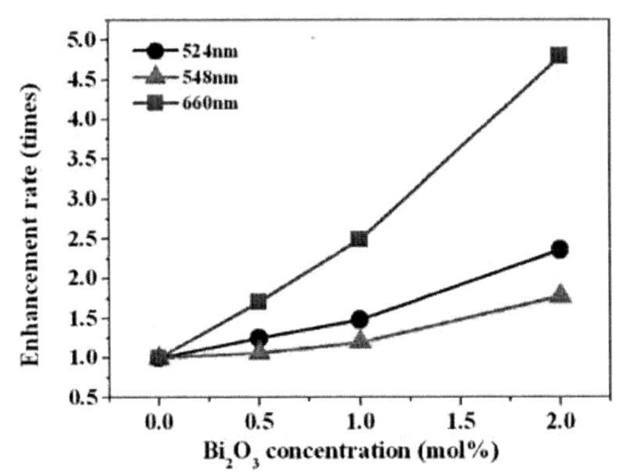

Figure 5. Enhancement rate of the peak intensities of the emission bands at 524, 548, and 660nm with bismuth concentration in the Bi-Er-Yb triply doped glasses (SBA-A, B, C, D).

CONCLUSIONS

Spectral properties of the Bi-Er-Yb triply doped borosilicate glasses were investigated to understand energy transfer process between Er^{3+}, Yb^{3+}, and bismuth ions. Borosilicate glasses codoped with Er^{3+}/Yb^{3+} ions and different concentrations of bismuth were fabricated by conventional glass melting process. Absorption bands appeared near 470, 700, and 800nm by bismuth ions and NIR and visible emissions were found upon excitation at 805nm using the cw-Ti:Sapphire laser.

While the emission intensities near 1540nm of Er^{3+} ions and 984 and 1020nm of Yb^{3+} ions were considerably enhanced by bismuth incorporation, the intensity near 1230nm of bismuth ions was found to decrease by Er^{3+} and Yb^{3+} ions. The intensities of the visible emission bands at 524, 548, and 660nm of Er^{3+} ions was also found to increase. The emission enhancement of the rare earth ions in NIR and visible range by bismuth incorporation was well explained by several energy transfer processes from bismuth ions to Er^{3+} and Yb^{3+} ions.

ACKNOWLEDGMENTS

This work was partially supported by New Growth Engine Industry Project by the Ministry of Knowledge Economy, NRF through the research programs (No. 2008-0061843 and No. 20100020794), the Gwangju Institute of Science and Technology Top Brand Project (Photonics 2020), the Brain Korea-21 Information Technology project, and the Asian laser center program of GIST, the Ministry of Education, Science and Technology, South Korea.

REFERENCES

[1]D. Matsuura, Red, green, and blue upconversion luminescence of trivalent-rare-earth ion-doped Y_2O_3 nanocrystals, *Appl. Phys. Lett.*, **81**, 4526-4528 (2002).

[2]H. Lin, D. Yang, G. Liu, T. Ma, B. Zhai, Q. An, J. Yu, X. Wang, X. Liu, and E. Y. B. Pun, Optical absorption and photoluminescence in Sm^{3+} - and Eu^{3+} -doped rare-earth borate glasses, *J. Lumin.*, **113,** 121-128 (2005).

[3]S. Tanabe, N. Sugimoto, S. Ito, and T. Hanada, Broad-band 1.5μm emission of Er^{3+} ions in bismuth-based oxide glasses for potential WDM amplifier, *J. Lumin.*, **87-89,** 670-672 (2000).

[4]N.Q. Wang, X. Zhao, C.M. Li, E.Y.B. Pun, and H. Lin, Upconversion and color tunability in $Tm^{3+}/Ho^{3+}/Yb^{3+}$ doped low phonon energy bismuth tellurite glasses, *J. Lumin.*, **130**, 1044-1047 (2010).

[5]Y. Chen, Y. Huang, and Z. Luo, Spectroscopic properties of Yb^{3+} in bismuth borate glasses, *Chem. Phys. Lett.*, **382**, 481-488 (2003).

[6]X. Shen, Q. Nie, T. Xu, and Y. Gao, Optical transitions of Er^{3+}/Yb^{3+} codoped TeO_2-WO_3-Bi_2O_3 glass, *J. Spec. Act. Part A*, **61**, 2827-2831 (2005).

[7]L. Zhang, H. Hu, C. Qi, and F. Lin, Spectroscopic properties and energy transfer in Yb^{3+}/Er^{3+} -doped phosphate glasses, *Opt. Mat.*, **17**, 371-377 (2001).

[8]M. Tsuda, K. Soga, H. Inoue, S. Inoue, and A. Makishima, Effect of Yb^{3+} doping on upconversion emission intensity and mechanism in Er^{3+}/Yb^{3+}-codoped fluorozirconate glasses under 800nm excitation, *J. Appl. Phys.*, **86**, 6143-6149 (1999).

[9]Y. Fujimoto, and M. Nakatsuka, Infrared Luminescence from Bismuth-Doped Silica Glass, *Jpn. J. Appl. Phys.*, **40**, L279-L281 (2001).

[10]M. Peng, J. Qiu, D. Chen, X. Meng, I. Yang, X. Jiang, and C. Zhu, Bismuth- and aluminium-codoped germanium oxide glasses for super-broadband optical amplification, *Opt. Lett.*, **29**, 1998-2000 (2004).

[11]X. Meng, J. Qiu, M. Peng, D. Chen, Q. Zhao, X. Jiang, and C. Zhu, Near infrared broadband emission of bismuth-doped aluminophosphate glass, *Opt. exp.*, **13**, 1628-1634 (2005).

[12]M. A. Hughes, T. Akada, T. Suzuki, Y. Ohishi, and D.W. Hewak, Ultrabroad emission from a bismuth doped chalcogenide glass, *Opt. Exp.*, **17**, 19345-19355 (2009).

[13]X. Meng, J. Qiu, M. Peng, D. Chen, Q. Zhao, X. Jiang, and C. Zhu, Infrared broadband emission of bismuth-doped barium-aluminum-borate glasses, *Opt. exp.*, **13**, 1635-1642 (2005).

[14]J. Ruan, Y. Chi, X. Liu, G. Dong, G. Lin, D. Chen, E. Wu, and J. Qiu, Enhanced near-infrared emission and broadband optical amplification in Yb–Bi co-doped germanosilicate glasses, *J. Phys. D: Appl. Phys.*, **42**, 155102 (2009).

[15]N. Dai, B. Xu, Z. Jiang, J. Peng, H. Li, H. Luan, L. Yang, and J. Li, Effect of Yb^{3+} concentration on the broadband emission intensity and peak wavelength shift in Yb/Bi ions co-doped silica-based glasses, *Opt. Exp.*, **18**, 18642-18648 (2010).

[16]Z. Bai, H. Sun, T. Hasegawa, M. Fujii, F. Shimaoka, Y. Miwa, M. Mizuhata, and S. Hayashi, Efficient near-infrared luminescence and energy transfer in erbium/bismuth codoped zeolites, *Opt. Lett.*, **35**, 1926-1928 (2010).

[17]R. Yang, M. Mao, Y. Zhang, Y. Zhuang, K. Zhang, and J. Qiu, Broadband near-infrared emission from Bi-Er-Tm co-doped germinate glasses, *J. Non-Cryst. Solids*, **357**, 2396-2399 (2011).

[18]J. Yang, S. Dai, N. Dai, S. Xu, L. Wen, L. Hu, and Z. Jiang, Effect of Bi_2O_3 on the spectroscopic properties of erbium-doped bismuth silicate glasses, *J. Opt. Soc. Am. B*, 20, 810-815 (2003).

[19]W.T. Carnall, P.R. Fields, and K. Rajnak, Electronic Energy Levels in the Trivalent Lanthanide Aquo Ions, I. Pr^{3+}, Nd^{3+}, Pm^{3+}, Sm^{3+}, Dy^{3+}, Ho^{3+}, Er^{3+}, and Tm^{3+}, *J. Chem. Phys.*, **49**, 4424-4442 (1968).

FABRICATION AND ESTIMATION OF DIFFUSION COEFFICIENT OF Pb IN PbO/GeO_2-CODOPED OPTICAL FIBER WITH THERMALLY EXPANDED CORE

Seongmin Ju,[1] Pramod R. Watekar,[2] Dong Hoon Son,[1] Taejin Hwang,[3] and Won-Taek Han[1]

[1]Department of Information and Communications/Department of Photonics and Applied Physics, Gwangju Institute of Science and Technology, 261 Cheomdan-gwagiro, Buk-Gu, Gwangju, 500712, Korea; [2]Sterlite Optical Technologies, E1, E2, E3, MIDC Waluj, Aurangabad 431136, India; [3]Production Technology R&D Department, Korea Institute of Industrial Technology, 7-47, Songdo-dong, Yeonsu-Gu, Incheon, 406840, Korea

ABSTRACT

Diffusion coefficient of PbO in PbO/GeO_2-codoped optical fibers was estimated by using the change in mode field diameter of the fiber upon heat treatment for core expansion. The fiber showed the highest expansion in the MFD by about two times after heat treatment at 1200°C for 1 hour and the diffusion coefficient of Pb ions was found to be about 1.60×10^{-14} m^2/s at 1200°C.

INTRODUCTION

Fibers with thermally diffused expanded core (TEC) have been developed to reduce the splice loss between the optical fiber and other fibers or components and to increase the margin of misalignment. TEC fiber is simply made by using diffusion of large refractive index dopant ions in fiber core into cladding region radially.[1-6] Although several methods such as butt-joint coupling,[7] bulk lens systems,[8-10] and lensed fibers[11] are available to minimize optical coupling loss occurring due to modal profile mismatch, drawback of them is very small tolerance against misalignment. Moreover, the extent of core expansion or mode field diameter (MFD) due to diffusion of GeO_2 in optical fiber core is practically limited because of very small diffusion coefficient of GeO_2, 2.46×10^{-17} m^2/s at 1200°C in silica glass fiber.[12] In addition, conventional thermal diffusion method using a micro-burner lacks controllability in temperature and it induces a dimensional variation of fiber after heat treatment. Thus, TEC fibers with fast diffusion of dopant ions can be a solution and seeking such dopants is important to shorten the time of heat treatment for expansion of the fiber core maintaining the single mode condition.[4,13]

In this paper, we report a new optical fiber incorporated with PbO in addition to GeO_2 in the fiber core to enhance the thermal diffusion because diffusion coefficient of PbO is much larger than that of GeO_2. We have also demonstrated a new method to estimate diffusion coefficient of dopant in the core region of the optical fiber using MFD of expanded core fiber by use of halogen lamp as a heat source.[14] Optical properties of PbO/GeO_2-codoped TEC fibers have been also investigated by using the comparative modeling method.

This study has been arranged as follows by answering the following questions: (a) How can the MFD method be used and justified to determine the diffusion coefficient of ions? To answer this query, we have performed experimental investigation to determine core diameter change of the TEC fiber by the MFD analysis and by the electron probe micro-analyzer (EPMA) method, where we found that there was a very good match between two results. (b) How can the addition of Pb ions be justified to increase the core size? To prove it, we have done several experimental investigations as described in the subsequent sections and justified that Pb ions are one of the best options to increase the core size and thereby to develop a highly expanded core optical fiber. And lastly, (c) How can contributions of Pb ions and Ge ions be separated for enhanced core size upon heat treatment? We tried to differentiate the contributions of Ge-ions and Pb-ions.

EXPERIMENTAL PROCEDURE

To fabricate PbO/GeO_2-codoped optical fibers, PbO/GeO_2-codoped fiber preforms were made by using the modified chemical vapor deposition (MCVD) process. PbO was doped in the core by soaking the silica glass tube deposited inside with GeO_2-doped core layers in doping solution for two hours. The solution was prepared by dissolving reagent grade PbO powder in nitric acid solution to maintain the PbO concentration of 0.0265 mole. Then the tube was sintered and sealed to a fiber preform. The preform was drawn into a fiber using the Draw Tower (DT) at 2150°C. To estimate the diffusion coefficients of PbO, the PbO/GeO_2-codoped fibers of the different core sizes were fabricated. For the sake of comparison, optical fibers doped with only GeO_2 were also fabricated. Optical parameters of the fabricated fibers are listed in Table I. General structure of the fabricated fibers is illustrated in Fig. 1 and the cross-sectional images of the fibers are shown in Fig. 2.

Table I. Optical parameters of the fabricated optical fibers

Parameter	GeO_2-doped fiber (Fiber-1)	PbO/GeO_2-codoped fiber (Fiber-2)	GeO_2-doped fiber (Fiber-3)	PbO/GeO_2-codoped fiber (Fiber-4)
Core refractive index difference (Δn)	0.0369	0.0227	0.0067	0.0067
Core diameter (μm)	2.68	3.11	6.44	6.50
Effective mode field diameter (μm)	3.80	4.95	8.95	8.94

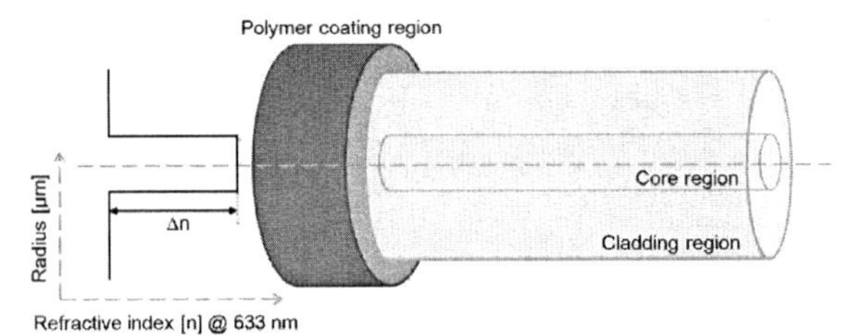

Figure 1. General structure of the fabricated fibers.

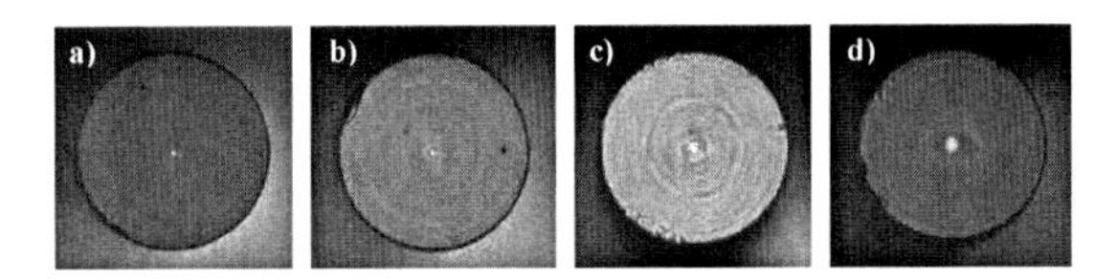

Figure 2. Cross-sectional images of (a) GeO_2-doped fiber (Fiber-1), (b) PbO/GeO_2-codoped fiber (Fiber-2), (c) GeO_2-doped fiber (Fiber-3), and (d) PbO/GeO_2-codoped fiber (Fiber-4).

To estimate diffusion coefficient of the dopant in the core of the optical fiber using the MFD of the expanded core fiber, the heat treatment was carried out at 1200°C for different time durations of 10 ~ 60 minutes by broadly focusing the light from the halogen lamp equipped in the image furnace onto the fiber. The radiation from the halogen lamp was measured to be at near infrared from 0.9 μm to 1.6 μm. After the heat treatment, the fiber was cut at the center of the heated section into two pieces and then MFD was measured at 1.55 μm using the far-field pattern method.[15] Diffusion coefficient of PbO and GeO_2 in the core expanded germano-silicate glass optical fiber was calculated using the measured MFD. The propagation loss in the fiber was obtained by comparing optical loss measured with the optical spectrum analyzer (OSA, Ando AQ6317B) before and after the heat treatment.

THEORY

The diffusion coefficient of dopant in core of a fiber can be calculated using a change in core diameter before and after thermal expansion by heat treatment. There are difficulties involved in determining exact core diameter after diffusion and one has to use the EPMA method to determine the dopant concentration and its profile and then determine the core diameter. Now if dopant concentration is small enough (for example, PbO concentration in this study) to be detected by the EPMA, the only way that remains is to use the expensive concentration determination equipments. We overcame this difficulty by using a simple MFD determination method where optical far field was measured and used to determine the change of core diameter due to application of heat.

In the case of a single mode optical fiber with any arbitrary profile, when there is the diffusion of Ge ions from core to cladding, total amount of GeO_2 is maintained while its refractive index profile shape changes. To estimate diffusion of ions from the core, we need a parameter that can address localized change in Ge concentration. A mode field diameter of the optical fiber depends on the profile shape and hence it can be a useful measurement to study diffusion of ions from core to cladding upon heat treatment. Far field MFD can be determined from the experimental measurement of the far-field power as given below,[11,16]

$$MFD = \frac{\lambda}{\pi}\left[\frac{2\int_0^{\pi} F^2(\theta)\sin(\theta)\cos(\theta)d\theta}{\int_0^{\pi} F^2(\theta)\sin^3(\theta)\cos(\theta)d\theta}\right]^{\frac{1}{2}} \quad (1)$$

where $F^2(\theta)$ and λ are the angular far-field power distribution and the mean wavelength of the light source, respectively. By using Eq. (1) and the well-known diffusion equation (Eq. (2)), we can determine diffusion coefficients, D [m^2/s], of GeO_2 and PbO in the optical fiber.

$$D = \frac{\langle [2(x_j - x_i)]^2 \rangle}{q_i t} \quad (2)$$

where x_i [m] and x_j [m] are diameters of the fiber core before and after heat treatment, respectively, t is heat treatment time, and q_i is numerical constant which depends on dimensionality: $q_i = 2$, 4, or 6, for 1, 2, or 3 dimensional diffusion, respectively.

Now coming to the first point in the introduction, i.e., validity of the MFD method to determine the change of core diameter, we directly measured the far field MFD of the GeO_2-doped fiber (Fiber-1) at 25°C. The fiber was then heat treated at 1200°C for 60 minutes and the far field MFD was again measured. The measured far field MFD before and after the heat treatment is listed in Table II, where the MFD change was about 1.08 μm. The same fiber samples were then taken for the EPMA and the results are shown in Table II and Fig. 3, where the core diameter change was about 1.04 μm before and after heat treatment at 1200°C for one hour. The error in estimation of the core diameter change by using the MFD method is below 4%, and this justifies the validity of MFD method to determine the diameter change of the core after the heat treatment.

Table II. Change in core diameter of the GeO_2-doped optical fiber (Fiber-1)

(a) Far field MFD method:	(b) EPMA method:
MFD at 25 °C = 3.60 μm MFD at 1200 °C (60 min) = 4.68 μm MFD change = 1.08 μm	Core diameter change = 1.04 μm
Error between the two methods = 3.85 %	

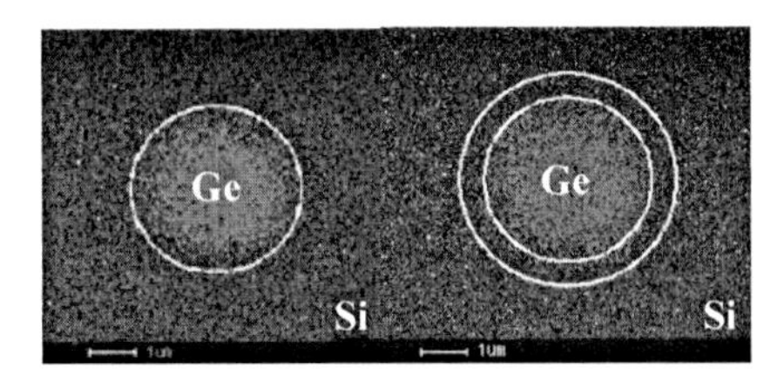

Figure 3. Concentration profiles of GeO_2 in the optical fiber (Fiber-1) before (left hand side) and after the heat treatment at 1200 °C for 60 minute (right hand side).

RESULTS AND DISCUSSION

Variations of the measured far field MFD of the different fibers (Fiber-1 to Fiber-4) heat treated at 1200°C at various heat treatment time are shown in Fig. 4, which is a linear curve fit to obtain the MFD at 1200°C (time = 0 min) for the various fibers. The temperature of the IR heater increased from 25°C to 1200°C and then it was kept constant at 1200°C.

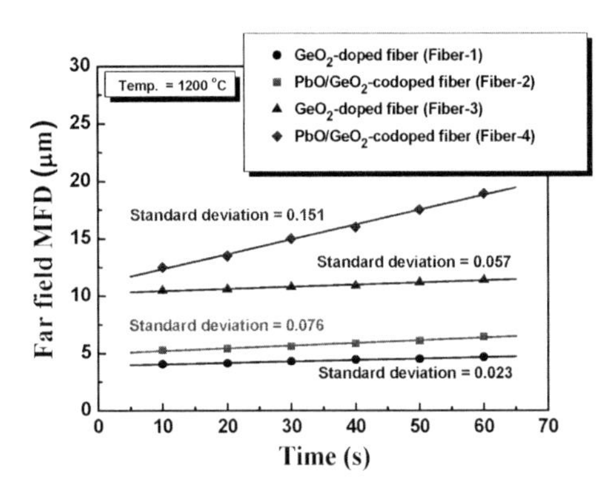

Figure 4. Measured far field MFDs of the fabricated fibers

The far field MFD of the GeO_2-doped fiber increased by about 18% (Fiber-1) and 10% (Fiber-3) after heat treatment at 1200°C for 0 to 60 minutes, that of the PbO/GeO_2-codoped fiber increased by about 27% (Fiber-2) and 69% (Fiber-4) by thermal treatment for 0 to 60 minutes at

1200°C. These results clearly explain the answer for the second question about justification of addition of Pb ions to enhance the core expansion. It is worth noting that the measured light propagation loss at 1550 nm of both the GeO_2-doped fiber and the PbO/GeO_2-codoped TEC fiber was found to be as low as 0.15 dB. The far field MFD at 1200°C was found to increase with the increase of time from 0 min to 60 min as shown in Table III and the diffusion coefficient was determined from Eq. (1) and Eq. (2) with the numerical constant, q_i of 4 (because the far-field MFD was obtained from the 2-dimensional diffusion in the cross-sectional plane). As shown in Table IV, especially the PbO/GeO_2-codoped fibers show a large change in the far field MFD at 1200°C for 1 hour and therefore a very large diffusion coefficient is expected. The PbO/GeO_2-codoped fiber, Fiber-4 (Fiber-2) showed nearly 49 times (3.5 times) larger diffusion coefficient than the GeO_2-doped fiber, Fiber-3 (Fiber-1). These results indicate that the diffusion coefficient of dopant was different with the core size of the fiber due to the difference in the dopant concentration and the dopant distribution in the core.

Table III. Effect of heat treatment time (at 1200°C) on the MFD of the optical fibers

	GeO_2-doped fiber (Fiber-1)	PbO/GeO_2-codoped fiber (Fiber-2)	GeO_2-doped fiber (Fiber-3)	PbO/GeO_2-codoped fiber (Fiber-4)
1200 °C (0 min)	MFD = 3.94 μm	MFD = 5.00 μm	MFD = 10.26 μm	MFD = 11.07 μm
1200 °C (60 min)	MFD = 4.68 μm	MFD = 6. 38 μm	MFD = 11.35 μm	MFD = 18.73 μm

Table IV. Diffusion coefficients of Pb and Ge ions (at 1200°C, 60 min) in optical fibers. The fabricated MFD difference was obtained between its values at 0 min and 60 min of heating time

	GeO_2-doped fiber (Fiber-1)	PbO/GeO_2-codoped fiber (Fiber-2)	GeO_2-doped fiber (Fiber-3)	PbO/GeO_2-codoped fiber (Fiber-4)
MFD difference (μm)	0.74	1.38	1.10	7.66
Diffusion coefficient (m^2/s)	1.51×10^{-16}	5.23×10^{-16}	3.33×10^{-16}	1.63×10^{-14}
Standard deviation	4.43×10^{-17}	2.14×10^{-17}	1.08×10^{-16}	1.80×10^{-15}

The diffusion coefficient of PbO with the high concentration gradient in the core is expected to be larger than that with the small concentration gradient in the same core area. In the case of different diffusion areas, however, the diffusion coefficient of dopant depends on the concentration gradient and the initial diffusion area. In this experiment, even though the optical fiber (Fiber-4) with 11.07 μm core diameter ($D = 1.63\times10^{-14}$ m^2/s) has larger diffusion coefficient of dopants than Fiber-2 with 5.00 μm core diameter ($D = 5.28\times10^{-16}$ m^2/s), the concentration of dopants in the fiber core region was small due to the difference of the initial area. Thus, it can be stated that the diffusion coefficient of dopant depends more on the initial diffusion area than the dopant concentration because the dopant in the fiber core diffuses out radially and the diffusion coefficient of PbO is larger than that of GeO_2.

As shown in Table IV, the diffusion coefficient of the PbO/GeO_2-codoped fibers was larger than that of the GeO_2-doped fiber and therefore we obtained a large expansion of core diameter after heating at 1200°C for 60 minutes, giving a very efficient thermally expanded core fiber. However, a question remains, whether the large diffusion coefficient and subsequent high expansion of the core is because of PbO or GeO_2 or both? In other words, what would be the diffusion coefficient of PbO in Fiber-2 and Fiber-4, if no GeO_2 is in the core? To address this issue, we need to estimate the contributions of each dopant in the optical fiber core to the measured diffusion coefficient by finding out relationships between mode field diameter, temperature, time and concentration. To make the thing simple, accurate and practicable, we chose the GeO_2-doped optical fibers where it was easy to calculate radial distributions of the mode fields by using the known refractive index profile parameters such as core diameter and core index peak at room temperature. Using the measured data of the far field MFDs at 1200°C (60 min) for GeO_2-doped fibers, we established an empirical relationship between the far field MFD and the radial mode field distribution given by:

$$\Delta MFD_{ff} = 0.47365 + 0.06938 \times 2\sqrt{\frac{2\pi\left(\int_0^\infty E^2 r dr\right)^2}{\left(\pi\int_0^\infty E^4 r dr\right)}} \qquad (3)$$

where ΔMFD_{ff} is the absolute difference of the far field MFDs in μm between 0 min (an instance of reaching temperature from 25°C to 1200°C) and 60 min of heat treatment at 1200°C, E is the radial electric field distribution (mode field, $W^{1/2}/m$) at 25°C and r is the radial parameter in m. Eq. (3) was directly used to estimate the far field MFD difference of the PbO/GeO_2-codoped fibers (if only GeO_2 effect is considered) heated for 60 min at 1200°C. By using Eq. (2) and Eq. (3), it was found that the contribution of GeO_2 to the diffusion coefficient of the PbO/GeO_2-codoped fiber (Fiber-2) was about 35%, while it was about just 2% for Fiber-4. Contribution of each dopant to the diffusion coefficient of PbO and GeO_2 in the fibers is listed in Table V. It is evident from Table V that the Pb ions exhibited very large diffusion coefficient, which contributed to the enhancement of core diameter after heat treatment at 1200°C in the TEC fiber. For instance, the diffusion coefficients of PbO in Fiber-2 and Fiber-4 were 3.42×10^{-16} m^2/s and 1.60×10^{-14} m^2/s at 1200 °C, respectively.

Table V. Effect of heat treatment time (at 1200°C) in the MFD of the optical fibers

	PbO/GeO_2 -codoped fiber (Fiber-2)	Equivalent GeO_2-doped fiber (Fiber-2)	PbO/GeO_2 -codoped fiber (Fiber-4)	Equivalent GeO_2-doped fiber (Fiber-4)
MFD difference 1200°C	1.38 μm	0.82 μm	7.66 μm	1.09 μm
Diffusion coefficient [m^2/s] of PbO and GeO_2 at 1200°C	5.28×10^{-16}	1.86×10^{-16}	1.63×10^{-14}	3.33×10^{-16}
Diffusion coefficient [m^2/s] of of PbO only at 1200°C	3.42×10^{-16}	--	1.60×10^{-14}	--

* Far field MFD after heat treatment at 1200°C for 1 hr

CONCLUSIONS

Diffusion characteristics of a new TEC fiber based on the PbO/GeO_2-codoped fiber upon heat treatment using the halogen lamp was investigated. The MFD of the PbO/GeO_2-codoped fibers were expanded from about 5.00 μm to 6.38 μm (Fiber-2) and from about 11.07 μm to 18.73 μm (Fiber-4) after heat treatment at 1200°C for 1 hour, far greater than those of the GeO_2-doped fibers.

The diffusion coefficient of PbO in the optical fiber core was found to be 3.42×10^{-16} m^2/s (for Fiber-2) and 1.60×10^{-14} m^2/s (for Fiber-2).

ACKNOWLEDGMENTS

This work was supported partially by the Ministry of Science and Technology, the NRF through the research programs (No. 2008-0061843 and No. 20100020794), the New Growth Engine Industry Project of the Ministry of Knowledge Economy, the Core Technology Development Program for Next-Generation Solar Cells of Research Institute of Solar and Sustainable Energies (RISE), the Brain Korea-21 Information Technology Project, and by the (Photonics 2020) research project through a grant provided by the Gwangju Institute of Science and Technology in 2012, South Korea.

REFERENCES

[1]K. Shiraishi, Y. Aizawa, and S. Kawakami, J. Lightwave Technol., **8**, 1151-1161 (1990).
[2]H. Hanafusa, M. Horiguchi, and J. Noda, Electron. Lett., **27**, 1968-1969 (1991).
[3]M. Kihara, M. Matsumoto, T. Haibara, and S. Tomita, J. Lightwave Technol., **14**, 2209-2214 (1996).
[4]Y. Ando and H. Hanafusa, IEEE Photonics Tech. Lett., **4**, 1028-1031 (1992).
[5]S. Savović, and A. Djordjevich, Optical Materials, **30**, 1427-1431 (2008).
[6]G.S. Kliros and N. Tsironikos, Optik: Int. J. Light Electron Opt., **116**, 365-374 (2005).
[7]Y. Ohtera, O. Hanaizumi, and S. Kawakami, J. Lightwave Technol., **17**, 2675-2681 (1999).
[8]E. Weidel, Electron. Lett., **11**, 436-437 (1975).
[9]Y. Odagir and K. Kobayashi, in Technical Digest, Ann. Meet. IECE, Tokyo, Japan, **891** (1980).
[10]M. Saruwatari and T. Sugie, IEEE J. Quantum Electron., **QE-17**, 1021-1027 (1981).
[11]H. Zhou, W. Liu, Y. Lin, S.K. Mondal, and F.G. Shi, IEEE Trans. Adv. Packag., **25**, 481-487 (2002).
[12]H. Yamada and H. Hanafusa, IEEE Photonic Tech. L., **6**, 531-533 (1994).
[13]J. DiMaio, B. Kokuoz, T.L. James, T. Harkey, D. Monofsky, and J. Ballato, Opt. Express, **16**, 11769-11775 (2008).
[14]S. Ju, P.R. Watekar, C.J. Kim, and W.-T. Han, J. Non-Cryst. Solids, **356**, 2273-2276 (2005).
[15]M. Artiglia, G. Coppa, P.D. Vita, H. Potenza, and A. Sharma, J. Lightwave Technol., **7**, 1139-1152 (1989).
[16]R. Tewari, M. Basu, and H.N. Acharya, Optics Communication, **174**, 405-411 (2000).

*Corresponding author: jusm@gist.ac.kr

EFFECTS OF REDUCING AGENT ON PHOTOLUMINESCENCE PROPERTIES OF COPPER ION DOPED ALKALI BOROSILICATE PHASE-SEPARATED GLASSES

Fumitake Tada,[1] Sayaka Yanagida[1] and Atsuo Yasumori[1]
[1]Department of Materials Science and Technology, Tokyo University of Science
2641 Yamazaki, Noda, 278-8510 Chiba, Japan

ABSTRACT

The Cu^+ clusters containing sodium-borosilicate phase-separated glasses were prepared by addition of SnO as a reducing agent. The effects of the amount of SnO on the formation of Cu^+ clusters and photoluminescence (PL) properties were investigated. The obtained glasses showed the yellow PL owing to Cu^+ clusters and its color coordinates were close to that of an incandescent lamp. By increasing the additive amount of SnO, the absorption due to Cu^+ clusters in the glasses increased, however, the excess increase of Cu^+ clusters resulted in the saturation of the yellow PL intensity because of the decrease of the penetration depth of the excitation light.

INTRODUCTION

Energy-savings is required for various lighting systems in order to control energy consumption. White light emitting diode (LED) is the most effective lighting system because the LED lighting has various advantages such as long lifetime, high emission efficiency, a small size, a light mass, a simple electric circuit and high shock-resistance. Recently, there are various emission systems to obtain white color light by use of a LED chip. "One-chip type" is the most typical one. This type of white LED consists of a blue LED chip and yellow light emitted phosphor which excited by blue emission light and the pseudo white light is made by mixture of blue and yellow lights.[1-3] However, a low thermal durability and a cool color light become problems in this type when it is applied to interior lighting, because the yellow light phosphors are embedded in an organic polymer matrix and the intensity of blue light is too strong to use as a warm white light.[2, 4]

It is known that the optical absorption and photoluminescence (PL) properties of Cu ions in silicate glasses depends on their valence state. The Cu^{2+} ion doped glass shows blue colors by their d-d transition,[5, 6] and the Cu^+ ion doped one is colorless but exhibits blue light emission under UV light irradiation.[7] The glass containing Cu_2O or Cu^0 colloids shows red color by their absorption or surface plasmon resonance (SPR) absorption, respectively.[8] On the other hand, the valence state represented as Cu^+ clusters exhibits yellow light emission under near-UV light irradiation of which color is expected to be very suitable for warm color lighting.[9] In order to form the Cu^+ clusters in the glass, the Cu^{2+} ions should be close to a reducing agent such as Sn^{2+} ions, the isolated Cu^+ ions should also be close to each other and the Cu ions must not proceed excessively to reduction state such as Cu^0 colloids.

In order to obtain a warm white light, we have prepared Cu ion doped alkali-borosilicate glasses which emitted a yellow light by irradiation of near-UV light in the range from 360-400nm.[10] This glass can have a high transparency in visible light range, a high thermal durability, a high chemical durability and a high mechanical strength. We controlled the valence state of Cu ions and formed Cu^+ clusters by use of phase separation of an alkali-borosilicate glass, because transition metal ions such as Cu were well known to concentrate in alkali-borate rich phase.[11] It was experimentally confirmed because the PL intensity from the Cu and Sn ions co-doped phase separated glasses in the

Na_2O-B_2O_3-SiO_2 system steeply decreased after the Na_2O-B_2O_3 rich phase was selectively leached out.[10] In order to obtain higher intensity of yellow light emission from the phase-separated glass containing Cu^+ clusters, it is necessary to examine the effect of the amount of a reducing agent, SnO, on the emission properties such as the intensity and its excited and emitted wavelength range. In this study, we investigated the effects of SnO on the formation of Cu^+ clusters and their emission properties by changing the SnO content in the Cu ion doped phase-separated sodium borosilicate glasses. The PL and the absorption properties of the glasses were examined.

EXPERIMENT

The composition of the mother glass was $6.6Na_2O$-$28.3B_2O_3$-$65.1SiO_2$ (mol%), which is located at the center of the metastable immiscibility region in this ternary system.[12] This composition was selected because the glass of this composition showed highest PL intensity in our previous study.[10] Cu_2O (0.2 mol%) and SnO (1-7 mol%) was externally added to the glass composition. The glasses were prepared by use of a conventional melt quenching method at 1500°C for 1h in air using an alumina crucible. The PL and the absorption properties of the glasses were examined for the polished plate samples (thickness 1.3-1.7 mm) by use of a fluorescence spectrophotometer (JASCO FP-6500) and an ultraviolet-visible (UV-vis.) spectrophotometer (SHIMADZU UV-2400PC), respectively.

RESULTS AND DISCUSSIONS

Absorption Properties

The prepared (quenched) glasses were colorless and transparent without obvious light scattering, but they are considered to have very fine phase separation texture.[10] The UV-vis. absorption spectra of the mother glass and the SnO doped one are shown in Figure 1. The absorption up to around 340 nm is due to the $S_0 \rightarrow T_1$ transition of Sn^{2+} ions.[13] Figure 2 shows the UV-vis. absorption spectra of Cu/Sn ions co-doped samples with Sn ion non-doped one. There are four valence states of Cu in the glass as mentioned above, which are Cu^+ isolated ions, Cu^+ clusters, Cu^{2+} ions and Cu_2O or Cu^0 colloids. The weak and broad absorption at around 600-800 nm which is attributed to the d-d transition

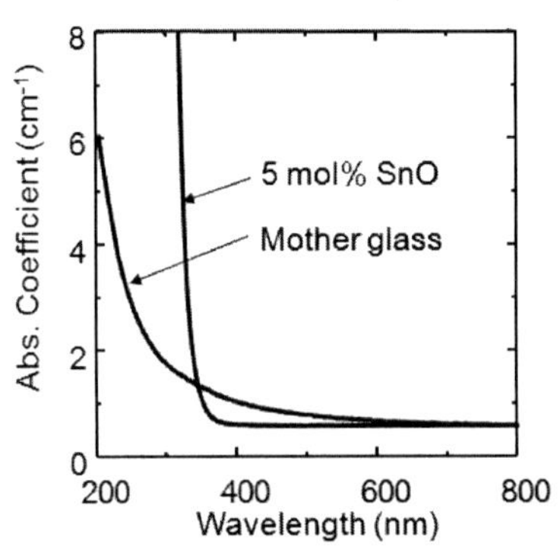

Figure 1. UV-vis. absorption spectra of the mother glass and the Sn ion doped one.

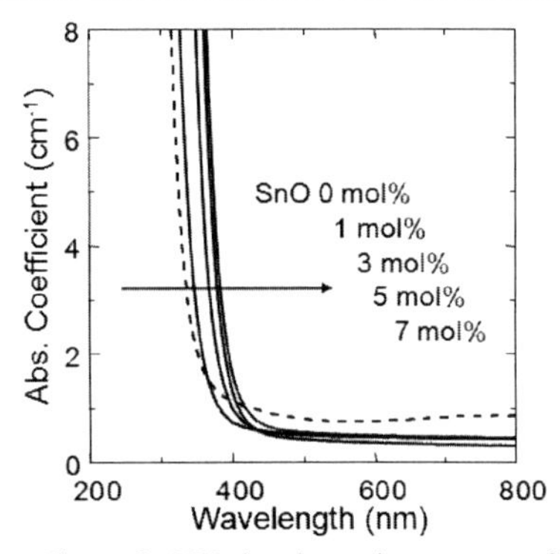

Figure 2. UV-vis. absorption spectra of Cu/Sn ions co-doped samples with Sn ion non-doped one.

of Cu^{2+} ions[5,6] disappeared by addition of 1 mol% of SnO. The absorption edge wavelength red shifted with the increase in SnO content, whereas the absorption due to the Cu_2O or Cu^0 colloids at around 570 nm did not appear on the spectra of the samples containing large amounts of SnO.[8]. These results of the absorption spectra indicate that Sn ions work as reducing agent for the reduction from Cu^{2+} to Cu^+ and that the absorption in the range of 300-400 nm is probably caused by Cu^+ isolated ions and/or Cu^+ clusters.

Photoluminescence Properties

Figure 3 shows the change of the excitation and the emission spectra (monitor wavelength 600 nm and excitation wavelength 365 nm) of the Cu/Sn ions co-doped samples with the increase in SnO content. The yellow emission of which peak top wavelength is at around 600 nm owing to Cu^+ clusters were observed in all samples. The emission intensity steeply increased with the increase in SnO content up to 3 mol% and then slightly decreased with further increase in SnO content. The peak wavelength of the excitation spectra red shifted with the increase in SnO content. This red-shift is coincident with that of the absorption edge shown in Figure 2. Therefore, these results indicate that the absorption in the range of 300-400 nm is attributed to Cu^+ clusters. In order to confirm these discussions, the excitation and emission spectra of the samples were measured by changing the excitation wavelength. Figure 4 shows the change of the excitation and emission spectra (monitor wavelength 635 nm and excitation wavelength 385 nm) of the samples with the increase in SnO content. The change of the excitation peak wavelength showed the same tendency as the result of the excitation wavelength at 365 nm, but the emission intensity increased with the increase in SnO content up to 5 mol%.

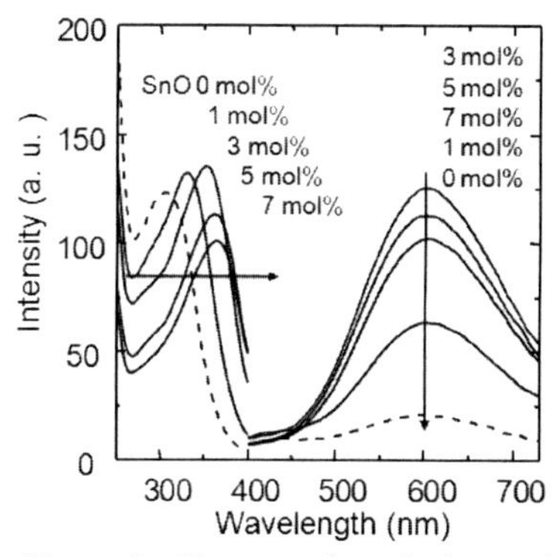

Figure 3. Changes of excitation and emission spectra of the samples with SnO contents. (Excitation wavelength 365 nm, Monitor wavelength 600 nm)

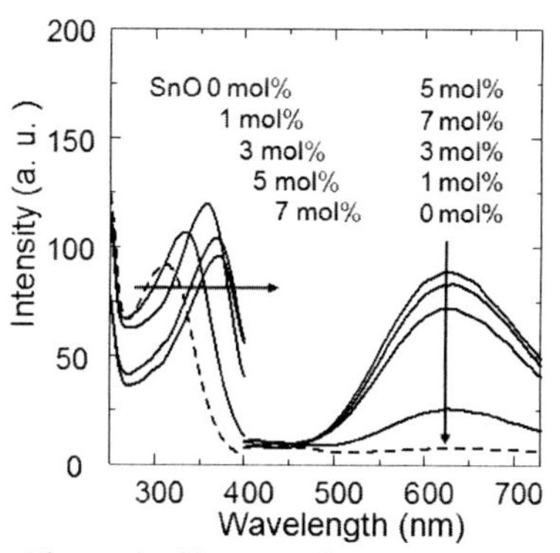

Figure 4. Changes of excitation and emission spectra of the samples with SnO contents. (Excitation wavelength 385 nm, Monitor wavelength 635 nm)

In order to examine the relation between the absorption wavelength and the amount of Cu^+ clusters in the glass, the emission peak intensity and the absorption coefficient at excitation wavelength 365 nm or 385 nm were plotted against SnO content, and the results are shown in Figure 5 and Figure 6, respectively. Both the PL peak intensities and the absorption coefficients at 365 and 385 nm linearly increased with the increase in SnO content in the range of small content of SnO. These results confirm that the absorption in the range of 300-400 nm is attributed to Cu^+ clusters and their number increased with the increase in SnO content. However, the PL peak intensities saturated and/or slightly decreased by further addition of SnO more than 3 and 5 mol% for the excitation at 365 and 385 nm, respectively. For the explanation of these emission properties, the difference of the absorption coefficients between 365 nm and 385 nm of the excitation wavelength should be considered, which are equivalent to the penetration depth of the excitation light into the glasses. Figure 7 shows the schematic models of the explanation of the emission properties. The amount of Cu^+ clusters in sodium-borosilicate glasses increased with the increase in SnO content. However, the excess increase in the number of Cu^+ clusters resulted in the decrease in the penetration depth of the excitation light. The absorption coefficient at

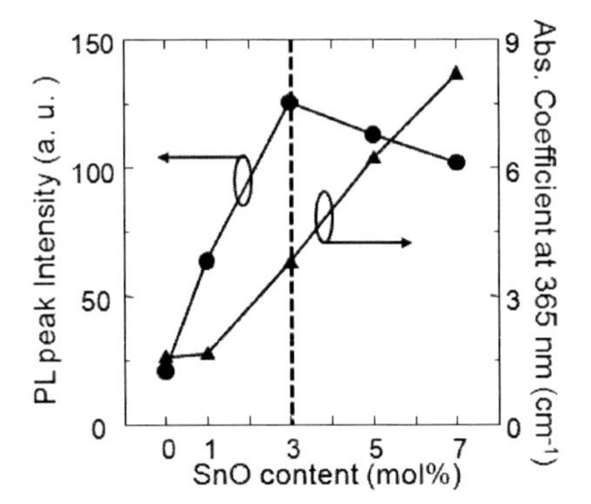

Figure 5. Changes of the emission peak intensity and the absorption coefficient at 365 nm with the increase of SnO content.

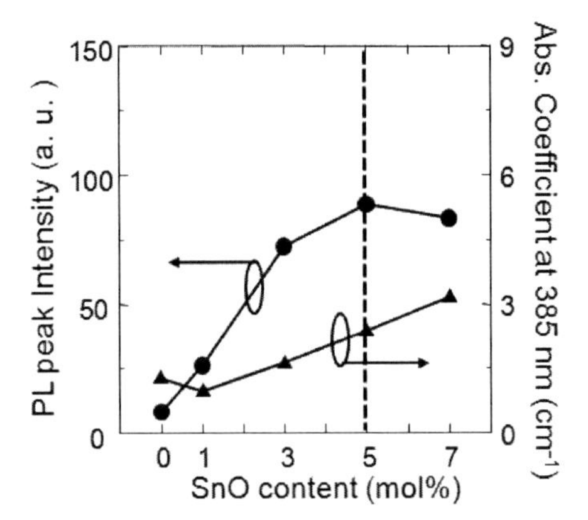

Figure 6. Changes of the emission peak intensity and the absorption coefficient at 385 nm with the increase of SnO content.

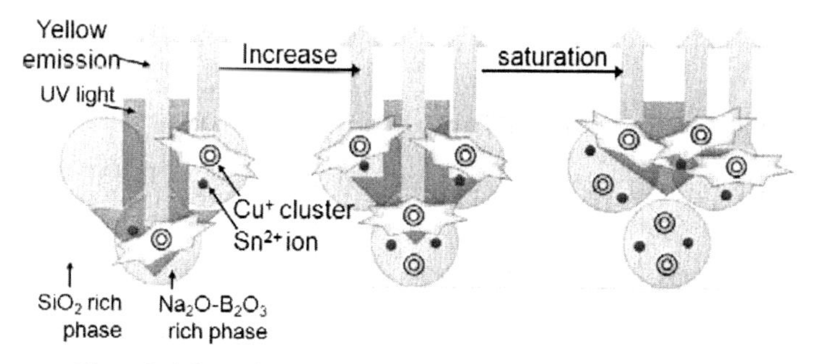

Figure 7. Schematic models of the absorption and the emission due to Cu^+ clusters in the glasses containing different SnO contents.

365 nm was almost three times larger than that at 385 nm. Therefore, the emission intensity excited at 365 nm saturated at 3 mol%, whereas that excited at 385 nm saturated at larger content, 5 mol%. These results and discussions suggest that it is necessary to control the thickness and SnO content of the sample appropriately in order to obtain the highest emission intensity, and the effect of the excited light intensity on the emission intensity should be also examined.

Color Temperature

Figure 8 shows the CIE chromaticity diagram with the plots of the color coordinates of the samples which were calculated from their emission spectra. The x and y values of the coordinates in the CIE chromaticity diagram for each sample excited at 365 nm or 385 nm are shown in Table I. The color coordinates of the samples containing 1-7 mol% of SnO excited at 365 nm locates near the points of 3000 K of the color temperature, which is close to that of an incandescent lamp of 2800 K in general. Moreover, the coordinates of the samples excited at 385 nm shifted toward those of a red color point, which is further close to a warm color and more suitable for the interior lighting

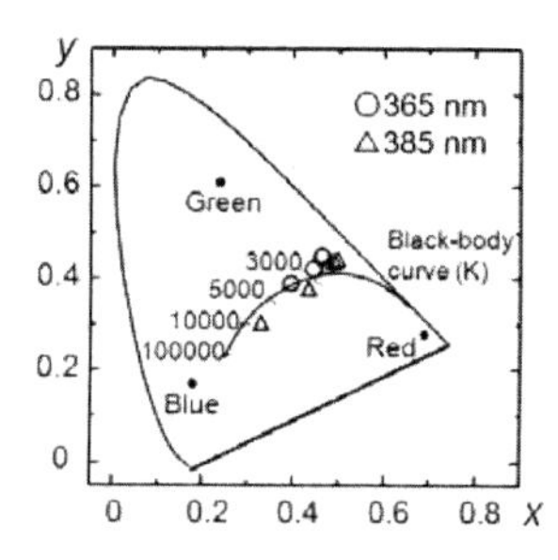

Figure 8. Color coordinates of the samples excited at 365 nm or 385 nm on CIE chromaticity diagram.

Table I. Coordinate values of the emission colors of the samples excited at 365 nm or 385 nm.

Ex. Wavelength (nm)		365		385	
Coordinate Value		x	y	x	y
	0	0.40	0.38	0.33	0.30
SnO (mol%)	1	0.44	0.42	0.43	0.37
	3	0.46	0.45	0.48	0.42
	5	0.46	0.45	0.49	0.43
	7	0.46	0.45	0.49	0.43

SUMMARY

The Cu^+ clusters containing sodium-borosilicate glasses were prepared by use of SnO as a reducing agent in the metastable immiscibility region. The obtained phase-separated glasses showed the yellow photoluminescence owing to Cu^+ clusters by the irradiation of near-UV lights from 300-400 nm. The color coordinates of the yellow emission from the samples located at the near 3000 K of the color temperature, which is close to that of an incandescent lamp. The number of Cu^+ clusters in the sodium-borosilicate glasses increased with the increase in the additive amount of SnO. However, the excess increase in the number of Cu^+ clusters resulted in the saturation of the yellow emission intensity because of the decrease in the penetration depth of the excitation light. Therefore, the appropriate controls of the thickness and/or Cu/Sn ions contents of the glasses are necessary to achieve the high warm color emission especially for the interior lighting system.

ACKNOWLEDGMENT

This research was supported by the Japan Society for the Promotion of Science (JSPS), Grant-in-Aid for Scientific Research (B), 21360326.

REFERENCES

[1]P. Schlotter, J. Baur, Ch. Hielscher, M. Kunzer, H. Obloh, R. Schmidt, and J. Schneider, *Materials Science and Engineering B*, **59**, 390–394 (1999).

[2]N. Narendran, Y. Gu, J. P. Freyssinier, H. Yu, and L. Deng, *J. Crystal Growth*, **268**, 449–456 (2004).

[3]T. Taguchi, *IEEJ Transactions on Electrical and Electronic Engineering*, **3**, 21-26 (2007).

[4]S. Tanabe, S. Fujita, S. Yoshihara, A. Sakamoto, and S. Yamamoto, *5th International Conference on Solid State Lighting*, **5941**, 594112-1-6 (2005).

[5]H. Hosono, H. Kawazoe, and T. Kanazawa, *J. Non-Crystalline Solids*, **33**, 103-115 (1979)

[6]G. Ramadevudu, Md. Shareefuddin, N. Sunitha Bai, and M. Lakshmipathi Rao, *J. Non-Crystalline Solids*, **278**, 205-212 (2000).

[7]R. Debnath, and S. K. Das, *Chemical Physics Letters*, **155**, 52-58 (1989).

[8]R. H. Magruder III, J. E. Wittig, and R. A. Zuhr, *J. Non-Crystalline Solids*, **163**, 162-168 (1993).

[9]C. Parent, P. Boutinaud, G. Le Flem, B. Moine, C. Pedrini, D. Garcia, and M. Faucher, *Optical Materials 4*, 107-113 (1994).

[10]A. Yasumori, F. Tada, S. Yanagida, and T. Kishi, *J. Electrochemical Society*, **159**, J1-J5 (2012) to be published.

[11]W. Vogel, *J. Non-Crystalline Solids*, **25**, 170-214 (1977).

[12]W. Haller, D. H. Blackburn, F. E. Wagstaff, and R. J. Charles, *J. American Ceramic Society*, **53**, 34-39 (1970).

[13]L. Skuja, *J. Non-Crystalline Solids*, **149**, 77-95 (1992).

*Corresponding author: t_black_label@yahoo.co.jp

THE ROLE OF CHEMICAL COMPOSITION AND MEAN COORDINATION NUMBER IN Ge-As-Se TERNARY GLASSES

Rong-Ping Wang, Duk Yong Choi, Steve Madden, and Barry Luther-Davies

Centre for Ultrahigh-bandwidth Devices for Optical Systems, Laser Physics Centre
The Australian National University
Canberra, Australia

ABSTRACT

We prepared a series of Ge-As-Se glasses with different chemical compositions and mean coordination numbers (MCN) from 2.2 to 2.94. We further measured their physical properties in order to understand the effect of MCN and chemical compositions on the physical properties of the glasses. It was found that, while glass transition temperatures, T_g, in the glasses with stoichiometric compositions generally increase linearly with increasing MCN, they depart from the linear behaviour in the glasses with high MCN >2.6 and low MCN<2.4. The fluctuation of T_g is less than 5% in the glasses with the same MCN of 2.5 but different chemical compositions. On the other hand, the density and elastic moduli of the glasses show two transition thresholds at MCN=2.45 and 2.65, respectively. The glasses with the same MCN of 2.5 but different chemical compositions show a change of 3% in density and 5% in elastic moduli, respectively. All these results suggest that, in a region from MCN=2.4 to 2.55, MCN could be a primary factor to determine the physical properties of the ternary glasses.

INTRODUCTION

It has been argued that the physical properties of chalcogenide glasses would be predominantly controlled by the mean coordination number (MCN, which is the sum of the products of the individual abundance times the valency of the constituent atoms) on the basis of the theory of constraint counting, irrespective of their actual chemical compositions.[1,2] A simple counting of the bond length and angle constraints on the total number of degrees of freedom available to a mole of three-dimensionally connected atoms indicated the existence of a phase transition at MCN=2.4 from an under-constrained "floppy" network to an over-constrained "rigid" phase.[1,2] Tanaka suggested that a second phase transition existed at MCN=2.67 which represents a topological change from a 2-D to 3-D "stressed rigid" phase.[3] Increasingly experimental evidence has shown that many chemical and physical properties change abruptly at these MCN values.[4-9]

While it is easily understood that the structure and physical properties of chalcogenide glasses can be tuned by chemical compositions, significant role of MCN in determining physical properties of chalcogenide glasses is somehow confused, particularly in ternary glasses since one can prepare the glasses with different compositions but the same MCN. On the other hand, emerging applications like integrated waveguide devices require the fabrication of chalcogenide glasses in the form of thin films and it has generally been found that such films, unlike the bulk glasses, have unstable physical properties which usually degrade the device performance and reliability.[10] While these instabilities generally arise because the films are prepared in non-equilibrium conditions and this leads to different bond configurations from the bulk materials,[10] searching the glasses with the best chemical compositions in so-called intermediate phase, where the glasses are thought to be stable, could be the best solution to achieve the best films against any degradation. Therefore to elucidate the different role of MCN and chemical compositions in tuning physical properties of the glasses, and to search the best compositions in ternary glasses, in this paper we prepared numbers of Ge-As-Se glasses with specially designed compositions, and characterized them using various tools in order to elucidate the correlation between MCN, chemical compositions and the structural and physical properties of the glasses.

EXPERIMENTS

The details for the glasses preparation can be found in our previous paper.[7,8] In total, we prepared and measured 3 groups of glasses with different compositions, including (1) 15 pieces of GeAsSe glasses with different chemical compositions which randomly occupy the glass forming region; (2) five different stoichiometric compositions, $Ge_{7.5}As_{35}Se_{57.5}$, $Ge_{10}As_{30}Se_{60}$, $Ge_{12.5}As_{25}Se_{62.5}$, $Ge_{15}As_{20}Se_{65}$, and $Ge_{20}As_{10}Se_{70}$, with same MCN of 2.5; (3) five stoichiometric compositions with different MCN including $Ge_{6.25}As_{32.5}Se_{61.25}$ (MCN=2.45), $Ge_{12.5}As_{25}Se_{62.5}$ (MCN=2.5); $Ge_{18.75}As_{17.5}Se_{63.75}$ (MCN=2.55), $Ge_{25}As_{10}Se_{65}$ (MCN=2.6) and $Ge_{31.25}As_{2.5}Se_{66.25}$ (MCN=2.65). All these glass compositions have been indicated in Figure 1.

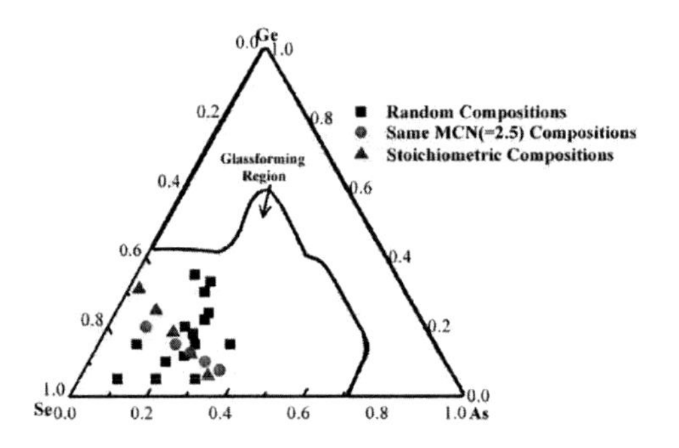

Figure 1. Glass compositions in Ge-As-Se glass-forming region.

Glass transition temperatures, T_g, were measured using a differential scanning calorimeter (Shimadzu DSC-50) with a scanning rate of 10K/min in a nitrogen gas flow of 30ml/min, and no devitrification or crystallization was observed for the glasses in this study. The density of the glasses was measured using a Mettler H20 balance (Mettler-Toledo Ltd., Switzerland) with a MgO crystal as a reference. Ultrasonic pulse interferometry was employed to measure both shear and compressional wave velocities from which elastic moduli were calculated.

RESULTS AND DISCUSSION

T_g as a function of MCN is shown in Figure 2. First we consider a group of the glasses with stoichiometric compositions as indicated by blue triangles. These stoichiometric glasses can only be formed at a narrow region with MCN values from 2.4 to 2.67. It is generally accepted that these glasses consist of perfectly tetrahedral $GeSe_{4/2}$ and pyramidal $AsSe_{3/2}$ units with negligible amount of other structures. It has been well established that, the value of T_g is closely correlated with the connectivity and rigidity of the vitreous network.[11] Therefore increasing concentration of the network forming elements such as Ge and As enhances the rigidity of the network, leading to a high T_g. Moreover, a linear correlation between logistic T_g and MCN was found for the chemically stoichiometric glasses. On the basis of an empirical relationship discovered by Tanaka for a wide variety of stoichiometric molecular glasses,[12] T_g has a correlation with M (MCN) as $\ln T_g \sim 1.6M+2.3$. The fitting of T_g for our samples reveals a linear relation of $\ln T_g \sim (1.49\pm0.10)M+(2.53\pm0.26)$, which is in good agreement with Tanaka's results.

T_g for the glasses with random compositions is shown in Fig. 2 as square dots. We note that, when the fitting line of T_g for the stoichiometric glasses is extended to lower and higher MCN values, the experimental values of T_g for those glasses with MCN less than 2.55 are very close to fitting the line while those with MCN more than 2.6 are obviously below the line. On the other hand, five pieces of glasses with same MCN of 2.5 but different chemical compositions show T_g at 512.5K, 516K, 520.9K, 524.8K, and 518K, for $Ge_{7.5}As_{35}Se_{57.5}$, $Ge_{10}As_{30}Se_{60}$, $Ge_{12.5}As_{25}Se_{62.5}$, $Ge_{15}As_{20}Se_{65}$, and $Ge_{20}As_{10}Se_{70}$, respectively. In spite of the large change in Ge content, T_g values for these five pieces of glasses are almost the same with a fluctuation of less than 5%.

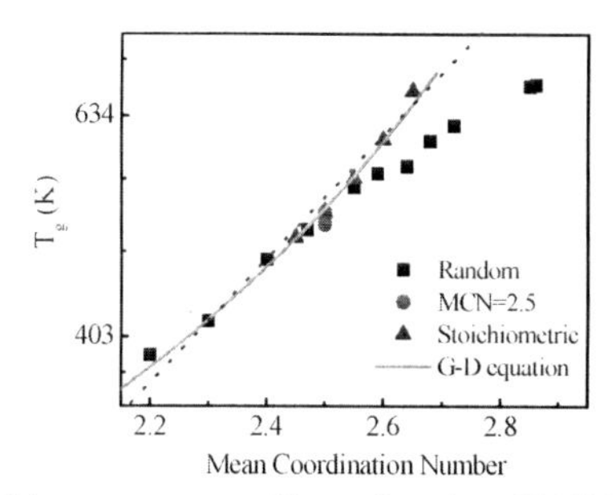

Figure 2. Glass transitions temperatures T_g as a function of MCN for $Ge_xAs_ySe_{1-x-y}$ ternary glasses. The black squares indicate that the glasses have random compositions in the glass-forming region, the red dots indicate that the glasses have a same MCN of 2.5 but different chemical compositions, and the blue triangles are those glasses with stoichiometric compositions. The blue line is a linear fitting of T_g for the stoichiometric glasses, and the green curve is a fitting of T_g using modified Gibbs-DiMarzio equation.

As mentioned above, T_g is closely correlated with the connectivity and rigidity of the glass network.[11] For the glasses with MCN>2.6, increasing concentrations of Ge and As induce a large amount of the homopolar bonds like Ge-Ge and As-As, that could cut the network connectivity, leading to the low T_g values compared with the blue line (stoichiometric compositions) in Fig.2. Here we have to emphasize that quantitative analysis of T_g based on chemical bonding energy is difficult due to two reasons: (1) it is hard to get the exact percentage of each kind of chemical bond; (2) the change of MCN will result in the topological transitions of the glass structure, therefore continuous random network in the glass does not exist for all the compositions.[13] Nevertheless, we observed a narrow region from MCN of 2.45 to 2.55 where T_g increases with increasing MCN but less relevant to the chemical compositions. The investigation of the correlation of the region with the intermediate phase is still in progress. Certainly the present results suggest that, in this region the physical properties of the ternary glasses are mainly determined by MCN.

We also noted that, Varshneya et.al. used a modified Gibbs-Dimarzio equation to describe the glass transition temperature trends in multicomponent chalcogenide glasses,[14] where T_g was found to be correlated with MCN as $T_g=T_0/[1-\beta(MCN-2)]$. Here T_0 is the glass transition temperature of the non-cross-linked parent of Se and β is a constant. They found that, for the chalcogen-rich Ge-Sb-Se glasses, T_g can be well fitted as $T_g=310/[1-0.75(MCN-2)]$. While the same approach was applied for our data, a correlation of $T_g=320/[1-0.78(MCN-2)]$ was obtained for the chalcogen-rich Ge-As-Se glasses as shown in Figure 2 as a green curve. Basically the fitting parameters of T_0 and β agree with each other although they are in the different glass systems. However, we found that T_g of the chalcogen-deficient glasses does not fall on the green curve due to the fact that the chalcogens are too few to form continuous chains in these glasses.[14]

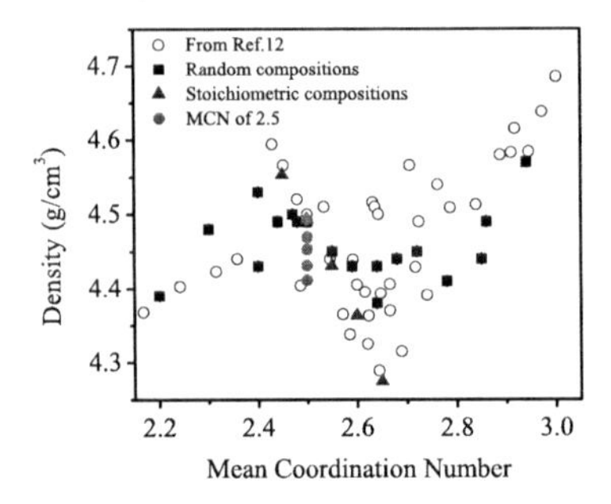

Figure 3. Density as a function of MCN.

Figure 3 shows the density of $Ge_xAs_ySe_{1-x-y}$ glasses as a function of MCN. The data from Ref.15 have also been included as open circles. The samples from two different sources exhibit similar behavior. The density first increases with increasing MCN exhibiting a maximum at MCN≈2.45 before decreasing to a minimum at MCN around 2.65. For MCN<2.4 corresponding, for example, to glass samples such as $Ge_5As_{10}Se_{85}$, $Ge_5As_{20}Se_{75}$ and $Ge_5As_{30}Se_{65}$, if there were no change in the arrangement of the atoms in the glasses, progressively replacing Se by As atoms should reduce the glass density because Se is heavier than As. Therefore, the observed opposite trend of increasing density with increasing MCN for MCN<2.4 suggests that the addition of As (or Ge) could result in significant changes to the topological structure of the glasses.

For the glasses with MCN values between 2.45 and 2.65, the density decreases with increasing MCN, indicating that these is no significant atomic rearrangement in the glasses. The stoichiometric glasses as indicated by triangles in Fig.3 show a linear decrease in the density with increasing MCN while the fluctuation of the density for the glasses with same MCN but different compositions is less than 3%, suggesting that MCN and chemical compositions are first and second factors to determine the density of the glasses, respectively, in this region.

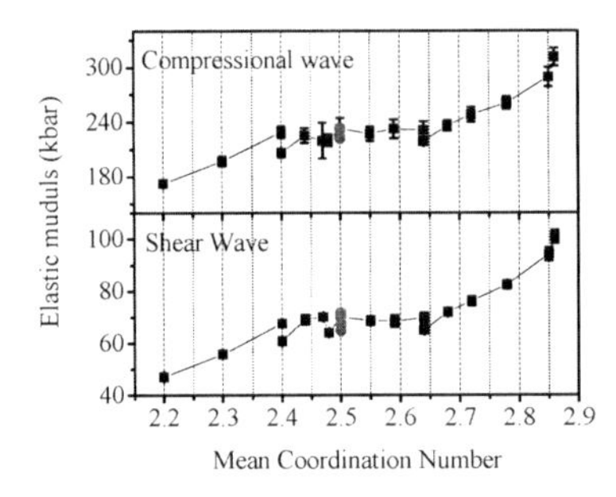

Figure 4. Shear and compressional elastic moduli as a function of MCN. The black squares correspond to the glasses with random compositions, while the red dots to those with same MCN of 2.5. The eyes are guided to the lines.

The evolution of shear and compressional elastic moduli as a function of MCN is plotted in Fig.4. The key result is the simultaneous appearance of two transition thresholds at MCN≈2.45 and 2.65, respectively. For MCN values below 2.4, we observed increasing moduli with increasing MCN for Se-rich samples such as $Ge_5As_{10}Se_{85}$, $Ge_5As_{20}Se_{75}$ and $Ge_5As_{30}Se_{65}$ glasses, which is in contrast with the results reported in Refs. 16 and 17 where the elastic moduli below MCN=2.4 were almost constant. The differences could be caused by the complete neglect of van de Waals interactions and dihedral angle forces in Ref.16, and the presence of 6-8% oxygen impurity in their samples used in Ref.17. Since the van de Waals interaction decreases with decreasing Se concentration in the three Se-rich samples, an increase in the elastic moduli becomes a natural result of introducing Ge and As.

For MCN between 2.45 and 2.65, shear and compressional elastic moduli are almost constant at ≈68 and ≈230 kbar, respectively. Between 2.4 and 2.65, several experimental results have confirmed that chalcogenide glasses have a layer-type structure which contains flexible segments.[18-20] Thus, when compressing the samples, the van der Waals forces between the segments are mainly responsible for the moduli, whilst the covalent bonds within an individual segment are almost unperturbed by pressure. Therefore all the glasses have moduli of similar magnitude, irrespective of their detailed microscopic structure. The glasses with same MCN of 2.5 as indicated by the red dots also show negligible change in their elastic modulus even though the chemical compositions of the glasses are different.

For MCN values above 2.65, the segments will be cross-linked by increasing Ge and As contents forming a stressed-rigid phase. Therefore the elastic moduli increase again with increasing MCN. Tanaka reported that the constraint-counting formalism yields a value of MCN~2.67 for transition to a rigid state if a two-dimensional layer structure is embedded in a 3-dimensional space.[3] The transition we observed at 2.65 is, therefore, very close to that threshold.

CONCLUSIONS

In summary, we found two transition thresholds at MCN=2.45 and 2.65, respectively, from the data of the density and elastic moduli of the glasses. The analysis of the experimental results on the glasses with same MCN but different chemical compositions suggest that, in a region from MCN= 2.45 to 2.55, the physical properties of the ternary glasses are dominated by MCN, and the chemical compositions could be second factor to tune the physical properties.

ACKNOWLEDGEMENTS

This research was partly supported by the Australian Research Council through its Centres of Excellence and Discovery projects.

REFERENCES

[1]J.C. Phillips, Topology of covalent non-crystalline solids I: Short-range order in chalcogenide alloys, *J.Non-Cryst.Solids* 34, 153-181(1979).

[2]M.F. Thorpe, Continuous deformations in random networks, *J.Non-Cryst.Solids* 57, 355-370(1983).

[3]K.Tanaka, Structure phase transitions in chalcogenide glasses, *Phys.Rev.B* 39, 1270-1279(1989).

[4]S. Mahadevan and A. Giridhar, Coexistence of topological and chemical ordering effects in Ge-Ga-Se glasses, *J.Non-Cryst.Solids* 152, 42-49(1993).

[5]A. Srinivasan, K.N. Madhusoodanan, E.S.R. Gopal, and J. Philip, Observation of a threshold behaviour in the optical band gap and thermal diffusivity of Ge-Sb-Se glasses, *Phys.Rev.B* 45, 8112-8115(1992).

[6]D. Arsova, E. Skordeva, and E. Vateva, Topological threshold in GeAsSe glasses and thin films, *Solid State Communications* 90, 299-302(1994).
[7]D.A.P Bulla, R.P. Wang, A. Prasad, A.V. Rode, S.J. Madden, and B. Luther-Davies, On the properties and stability of thermally evaporated Ge-As-Se thin films, *Applied Physics A: Material Sciences and Processing* 96, 615-625(2009).
[8]R.P. Wang, A. Smith, B. Luther-Davies, H. Kokkonen, I. Jackson, Observation of two elastic thresholds in $Ge_xAs_ySe_{1-x-y}$ glasses, *Journal of Applied Physics* 105, 056109 (2009).
[9]Prasad, C.J Zha, R.P. Wang, A. Smith, S. Madden and B. Luther-Davies, Properties of $Ge_xAs_ySe_{1-x-y}$ glasses for all-optical signal processing, *Optics Express* 16, 2804-2815(2008).
[10]R.P. Wang, A.V. Rode, S.J. Madden, C.J. Zha, R.A. Javis, and B. Luther-Davies, Structural relaxation and optical properties in amorphous $Ge_{33}As_{12}Se_{55}$ films, *Journal of Non-crystalline Solids* 353, 950-952(2007).
[11]R.P. Wang, A.V. Rode, S.J. Madden, and B. Luther-Davies, Thermal characterization of Ge-As-Se glass by the differential scanning calorimetry, *Journal of Materials Science: Materials in Electronics 18*, s419-423(2007).
[12]K. Tanaka, Glass Transition of covalent glasses, *Solid State Commun.* 54, 867-869(1985).
[13]Ping Chen, C. Holbrook, P. Boolchand, D. G. Georgiev, K. A. Jackson and M. Micoulaut, Intermediate phase, network demixing, boson and floppy modes, and compositional trends in glass transition temperatures of binary As_xS_{1-x} system, *Phys. Rev. B* 78, 224208 (2008).
[14]A.N. Sreeram, D.R. Swiler and A.K. Varshneya, Gibbs-DiMarzio equation to describe the glass transition temperature trends in multicomponent chalcogenide glasses, *Journal of Non-Crystalline Solids 127*, 287-297(1991).
[15]Z.U. Borisova, *Glassy Semiconductors*, New York: Plenum, 1981, pp.263-267.
[16]H. He and M.F. Thorpe, Elastic properties of glasses, *Phys.Rev.Lett.*54, 2107-2110 (1985).
[17]B.L. Halfpap and S.M. Lindsay, Rigidity percolation in the Ge-As-Se alloy system, *Phys.Rev.Lett.* 57, 847-850 (1986).
[18]K. Tanaka, Elastic Properties of covalent glasses, *Solid State Commun.* 60, 295-297(1986).
[19]C.Lin, L.E. Busse and S.R. Nagel, Temperature dependence of the structure factor of GeS_2 glass, *Phys.Rev.B.* 29, 5060-5062(1984).
[20]K. Tanaka, High-pressure structural changes in chalcogenide glasses, *Solid State Commun.* 58, 469-471(1986).

*Corresponding author: rpw111@rsphysse.anu.edu.au

Author Index